MATHEMATICS AND SOCIETY

Wolfgang König
Editor

European Mathematical Society

Editor:

Prof. Wolfgang König
Weierstrass Institute for Applied Analysis
and Stochastics (WIAS)
Mohrenstr. 39, office 403
10117 Berlin
Germany

E-mail: koenig@wias-berlin.de

and
Institut für Mathematik
Technische Universität Berlin
Straße des 17. Juni 136
10623 Berlin
Germany

2010 Mathematics Subject Classification: 00-XX

Key words: Mathematics in the public; mathematics in architecture, biology, climate, encryption, engineering, finance, industry, nature shapes, physics, telecommunication, and voting systems; experimental mathematics, mathematics museums, mathematics for complex data

ISBN 978-3-03719-164-4

The Swiss National Library lists this publication in The Swiss Book, the Swiss national bibliography, and the detailed bibliographic data are available on the Internet at http://www.helveticat.ch.

Contact address:

European Mathematical Society Publishing House
Seminar for Applied Mathematics
ETH-Zentrum SEW A27
CH-8092 Zürich
Switzerland

Phone: +41 (0)44 632 34 36
Email: info@ems-ph.org
Homepage: www.ems-ph.org

Typeset using the authors' TEX files: Christoph Eyrich, Berlin
Printing and binding: Beltz Bad Langensalza GmbH, Bad Langensalza, Germany
∞ Printed on acid free paper
9 8 7 6 5 4 3 2 1

Mathematics possesses a central importance in our society. It shapes and influences many areas of our daily life, from education and culture via technology and industry to physics and information science, and more. Most obvious and most closest examples to our experiences comprise the design and the evaluation of insurance products and financing schemes, mobile wireless communication, electronic devices for a global positioning system, and much more. In many of these examples, the contribution of mathematics to the development of the functionality was decisive, and there are billions of users, but hardly anybody notices the mathematics involved anywhere in the final product. Indeed, this phenomenon seems to be even characteristic for mathematics: its rôle in the development of a new technology or device is vital, but cannot be seen in the product itself. This is even more distinctive in intangible concepts and procedures that have been shaped by mathematics, like encryption procedures that make possible online banking, e-mail and the internet, or finance products that are adapted to special market situations. Mathematics is important for the development of technology and industry for many centuries and will continue so for ever, and real-world problems inspire and accelerate mathematical research in many ways.

But there are also further, immaterial, impacts of mathematics on the society. For example, some exciting developments in mathematics raised interesting philosophic questions that are at times discussed in public, like the question about the validity of proofs: how well must a proof be checked, such that one can agree that the assertion has been proved? This is even more difficult to decide for computer-aided proofs, the existence of which has gained quite some popularity, at least within the mathematical community (think of the famous four-colour problem, which says that four colours suffice for colouring a map without giving the same colour to any two neighbouring areas). Popularity even outside the community is enjoyed by questions about the quest for solutions to famous open problems like the Clay Millennium Problems (www.claymath.org/millennium-problems). Another much-discussed aspect is the general rôle of mathematics in education, as it inspires many people, including pupils, as a freetime-occupation and stimulates them to further investigation of their own.

When we talk about mathematics in society, then we mean mathematics that has an influence on the daily life of a substantial part of the people, or mathematics that is an integral part of our society in one or another way. Furthermore, we mean such kind of mathematics that can be done only by a professional mathematical expert. We talk about the search of new concepts, methods and proofs on a level that necessitates investigation by mathematical researchers (jointly with the efforts of specialists of the application field, if necessary).

As we said at the beginning, much mathematical research whose results are decisive for the wealth of the society and for solutions to its problems is so well hidden that the public does not know about it, often not even the scientifically interested layman. Likewise, several application fields are generally known, but not the way in which mathematics makes an important contribution. There are many reasons for this unfortunate situation. One is of course that an understanding of the problem requires some least level of preparatory training in the application field, which the layman typically does not have. Another one, and may be the decisive one, is that the mathematics involved is quite abstract and specialised and needs even more training in mathematics. And a third main reason is the lack of expert people who are willing to take time for explaining things in a way that the scientifically interested layman can understand.

This volume is an attempt to rectify the situation. It tells success stories about some important examples of mathematics in society: a real-world problem to be solved, mathematical difficulties that the problem poses, and the rôle that mathematics plays in the solution. All these stories are written by eminent and experienced mathematicians, who shaped, and continue to shape, the application field that they describe. The goal was to cover a great diversity of application areas, and to explain things on a non-technical level, such that the scientifically interested layman can take home quite some message. Certainly also professionals, both those in mathematics and those in the application area, will greatly benefit. But, given that fifteen (rather diverse!) subjects are highlighted in this collection, nobody can be an expert in all of them, but every scientifically interested will find something that she or he is able to follow and to enjoy.

It is always difficult to talk about mathematics and its applications to real-world problems; even more, if readers without specialised education are addressed. The matter is generally difficult, the author's and the reader's background differ a lot, and also the readership is rather diverse: from expert mathematicians to mathematicians that work in other areas, and from experts on the application field to the layman who is just interested in science. All the authors of this volume have tried their best to address at least more than one of these groups, and some of them even aimed at the latter type of reader, which is definitely the most difficult task.

Being passionate mathematicians, the authors of this collection certainly present also a great deal of mathematical material; how else can you demonstrate its necessity and its benefits better? It may be the a exciting aspect of each of the essays that the authors' personal tastes shine through the choices of the subjects, through the ways of presentation and through the comments. It is always interesting to learn about the personal view of the expert at the things, even more as the editorial board encouraged the contributors to reveal some of that, and many of the authors gratefully used the opportunity to do so.

What are these application fields that we decided to present here? It was our purpose to present a most diverse collection of aspects and areas of mathematics that have shaped and continue to shape our society. Here is a survey.

Let us begin with a fundamental question about the interaction between mathematics and the society: the question about how a mathematical expert should write about his field for the public, such that (s)he will be understood and reaches her/his readership – precisely the situation in which all the authors of this book are! Read the advices and entertaining examples of how to do that and how one should not do that by George Szpiro! – We proceed with more abstract approaches to the question how mathematics enters and influences society. The enormous computer power that mankind is now able to use produced new ways of production of new mathematics, namely by computer-assisted proofs and by automated verification of existing, complex proofs and falsification of conjectures. Donald Bailey and Jonathan Borwein present deep and recent thoughts about this new field called *experimental mathematics*, and touch also some fundamental questions like the one about when a proof can be considered valid.

One of the reasons that mathematics is esteemed so high in our society is presumably its rôle in the education as a science that is most amenable to teaching methodologies that appeal to the natural wish of humans to find out by own doing. The conscience for this in our society is significantly increasing only for a few years, but it has lead to (and has been increased by) the foundation of science centers, hands-on museums, or even special mathematics museums. One of the most well-known educational centers of this kind (at least in Germany), the *Mathematikum*, was founded some ten years ago in Gießen by our author Albrecht Beutelspacher. In his contribution, he describes its concept and the reasons for its success. Furthermore, he gives an account on the highly interesting question, whether or not the institution of all these mathematics museums was worthwhile and how they changed the attitude of the population towards mathematics.

Let us proceed with application areas that have a notable random component. In finance, some few years ago, a big global crisis moved the world, and still the unspoken question remains in the room, whether or not the crisis happened *because* of the work of mathematicians or *in spite of* it. Read Walter Schachermayer's opinion about that! He also reports on the interesting history of the first applications of the Brownian motion to finance some hundred years ago. – One of the ubiquitous parts of applied mathematics is the field of statistics, whose goal is the probabilistic description of real-world phenomena. In the research of cancer, nowadays many approaches make use of an enormously huge amount of complex data. A solution to the problem how to extract useful information from these big data lies by no means in a strategy of brute-force computing with stronger and stronger computer power, but in the development of more and more sophisticated statistical methods, notably handling the high-dimensionality of the structure in the data. Aad van der Vaart and Wessel van Wieringen give some insight in the mathematical aspects of these methods. – Another application field of statistics is filtering theory within engineering. The story of this method and its historical successes is told by Ofer Zeitouni. – Probability plays also an important rôle in population models with application in biology. Especially in recent years, new models have been introduced that take care of the latest state of understanding of the evolution of biological populations. Jean Bertoin presents a pedagogical summary of some of the models that mathematical biologists work on most actively at the moment.

One of the areas with the oldest, most immediate, and most discussed connections with mathematics is physics. Already the ancient Greeks deeply thought about the laws that underly all the matter and the mathematical way to describe them. By example of the Second Law of Thermodynamics and the concept of entropy, Jürg Fröhlich brings to the reader the spirit of the old quest for the understanding of fundamental laws that control a great deal of actions around us.

Another large and classical part of mathematics, even the oldest, but still very modern and, one can say, ubiquitous field is geometry, which is more up-to-date than one might think. Actually, in architecture it plays nowadays an equally active rôle as thousands of years ago. More specifically, Hellmut Pottmann and Johannes Wallner concentrate on problems from discrete differential geometry that an architect has to solve if (s)he plans to design a piece of freeform architecture. Many pictorial examples from the practice illustrate the interplay between the mathematical theory and the intended form of the building. – While this article is about the geometry of objects that humans want to shape on their own, Christiane Rousseau concentrates on those shapes that appear in natural morphologies, for example animals and plants, and follow rules of geometry that lead to beautiful and functional solutions. She brings the rules underlying the natural shaping forces to the surface and explains why similar geometric forms appear in most diverse connections in the nature. Striking relations with fascinating mathematical objects like the Koch snowflake appear in a new context.

Let us come to applications of mathematics in industry. First of all, does "industrial mathematics" exist at all? One of the most experienced experts in this field, Helmut Neunzert, raises this question and extends it to the question "industrial mathematics versus academic mathematics". This is only a starting point for a large-scale survey on the history and the current situation in the relationship of mathematics, as is carried out in applied research, and industrial mathematics that is meant to solve explicit tasks, including a lot of philosophical considerations and personal statements! – Telecommunication is a concrete industrial field in which mathematics has much to do and to say, and its contributions are enormously diverse and versatile. Holger Boche and Ezra Tampubolon concentrate on a particular question that is ubiquitous in the theory and praxis of data transmission and can be resolved only by an ingenious use of highly developed mathematics: how can one handle the huge differences in the amplitude of the transmission of a signal by means of an orthogonal transmission scheme? They demonstrate that this annoying problem can be settled by use of some parts of mathematics that are considered to be quite pure, like additive combinatorics, but also more applied disciplines like functional and harmonic analysis.

Information security poses severe challenges for the society of the future. Cryptography now provides ramified techniques to deal with these challenges. It constitutes a wide field which is now heavily based on methods from a variety of mathematical disciplines, notably complexity theory and number theory. Claus Diem explores these connections. Also, the limitations of the mathematical approach to real life security are critically addressed.

A field that would be not guessed as a field of mathematical application is the field of politics, more precisely, voting systems. Actually, there are hardly any two votes that are carried out under precisely the same set of rules, and slight changes in the voting rules can lead to surprising changes in the result, and not only theoretically. Werner Kirsch gives a flavour of a mathematical concept of voting systems, its benefits, the effects that it contains and the conclusions that one can draw within the concept. The consequences of some theoretical results for the society can be pretty immediate, as he illustrates by means of historical examples.

The reader might have noticed that a field that is generally thought to have high affinity to mathematics has not yet (or only once, see the above mentioned contribution to statistics) been addressed: handling big data. Last, but not least, there is also one essay devoted to this important subject, in the connection of investigation of the climate and the weather. Here it is not possible to make experiments, one has to rely on mathematical descriptions and predictions of the reality. Obviously, huge amounts of data are available, but the biggest problems come from the enormous span of scales of the meaning of the data. Jörn Behrens describes the established mathematical models and methods in geoscience; in particular the rôle of the important field of uncertainty: what can mathematics do if we not even know the probability distribution of the unknown quantities in our equation?

This ends our small survey of the articles contained in this volume. Let me express my sincere thanks to the inspiring support of my colleagues that formed a awesome editorial board: Jochen Brüning, Hans Föllmer, Michael Hintermüller, Dietmar Hömberg, Rupert Klein, Gitta Kutyniok, and Konrad Polthier. Their broad expertise and overview helped a lot to identify a good choice of areas that should be contained in such a collection and experts that should be approached as authors for a book with such an intention. Let me also thank Claus Diem for careful reading and numerous hints and proposals, which led to a substantially better readability of several of the contributions.

I am also very grateful to the *European Mathematics Society Publishing House* and its staff, which was most helpful and flexible at every stage of the production. This book is part of the official documents for the *7th European Congress of Mathematics (7ECM)*, the quadrennial Congress of the European Mathematical Society, which is organised at Technische Universität Berlin on July 18-22, 2016.

Berlin, May 2016   Wolfgang König

Contents

CONTENTS

The truth, the whole truth and nothing but the truth: The challenges of reporting on mathematics

George G. Szpiro

I would like to address the role and responsibility of the journalist who writes about mathematics, and also share some personal experiences with you.

In recent years, the attitude of research mathematicians towards the popularization of their subject has undergone a change. While they previously jealously, and sometimes haughtily, guarded their work nearly as a secret science, they have now become more attuned to the demands of the general public. After all, even if mathematicians require less funds than scientists working in other fields, they still depend on the taxpayers' support. Hence, the general public deserves to know what it is that mathematicians do. As Jochen Brüning, one of the editors of this collection, pointed out, the change in attitude is apparent, for example, in the *Notices of the AMS*. While it used to be a rather dry newssheet, which reported mainly on meetings and job openings, it has become a highly interesting and entertaining source of information. But in contrast to mathematicians reporting enthusiastically about their or their colleagues' work, journalists who write for mainstream publications have an important task as interpreters of the works' significance. They must, however, always remain aware that they are no experts in the fields they write about. A certain measure of humbleness is required on their part and they must ask for help, even if their interlocutors may be unwilling, because they still prefer the protectedness of their ivory tower.

Before I can even begin to discuss how mathematics should be reported on in newspapers or magazines, I must first address a hurdle that needs to be overcome. This hurdle presents itself not in the form of reluctant readers, but rather in that of sceptical editors. When I proposed a monthly column on mathematics to the *Neue Zürcher Zeitung (NZZ) am Sonntag*, over a dozen years ago, I was unsure whether this idea would be accepted. I asked a prominent American colleague for his advice. He (if I write "he", I may also refer to a woman) answered that newspaper editors are generally very creative people who have turned to writing (instead of, say, the sciences) because they were often at odds with mathematics in school. This was thus the basis of their scepticism of everything mathematical. Now, I was fortunate and had willing editors. And, to my pleasure, it turned out that many more readers were interested in a column about mathematics than either I or my editors had expected.

The question that a mathematics journalist must ask himself concerns a simple but powerful formula: *the truth, the whole truth and nothing but the truth*.

Must a journalist who writes about mathematics follow this principle? At first glance, the question seems irrelevant, almost provocative. Of course the truth is a common good that should be a guiding principle for everyone – not just journalists – in all areas of life. In political reporting, for example, it is undisputable that journalists are absolutely obligated to the truth, the whole truth and nothing but the truth. When I wear my other hat, that of a foreign correspondent for the NZZ, it goes without saying that I always make a strong effort to rigorously observe all three parts of this maxim.

But is that which is self-evident for journalists in general also applicable to the field of mathematics?

Before I try to answer this question, I will first pose another one. Assuming that the work of mathematics journalists is different from that of other reporters, should they perhaps let themselves be inspired by articles in mathematics journals? In this case, the answer is a simple one. It is a clear No – the requirements of an article written for a daily newspaper or weekly magazine are much different from those of an article written for a mathematics journal.

- An academic author of a journal article expects that the reader will carefully work his way through the article, even if this requires effort. The journalist, on the other hand, has to make it easy for the reader.
- The academic aims at announcing results that have been proven in a strict, rigorous way; the journalist wants to inform, but also to amuse the reader.
- A journal article will be cited many years or even decades later, and a mistake in an argument may be discovered even long after the article was written. Newspaper articles are usually forgotten after a day, and all of their flaws with them.
- Scientists only submit a manuscript for publication when they are sure that they cannot improve it further. Journalists usually have a publication date thrust upon them and are under pressure to meet editorial deadlines.
- Usually, in a scientific journal, an author has as much space as needed, within reasonable limits. In the newspaper, only limited space is available.

In this limited space, an article on mathematics should include the following: a good title, an explanation of the problem, the history of the problem, the background of the mathematician who popularized it, unsuccessful attempts to solve the problem, the personality of the mathematician who produced the proof, the procedure used to arrive at the successful proof or at least the idea behind it, the implication of the theorem and some applications.

This all has to be done within the space of 600, 1200, or, under the best of circumstances, 2500 words. Thus we arrive at the question of whether, under these circumstances the journalist can – or should – allow himself to be led by the principle "the truth, the whole truth and nothing but the truth."

I will make it clear from the beginning: as a mathematics journalist, it is often not possible to completely live up to this standard; quite frequently, it is not even possible to come close. In my newspaper articles about mathematics, I am myself guilty of transgressing against all three parts of this principle to a certain degree. Hence, the mathematics journalist is caught between two worlds: the world of everyday reporting and the world of scientific journals.

Let us begin with the expectations of readers who are experts in the field.

One example: In the NZZ, I once wrote that "the Navier-Stokes equation can be exactly solved only in special cases [...] although approximations can be calculated using computer models, the numerical methods are very problematic." This led a reader to send a letter to the editor:

> This could have been written in the *Bild-Zeitung* [a German tabloid newspaper]. Every mathematics student knows that, in general, differential equations are not solvable in closed form etc., etc.

Another professor complained that my article was somewhat frivolous and superficial, and readers who are experts in the field regularly get worked up when I bring attention to ugly behind the scenes disputes or stupid mistakes in a proof. Instead of hanging out this dirty laundry for everyone to see, they say, I should limit myself to the beauty of mathematics, otherwise the readers will get a false impression of this discipline. Really? Is everything always beautiful in the mathematics world?

A few years ago I received a tip about an article in one of the most well-regarded mathematics journals, the *Annals of Mathematics*, that contained a mistake. The article's author was a young mathematician who was well on his way to a brilliant career. The publisher had not recognized the mistake until after publication. Now, the right thing to do would have been to publish a correction immediately in the Internet or in the publication itself, which would have informed the reader of the mistake and perhaps corrected it. But someone found it difficult to do this. Perhaps there was hope that the author could repair the mistake himself? Perhaps someone wanted to spare him the humiliation? In any case, the fact is that for years, no correction appeared. The author had long abandoned his work on the problem and had turned his back on mathematics completely – today he is a local government politician in the United States. This meant that, during all of that time, the dangerous possibility existed that other mathematicians who depended upon the reputation of the *Annals of Mathematics* would base their work on an erroneous result. The long overdue correction was not published until the present author sent the editors an inquiry about the matter.

Another example: For many years, the mathematician Wu-Yi Hsiang, professor at the University of California, Berkeley, insisted on the accuracy of his geometric proof of the Kepler conjecture, which concerns the densest way to pack spheres in a three-dimensional space. However, as it turned out, his supposed proof was wrong for several reasons. Even after Thomas Hales presented a (computer-supported) proof, Hsiang did not want to back down from his incorrect proof.

And yet another example: After Grigory Perelman published his famous proof of the Poincaré conjecture (now Poincaré's theorem) in the Internet, the Harvard professor Shing-Tung Yau began a nasty priority dispute with Perelman. He maintained that two of his protégés had closed gaps in Perelman's proof, and that therefore they should be given credit for the proof. The dispute, which carried nationalistic undertones, took a bizarre turn. When an article was published about the controversy in the American magazine *The New Yorker*, Yau accused it of defamation of character and threatened legal action. As a matter of fact, this article was critized as unfair by quite a few prominent mathematicians.

Should we not report on such events, to ensure that the readers do not get the "wrong" impression of mathematics work? Journalists are not the mathematics community's PR agents, nor are they the advocates of mathematics. It is not their job to merely report on the "beautiful aspects" of mathematics and leave out everything that is unattractive. When dealing with the sociology and politics of mathematics, journalists orient their work on that of the political journalist: no one would accept it if a journalist in Washington, Moscow or Berlin were not criticizing a government official, only to avoid creating a "wrong" image of the government. To the contrary, the readers expect correspondents to report to the best of their knowledge and belief and not keep anything secret.

Let's now leave aside the politics and sociology of mathematics and return to the question of whether the journalist – when we are dealing purely with mathematics – should be guided by the principle "the truth, the whole truth and nothing but the truth."

"The truth" means not knowingly writing something that is incorrect.

Mathematics is abstract (in contrast to biology, medicine or physics). In order to explain an abstract context to someone who is not an expert in the field, it is often necessary to use analogies with illustrative examples, even if they are sometimes a little off the mark. For example, spiders' webs can be used to illustrate graph theory, and a random walk can be illustrated by the stumbling of someone who is intoxicated. The example used is often not a perfect fit, and often the analogy is a little bit of a stretch. But examples and analogies often give laypeople at least an impression of the meaning of a mathematical theorem or the idea behind a proof.

"The whole truth" means not spreading half-truths.

Mathematics must be rigorous; all special cases have to be worked through. Every little gap has to be closed. In the newspaper, by contrast, a theorem often has to be reduced to an example, or it may only be possible to explain one single special case. Sometimes one must restrict to explaining a theorem in terms of its consequences for daily life or its applications. In their report, journalists can, at most, provide an idea of how a mathematician went about his work or what a proof has to look like in order to be valid. The unavoidable fact is that, unfortunately, this requires a lot of hand waving, something which is rightfully frowned upon by mathematicians. Once, when writing an article about the life's work of a prominent professor, I made the mistake of giving him the manuscript in advance. This unleashed days of correspondence in which every sentence was questioned, even though the changes would not have been noticed in the least by even the most attentive of readers – and they did nothing to make the article more understandable. Only the approaching delivery deadline brought an end to the back and forth between the professor and the author.

And thus, the journalist limits the explanation of Poincaré's conjecture to an explanation of a two-dimensional case, even though the famous conjecture actually relates to the three-dimensional case; the Radon transform is best explained using its application in medical imaging; and the fact that differential equations can also be solved using a numerical approach, well that is something that mathematics students know but that the general readership does not need to be distracted with.

"Nothing but the truth" means that truth should not be mixed with that which is irrelevant or misleading.

The journalist has to illustrate mathematics and bring it closer to those readers who are not necessarily interested in mathematics. To do this, the article must be interesting and not be reduced to "dry" mathematics. Readers are also interested in the human aspect and – as mentioned above – in the politics and sociology of mathematics. The journalist must respond to this desire. The reader is interested in the background and the personalities that are behind the theorems, even though they actually have nothing to do with the mathematics itself. In this respect, the fascinating personality of a person like Grigory Perelman is merely one example from the recent past. The fact that solutions of the so-called millennium problems will be awarded with prizes of one million dollars each has nothing to do with mathematics. Nevertheless, journalists cannot leave these facts out when writing newspaper articles about them. All of this means that scandals or information about erroneous proofs – including the *Schadenfreude* that goes along with them – occasionally make their way into the papers.

In summary, journalists do not produce new mathematic discoveries. For mathematicians, the best service that they can offer is to open the eyes of experts in one special field to another special field. For example, once, in an article about knot theory, I mentioned the work of two mathematicians – one is a Pole who is conducting research in Lausanne, the other is Chinese and teaches in North Carolina. They did not know each other before the article, but since it was published they have been collaborating with each other. On the other hand, many mathematicians may not quite realize what role they and their work play in society at large. It is through reports in the media that they obtain the necessary feedback, to appreciate more fully the social relevance of mathematicians and their work.

But journalists write for the general readership first and foremost. These readers need to be drawn in, informed, and also entertained. This means that, on the one hand, it is not always possible to write the truth, and certainly not the whole truth, but, on the other hand, sometimes somewhat more than the truth. At the same time, mathematics journalists must be aware that they bear a great responsibility to the general reader. In contrast to the daily news, which present a diversity of (usually) verified information to the public, most readers are not able to find out for themselves whether a mathematics report is accurate or not. The mathematics reporter is thus the shaper of the reader's opinions, with all of the obligations that this entails.

Experimental mathematics in the society of the future

David H. Bailey and Jonathan M. Borwein

Computer-based tools for mathematics are changing how mathematics is researched, taught and communicated to society. Future technology trends point to ever-more power-ful tools in the future. Computation in mathematics is thus giving rise to a new mode of mathematical research, where algorithms, datasets and public databases are as significant as the resulting theorems, and even the definition of what constitutes secure mathematical knowledge is seen in a new light.

1 Introduction

Like most other fields of scientific research, both pure and applied mathematics have been sig-nificantly affected in recent years by the introduction of modern computer technology. Just like their peers in other fields, mathematicians are using the computer as a "laboratory" to perform exploratory experiments and test conjectures, in a methodology that has been termed "experi-mental mathematics". The adjective "experimental" here is entirely appropriate, because, in a fundamental sense, there is little difference between a mathematician using a computer to ex-plore the mathematical universe and an astronomer using a large telescope facility to explore the physical universe.

By experimental mathematics we mean the following computationally-assisted approach to mathematical research [11]:

1. Gaining insight and intuition;
2. Visualizing mathematical principles;
3. Discovering new relationships;
4. Testing and especially falsifying conjectures;
5. Exploring a possible result to see if it merits formal proof;
6. Suggesting approaches for formal proof;
7. Replacing lengthy and error-prone hand derivations;
8. Confirming analytically derived results.

With regards to 6, we have often found that computer-based tools are useful to tentatively confirm preliminary lemmas; then we can proceed fairly safely to see where they lead. If, at the end of the day, this line of reasoning has not led to anything of significance, at least we have not expended large amounts of time attempting to formally prove these lemmas. And if computer tests falsify a conjecture, then no need to waste any time at all seeking a formal proof.

While computation is proving very useful in the exploration phase, computers are also being employed to produce formal proofs of mathematical results. One notable recent success, just concluded, is Thomas Hales' computer-based formal proof of the Kepler conjecture [16], a topic that we will revisit in Section 4.4.

2 Experimental methods in education

While some still resist, it is clear that computational tools are the wave of the future for mathematics instruction, certainly not replacing the instructor or hand computations and algebraic manipulations, but, instead, permitting mathematical principles to be taught with less pain and greater understanding.

To mention just one simple example, many of us recall using algebraic substitutions to rotate a geometric figure given by a formula. Nowadays such rotations can be performed easily using computer graphics tools. Indeed, modern computer technology places the cart before the horse – rather than using advanced algebra and calculus as tools to graph functions, instead we can now use computer-based graphics tools to learn principles of algebra and calculus.

As another simple example, we are taught in elementary calculus that a definite integral can be seen as the area under a curve, in particular the limit found by adding the areas of rectangles under a curve, subdividing the interval into finer and finer parts. While a few very simple examples of this sort can be done by hand algebra, it is arguably more instructive for students to let a computer program do the hard work. For example, by employing a simple trapezoidal approximation to evaluate the integral $\int_0^1 dx/(1 + x^2)$, with 10,000 subdivisions, one obtains the numerical result $0.78539815\ldots$, which is accurate enough for the Inverse Symbolic Calculator 2.0, available at http://isc.carma.newcastle.edu.au to identify as likely to be $\pi/4$ (it is, of course, equal to $\pi/4$).

In Section 4, we will present a few somewhat more sophisticated examples of experimental mathematics in action. Here we present two that require only a very modest background, and are exemplars of how computation can be incorporated into education even at the high school level.

2.1 A number theory example

Many high school students learn that the sum of the first n integers is $n(n + 1)/2$. Indeed, Gauss is reputed to have discovered this formula by himself in elementary school. What about sums of higher powers? The simple *Mathematica* command Sum[k^5,{k,1,n}] returns the formula

$$1^5 + 2^5 + \cdots + n^5 = \frac{1}{12}n^2(2n^2 + 2n - 1)(n + 1)^2. \tag{1}$$

Note that by typing the command F5[n_]:=n^2(2n^2+2n-1)(n+1)^2;Simplify[F5[m]-F5[m-1]], one can symbolically determine that the difference between formula (1) evaluated at an integer n and at $n - 1$ is n^5, which, since the formula is clearly valid for $n = 1$, constitutes a proof by induction that formula (1) is valid for all positive integers n.

In a similar vein, one can use the computer to explore sums of even higher powers. For example, using either *Maple* or *Mathematica*, one obtains the formula

$$\sum_{k=1}^{n} k^{10} = \frac{1}{66} n(2n+1)(n+1)(n^2+n-1) \cdot (3n^6 + 9n^5 + 2n^4 - 11n^3 + 3n^2 + 10n - 5), \quad (2)$$

so that, for example,

$$\sum_{k=1}^{10,000} k^{10} = 909590992424241424242434242424241924242425000. \quad (3)$$

Note the curious pattern of 42 repeated numerous times (except for the central 3) in the center of this number. What is the explanation? By using *Maple* or *Mathematica* to expand formula (2), one obtains

$$\sum_{k=1}^{n} k^{10} = \frac{1}{66} \left(6n^{11} + 33n^{10} + 55n^9 - 66n^7 + 66n^5 - 33n^3 + 5n \right). \quad (4)$$

Note that the fourth and fifth terms are -66 and 66, respectively, which, when divided by 66, are -1 and 1. Also note that without the $1/66$, the sum (3) above would be:

$$66 \sum_{k=1}^{10,000} k^{10} = 60033005499999993400000065999999967000000050000. \quad (5)$$

In this form, the correspondence between (4) and (5) is clear – by examining (5), one can literally read off the coefficients of (4) term by term (remembering that 999... is a key for a negative coefficient). When we evaluate the leading three leading terms of (4) for $n = 10,000$, it gives an integer that, when divided by 66, gives a decimal value that terminates in 4242424242..., which is the source of the 42s above. The central 3 in (3) is produced by the term $66n^5$ in (4), which, when divided by 66, is just n^5, adding one to the decimal digit 2 that is normally in this position.

2.2 An anomaly in computing pi

Gregory's series, discovered in the 17th century, is arguably the most elementary infinite series formula for π, although it converges rather slowly. It can be simply derived by simply noting that

$$\frac{\pi}{4} = \int_0^1 \frac{dx}{1+x^2} \quad (6)$$

$$= \int_0^1 \left(1 - x^2 + x^4 - x^6 + x^8 - \cdots \right) dx$$

$$= 1 - 1/3 + 1/5 - 1/7 + 1/9 - 1/11 + \cdots$$

In 1988, a colleague noted that Gregory's series, when evaluated to 5,000,000 terms by computer, gives a value that differs strangely from the true value of π. Here is the truncated Gregory value and the true value of π:

$$3.1415924535897932384646433832795027841971693993873058209749418223 0\ldots$$
$$3.1415926535897932384626433832795028841971693993751058209749445923 0\ldots$$

| 2 | −2 | 10 | −122 | 2770 |

The series value differs, as one might expect from a series truncated to 5,000,000 terms, in the seventh decimal place – a "4" where there should be a "6" (namely an error of 2). But the next 13 digits are correct! Then, following another erroneous digit, the sequence is once again correct for an additional 12 digits. In fact, of the first 46 digits, only four differ from the corresponding decimal digits of π. Further, the "error" digits appear to occur in positions that have a period of 14, as shown above. Why?

A great place to start is by enlisting the help of an excellent online resource for students and research mathematicians alike: Neil Sloane's Online Encyclopedia of Integer Sequences, available at www.oeis.org. This tool has no difficulty recognizing the sequence above, namely $(2, -2, 10, -122, 2770 \cdots)$, as "Euler numbers," which are coefficients E_{2k} in Taylor's series for the secant function:

$$\sec x = \sum_{k=0}^{\infty} \frac{(-1)^k E_{2k} x^{2k}}{(2k)!}. \tag{7}$$

Indeed, this discovery, made originally through the print version of the integer sequence recognition tool more than 25 years ago, led to a formal proof that the Euler numbers are indeed the "errors" here [11, p. 50-52].

3 Experimental methods in applied math

By many measures, the record of the field of modern high-performance, applied mathematical computation is one of remarkable success. Accelerated by relentless advances of Moore's law, this technology has enabled researchers in many fields to perform computations that would have been unthinkable in earlier times. Indeed, computation is rapidly becoming a third mode of scientific discovery, after theory and laboratory work.

The progress in performance over the past few decades is truly remarkable, arguably without peer in the history of modern science and technology. For example, in the November 2014 edition of the Top 500 list of the world's most powerful supercomputer (see Figure 1), the best system performs at over 30 Pflop/s (i.e., 30 "petaflops" or 30 quadrillion floating-point operations per second), a level that exceeds the sum of the top 500 performance figures approximately ten years earlier [21]. Note also that a 2014-era Apple MacPro workstation, which features approximately 7 Tflop/s (i.e., 7 "teraflops" or 7 trillion floating-point operations per second) peak performance, is roughly on a par with the #1 system of the Top 500 list from 15 years earlier (assuming that the MacPro's Linpack performance is at least 15% of its peak performance).

Just as importantly, advances in algorithms and parallel implementation techniques have, in many cases, outstripped the advance from raw hardware advances alone. To mention but a single well-known example, the fast Fourier transform ("FFT") algorithm reduces the number of operations required to evaluate the "discrete Fourier transform," a very important and very widely employed computation (used, for example, to process signals in cell phones), from $8n^2$ arithmetic operations to just $5n \log_2 n$, where n is the total size of the dataset. For large n, the

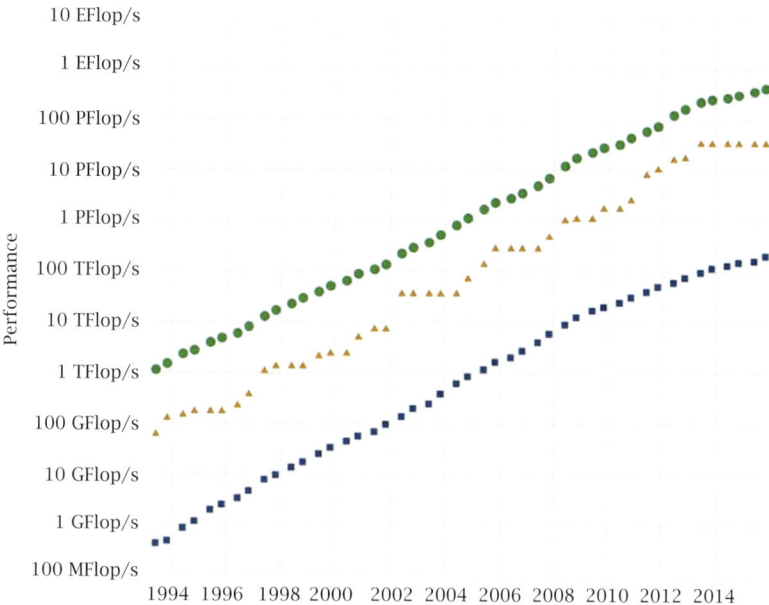

Figure 1. Performance of the Top 500 computers: Red = #1 system; orange = #500 system; blue = sum of #1 through #500

savings are enormous. For example, when n is one billion, the FFT algorithm is more than six million times more efficient.

Remarkable as these developments are, there is no indication that progress is slowing down. Moore's law, the informal rule in the semiconductor industry that the number of transistors on a chip roughly doubles every 18 months or so, continues apace, yielding a broad range of computer-based devices that continue to advance in aggregate power (even if power management concerns have limited increases in clock rate). And researchers continue to develop new and improved numerical algorithms, which, when coupled with improved hardware, will further improve the power of future systems. Thus we can look forward with some confidence to scientific computer systems in the 2025 year time frame that are roughly 100 times more power than systems available in 2015. Indeed, one's smartphone in the year 2025 may be comparable in power to the world's most powerful supercomputer in, say, 2005 or so.

In this section, we present a few relatively accessible examples of the experimental paradigm in action in applied mathematics. The intent here is certainly not to give an encyclopedic review of modern high-performance applied mathematical computing, but instead to illustrate, by a few examples (most of which require only modest computational resources), the truly exploratory nature of this work.

3.1 Gravitational boosting or "slingshot magic"

One interesting space-age example is the unexpected discovery of *gravitational boosting* by Michael Minovitch, who at the time (1961) was a student working on a summer project at the Jet Propulsion Laboratory in Pasadena, California. Minovitch found that *Hohmann transfer ellipses* were not, as then believed, the minimum-energy way to reach the outer planets. Instead, he discovered, by a combination of clever analytical derivations and heavy-duty computational experiments on IBM 7090 computers (which were the world's most powerful systems at the time), that spacecraft orbits which pass close by other planets could gain a "slingshot effect" substantial boost in speed, compensated by an extremely small change in the orbital velocity of the planet, on their way to a distant location [19]. Some of his earlier computation was not supported enthusiastically by NASA. As Minovitch later wrote,

> Prior to the innovation of gravity-propelled trajectories, it was taken for granted that the rocket engine, operating on the well-known reaction principle of Newton's Third Law of Motion, represented the basic, and for all practical purposes, the only means for propelling an interplanetary space vehicle through the Solar System.[1]

Without such a boost from Jupiter, Saturn, and Uranus, the Voyager mission would have taken more than 30 years to reach Neptune; instead, Voyager reached Neptune in only ten years. Indeed, without gravitational boosting, we would still be waiting!

A very similar type of "slingshot magic," deduced from computational simulations, was much more recently employed by the European Union's Rosetta spacecraft, which, in November 2014, orbited and then deployed a probe to land on a comet many millions of kilometers away. The spacecraft utilized gravity-boost swing-bys around the earth in 2005, 2007 and 2009, and around Mars in 2007. The spacecraft's final approach was very carefully orchestrated, starting with triangular-shaped paths and ending with an elliptical orbit tightly circling the comet, which has only a very feeble gravitational field. An animation of the Rosetta craft's path is available at www.esa.int/spaceinvideos/Videos/2014/01/Rosetta_s_orbit_around_the_comet.

Along this line, in December 2014 researchers at Princeton University and the University of Milan announced the discovery, aided by substantial computational simulations, of a new way to achieve Mars orbit, known as "ballistic capture." The idea of ballistic capture is instead of sending the spacecraft to where Mars will be in its orbit, as is done in missions to date, the spacecraft is instead sent to a spot somewhat ahead of the planet. As Mars slowly approaches the craft, it "snags" it into orbit about the planet. In this way most of the large rocket burn to slow the craft is avoided [14].

3.2   Iterative methods for protein conformation

The *method of alternating projections* (MAP), is a computational technique most often used in optimization applications. While a full mathematical treatment would require an excursion into Hilbert spaces and the like, the concept is fairly simple, to find a point in the intersect of several sets to iterate the following process: first "project" a point in a multidimensional space to its closest projection in on each of the sets (a process entirely analogous to the elementary geometry task of finding the closet point on a line to a point outside the line), and average these estimates. The *Douglas–Rachford method* (DR) "reflects," after projection using one of

---

1.   There are differing accounts of how this principle was discovered; we rely on the first-person account at www.gravityassist.com/IAF1/IAF1.pdf. Additional information on "slingshot magic" is given at www.gravityassist.com and www2.jpl.nasa.gov/basics/grav/primer.php.

Table 1. Average (maximum) errors from five replications with reflection methods of six proteins taken from a standard database

Protein	# Atoms	Rel. Error (dB)		RMSE		Max Error	
1PTQ	404	−83.6	(−83.7)	0.0200	(0.0219)	0.0802	(0.0923)
1HOE	581	−72.7	(−69.3)	0.191	(0.257)	2.88	(5.49)
1LFB	641	−47.6	(−45.3)	3.24	(3.53)	21.7	(24.0)
1PHT	988	−60.5	(−58.1)	1.03	(1.18)	12.7	(13.8)
1POA	1067	−49.3	(−48.1)	34.1	(34.3)	81.9	(87.6)
1AX8	1074	−46.7	(−43.5)	9.69	(10.36)	58.6	(62.6)

several reflection operations and then averages with the prior step. When the sets are convex, convergence is understood. In general, the sets are not convex, and yet the DR method often works amazing well (with non-convex sets) – see [12].

We illustrate the DR technique with the non-convex problem of reconstruction of protein structure (conformation) using only the short distances below about six Angstroms between atoms that can be measured by nondestructive magnetic resonance imaging (MRI) techniques (interatomic distances below 6Å typically constitute less than 8% of the total distances between atoms in a protein).

What do the reconstructions look like? We turn to graphic information for 1PTQ and 1POA, in Figure 2. These were respectively our initially most and least successful cases.

Note that the failure (and large mean or max error) is caused by a very few very large spurious distances. The remainder is near perfect.

While traditional numerical measures (relative error in decibels, root mean square error, and maximum error) of success held some information, graphics-based tools have been dramatically more helpful. It is visually obvious that this method has successfully reconstructed the protein whereas the MAP reconstruction method, shown below, has not. This difference is not evident if one compares the two methods in terms of decibel measurement (beloved of engineers). After 1000 steps or so, using the DR method, the protein shape is becoming apparent. After 2000 steps only minor detail is being fixed.

As shown in Figures 3 and 4, decibel measurement really does not discriminate this from the failure of the MAP method below which after 5000 steps has made less progress than DR after 1000. In Figures 3 and 4, we show the radical visual difference in the behavior of reflection and projection methods on IPTQ.

The first 3,000 steps of the 1PTQ reconstruction are available as a movie at http://carma.newcastle.edu.au/DRmethods/1PTQ.html.

A more robust stopping criterion. An optimized implementation suggested by the images above gave a ten-fold speed-up. This allowed for the experiment whose results are shown in Figure 5 to be performed. For less than 5,000 iterations, the error exhibits non-monotone oscillatory behavior. It then decreases sharply. Beyond this point progress is slower. This suggested that perhaps early termination was to blame, so we explored terminating when the error dropped below −100 dB.

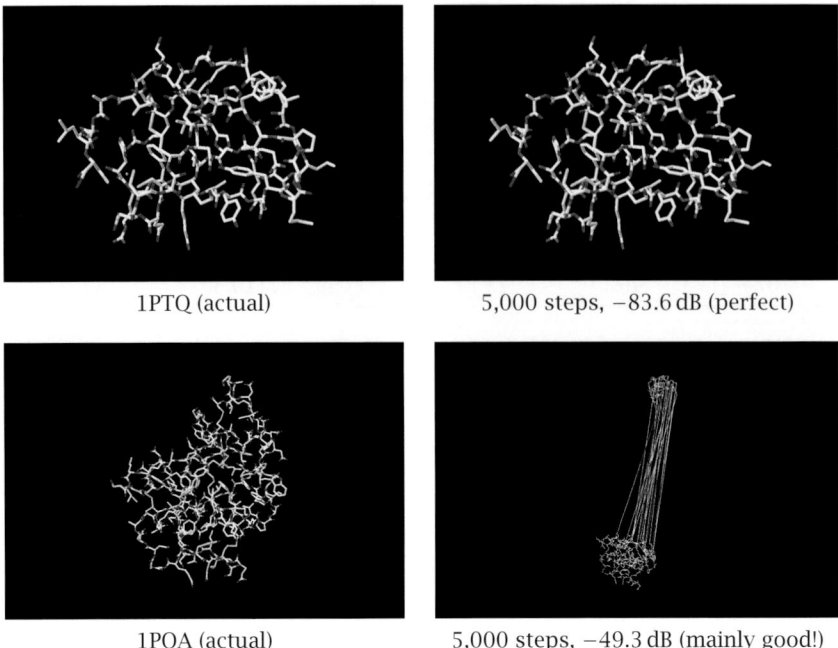

| 1PTQ (actual) | 5,000 steps, −83.6 dB (perfect) |

| 1POA (actual) | 5,000 steps, −49.3 dB (mainly good!) |

Figure 2. IPTQ and IPOA reconstructions

Similar results were observed for all the other test proteins. Nonetheless, MAP works very well for optical aberration correction (it was used to "fix" the Hubble telescope), and the method is now built in to software for some amateur telescopes.

4 New methods in mathematical research

Both of the present authors recall the time, earlier in their careers, when prominent mathematicians dismissed computation as of no relevance to mathematical research. "Real mathematicians don't compute" was the by-word. But times have changed. Nowadays it is not at all unusual for mathematicians, particularly relatively young mathematicians, to utilize computer-based tools to explore the mathematical universe, test conjectures and carry out difficult algebraic manipulations – see the list given in the introduction (Section 1).

In present-day mathematical research, the most widely used tools for experimental mathematics are the following:

- *Symbolic computing.* Symbolic computing, most often done using commercial packages such as *Maple* and *Mathematica*, is a mainstay of modern mathematical research, and is increasingly utilized in classroom instruction as well. Present-day symbolic computing tools are vastly improved over what was available even 10 years ago.
- *High-precision arithmetic.* Most work in scientific or engineering computing relies on either 32-bit IEEE floating-point arithmetic (roughly seven decimal digit precision) or 64-bit IEEE floating-point arithmetic (roughly 16 decimal digit precision). But in experimental mathemat-

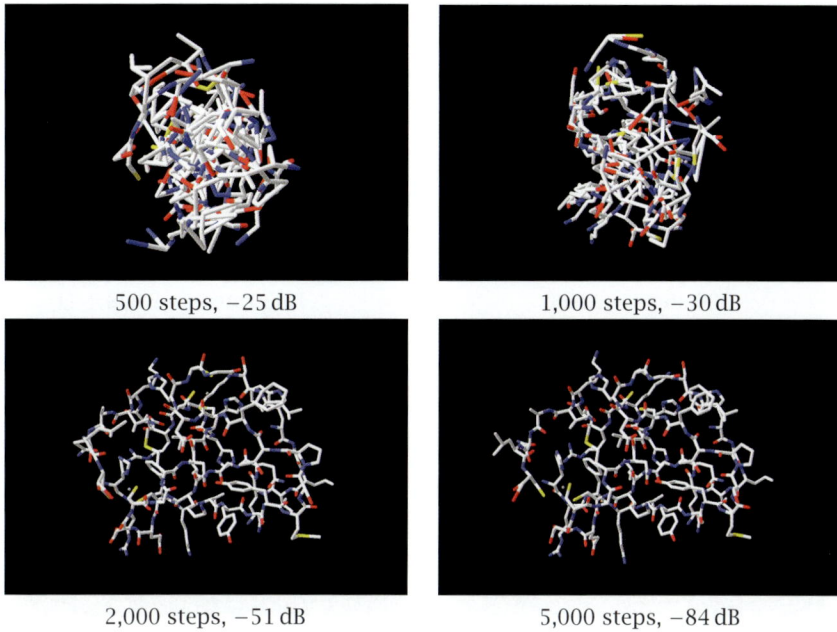

Figure 3. Decibel error by iterations for 1PTQ using DR

ics, studies often require very high precision – hundreds or thousands of digits. Fortunately software to perform such computations is widely available either in "freeware" or as a built-in feature of commercial packages such as *Maple* and *Mathematica*.

- *Integer relation detection.* Given a vector of real or complex numbers x_i, an *integer relation algorithm* attempts to find a nontrivial set of integers a_i such that $a_1x_1+a_2x_2+\cdots+a_nx_n = 0$. One common application of such an algorithm is to find new identities involving computed numeric constants. For example, suppose one suspects that some mathematical object x_1 (e.g., a definite integral, infinite series, etc.) might be given as a sum of terms $x_2, x_3, \cdots, x_n$, with unknown rational coefficients. One can then compute $x_1, x_2, \cdots, x_n$ to high precision (typically several hundred digits), and then apply an integer relation algorithm. If the computation produces an integer relation, then solving it for x_1 produces an experimental identity for the original integral, which then can be proven using conventional methods. The most commonly employed integer relation algorithm is the "PSLQ" algorithm of mathematician-sculptor Helaman Ferguson [11, 230–234]. In 2000, integer relation methods were named one of the top ten algorithms of the twentieth century by *Computing in Science and Engineering*.

4.1 The BBP formula for pi

As noted above, mathematicians have been fascinated by the mathematical constant $\pi = 3.1415926535\ldots$ since antiquity. Archimedes, in the third century BCE, was the first to provide a systematic scheme for calculating π, based on a sequence of inscribed and circumscribed

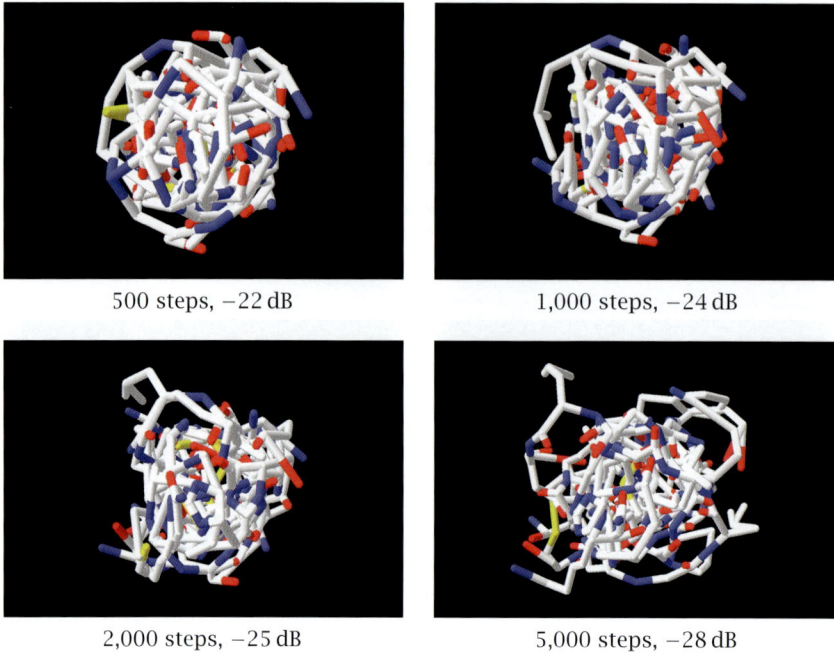

500 steps, −22 dB	1,000 steps, −24 dB
2,000 steps, −25 dB	5,000 steps, −28 dB

Figure 4. Decibel error by iterations for 1PTQ using MAP

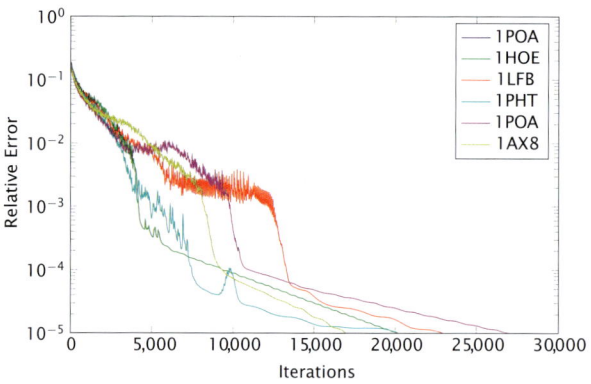

Figure 5. Relative error by iterations (vertical axis logarithmic)

polygons. The Chinese mathematician Tsu Ch'ung Chi, in roughly 480, computed seven correct digits; in 1665, Isaac Newton published 16 digits, but confessed "I am ashamed to tell you how many figures I carried these computations, having no other business at the time." These efforts culminated in the computation of π to 707 digits by William Shanks in 1873; alas, only the first 527 were correct. In the computer era, new techniques (such as FFT-based multiplication) and transistorized hardware resulted in vastly larger calculations – π was calculated to millions, then billions, and, as of October 2011, ten trillion decimal digits [4].

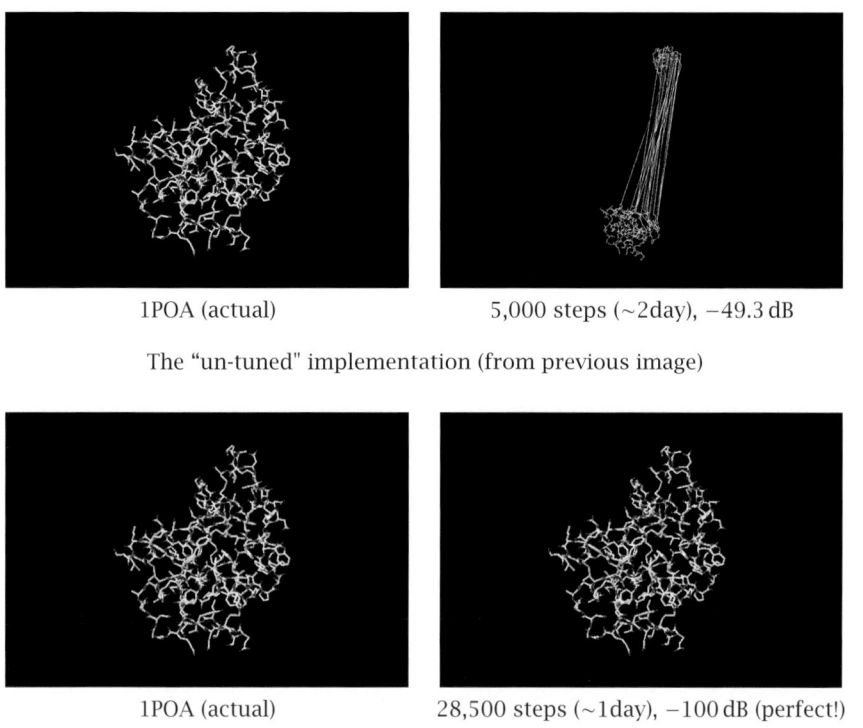

1POA (actual)   5,000 steps (~2day), −49.3 dB

The "un-tuned" implementation (from previous image)

1POA (actual)   28,500 steps (~1day), −100 dB (perfect!)

The optimised implementation

Figure 6. 1POA conformation before and after tuning

One motivation for such computations has been to see if the decimal expansion (or expansion in other number bases) of π repeats, which would suggest that π is a simple ratio of natural numbers. But in 1761, Lambert proved that π is irrational, and in 1882, Lindemann proved that π is transcendental, meaning that it is not the root of any algebraic polynomial with integer coefficients. Nonetheless, numerous questions still remain, notably the question of whether or not π is a *normal number* in a given number base. This will be discussed further in Section 4.2.

Thus it was with some interest researchers in 1996 announced the discovery of a new formula for π, together with a rather simple scheme for computing binary or hexadecimal digits of π, beginning at an arbitrary starting position, without needing to compute any of the preceding digits. The scheme is based on the following formula, now known as the "BBP formula" for π [9]:

$$\pi = \sum_{k=0}^{\infty} \frac{1}{16^k} \left(\frac{4}{8k+1} - \frac{2}{8k+4} - \frac{1}{8k+5} - \frac{1}{8k+6} \right). \qquad (8)$$

For our discussion here, perhaps the most relevant point is that *this formula was discovered using a high-precision computation (200 digits) together with the PSLQ algorithm.* Indeed, it was one of the earliest successes of what is now known as the experimental approach to mathematical research [4]. The proof of this formula (now known as the "BBP" formula for π)

is a relatively simple exercise in calculus. It is perhaps puzzling that it had not been discovered centuries before. But then no one was looking for such a formula.

The scheme to compute digits of π beginning at an arbitrary starting point is remarkably simple, but will not be given here; see [4] for details. This and similar algorithms have been implemented to compute hexadecimal digits of π beginning at stratospherically high positions. On March 14 (Pi Day), 2013, Ed Karrels of Santa Clara University announced the computation of 26 base-16 digits of π beginning at position one quadrillion [4]. His result: 8353CB3F7F0C9ACCFA9AA215F2. Here hexadecimal digits A, B, C, D, E, F denote 10, 11, 12, 13, 14, 15, respectively in base-16.

High-precision computations and PSLQ programs have been used to found numerous other "BBP-type" formulas, which permit arbitrary-digit calculation, for numerous other well-known mathematical constants. See [8] for some additional examples.

Certainly there is no need for computing π or other constants to millions or billions of digits in practical scientific or engineering work. There are certain scientific calculations that require intermediate calculations to be performed to higher than standard 16-digit precision (typically 32 or 64 digits may be required) [2], and certain computations in the field of experimental mathematics have required as high as 50,000 digits [7], but we are not aware of any "practical" applications beyond this level.

Computations of digits of π are, however, excellent tests of computer integrity—if even a single error occurs during a large computation, almost certainly the final result will be badly in error, disagreeing with a check calculation done with a different algorithm. For example, in 1986, a pair of π-calculating programs detected some obscure hardware problems in one of the original Cray-2 supercomputers. Also, some early research into efficient implementations of the fast Fourier transform on modern computer architectures had their origins in efforts to accelerate computations of π [11, p. 115].

4.2 Normal numbers

As we noted in the previous section, a long-standing unanswered question of pure mathematics (with some potential real-world applications) is whether or not π or other well-known mathematical constants are *normal*. A normal number (say in base ten) is a number whose decimal digit expansion satisfies the property that each of the ten digits $(0, 1, 2, \cdots, 9)$ appears, in the limit, $1/10$ of the time; every pair of digits, such as "27" or "83", appears, in the limit, $1/100$ of the time, and so on. Similar definitions apply for being normal base 2 or in other bases. Such questions have intrigued mathematicians for ages. But to date, no one has been able to prove (or disprove) any of these assertions – not for any well-known mathematical constant, not for any number base. For example, it is likely true that *every* irrational root of an integer polynomial (e.g., $\sqrt{2}$, $\sqrt[3]{10}$, $(1 + \sqrt{5})/2$, etc.) is normal to every integer base, but there are no proofs, even in the simplest cases.

Fortunately, modern computer technology has provided some new tools to deal with this age-old problem. One fruitful approach is to display the digits of π or other constants graphically, cast as a random walk [1]. For example, Figure 7 shows a walk based on one million base-4 pseudorandom digits, where at each step the graph moves one unit east, north, west or south, depending on the whether the pseudorandom base-4 digit at that position is $0, 1, 2$ or 3. The color indicates the path followed by the walk – shifted up the spectrum (red-orange-yellow-green-cyan-blue-purple-red) following a hue-saturation-value (HSV) model.

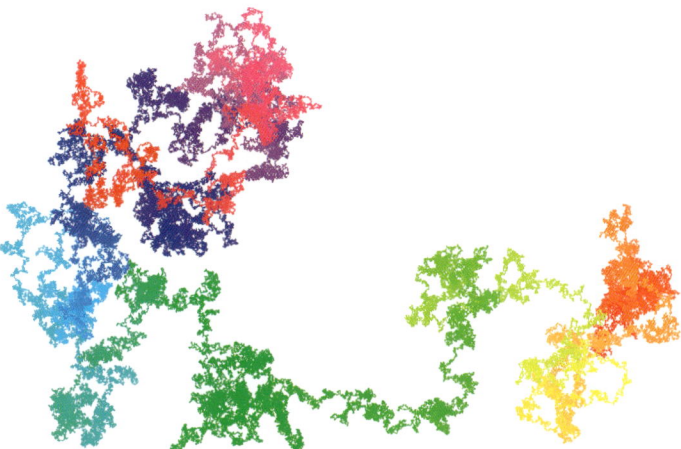

Figure 7. A uniform pseudorandom walk

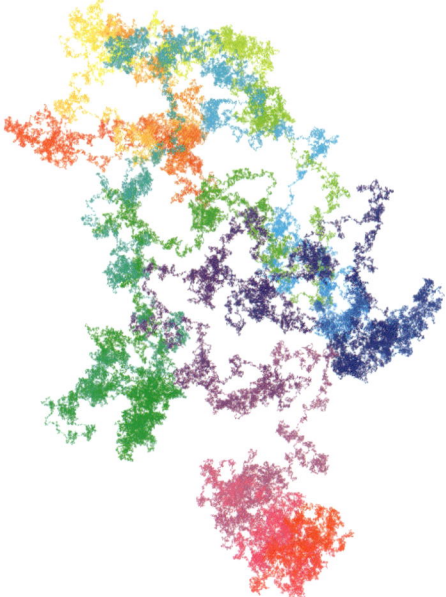

Figure 8. A walk on the first two billion base-4 digits of π

Figure 8 shows a walk on the first two billion base-4 digits of π. A hugely more detailed figure (based on 100 billion base-4 digits) is available to explore online at http://gigapan.org/gigapans/106803.

Although no rigorous inference regarding the normality of π can be drawn from these figures, it is plausible that π is 4-normal (and thus 2-normal), since the overall appearance of its graph is similar to that of the graph of the pseudorandomly generated base-4 digits.

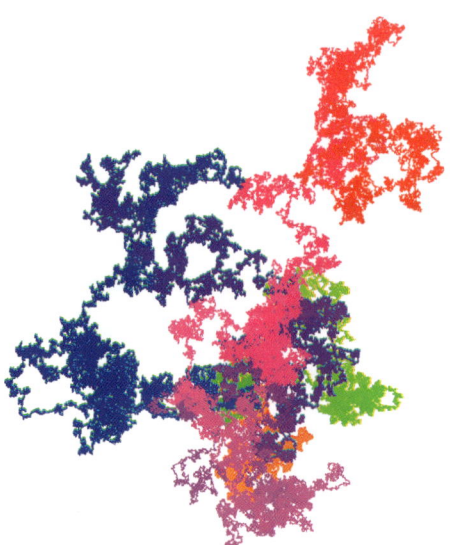

Figure 9.   A walk on the provably normal base-4 digits of Stoneham's constant

Figure 10.   A walk on the *nonnormal* base-6 digits of Stoneham's constant

This same tool can be employed to study the digits of Stoneham's constant, namely

$$\alpha_{2,3} = \frac{1}{2} + \frac{1}{3 \cdot 2^3} + \frac{1}{3^2 \cdot 2^{3^2}} + \cdots = 1.043478260869564531\ldots \tag{9}$$

This constant is one of the few that is provably normal base 2 (meaning that its binary expansion satisfies the normality property). What's more, it is provably *not* normal base 6, so that it is an explicit example of the fact that normality in one number base does not imply normality in another base [5]. For other number bases, including base 3, its normality is not yet known either way. Perhaps additional computer studies will clarify.

Figures 9 and 10 show walks generated from the base-4 and base-6 digit expansions, respectively, of $\alpha_{2,3}$. The base-4 digits are graphed using the same scheme mentioned above, with each step moving east, north, west or south according to whether the digit is $0, 1, 2$ or 3. Similarly, the base-6 graph is generated by moving at an angle of 0, 60, 120, 180, 240, or 300 degrees, respectively, according to whether the digit is 0, 1, 2, 3, 4 or 5.

From these three figures it is clear that while the base-4 graphs appear to be plausibly random (since they are similar in overall structure to Figures 7 and 8), the base-6 walk is vastly different, mostly a horizontal line. Indeed, we discovered the fact that $\alpha_{2,3}$ fails to be 6-normal by a similar empirical analysis of the base-6 digits – there are long stretches of zeroes in the base-6 expansion [5].

4.3 Ising integrals

High-precision computation and the PSLQ algorithm have been invaluable in analyses of definite integrals that arise in mathematical physics settings. For instance, consider this family of n-dimensional integrals [6]:

$$C_n = \frac{4}{n!} \int_0^\infty \cdots \int_0^\infty \frac{du_1 du_2, \cdots du_n}{\left(\sum_{j=1}^n (u_j + 1/u_j)\right)^2}. \tag{10}$$

Direct numerical computation of these integrals from (10) is very difficult, but it can be shown that

$$C_n = \frac{2^n}{n!} \int_0^\infty p K_0^n(p) \, dp, \tag{11}$$

where K_0 is the *modified Bessel function*, and in this form these integrals can be numerically computed. Indeed, it is in this form that C_n arises in quantum field theory, as we subsequently learned from David Broadhurst. Upon calculating these values, we quickly discovered that the C_n rapidly approach a limiting value. For example,

$$C_{1024} = 0.63047350337438679612204019271087890435458707 87\ldots$$

What is this number? When one copies the first 30 or 40 digits into the online Inverse Symbolic Calculator (ISC) at http://isc.carma.newcastle.edu.au, it quickly returns the result

$$\lim_{n \to \infty} C_n = 2e^{-2y}, \tag{12}$$

where y denotes Euler's constant $= 0.5772156649\ldots$ This fact was rigorously proven shortly after discovery. Numerous other results for related families of integrals have been discovered by computer in this fashion [6].

4.4 Formal verification of proof

In 1611, Kepler described the stacking of equal-sized spheres into the familiar arrangement we see for oranges in the grocery store. He asserted that this packing is the tightest possible. This assertion is now known as the Kepler conjecture, and has persisted for centuries without proof.

In 1994, Thomas Hales, now at the University of Pittsburgh, proposed a five-step program that would result in a proof: (a) treat maps that only have triangular faces; (b) show that the face-centered cubic and hexagonal-close packings are local maxima in the strong sense that they have a higher score than any Delaunay star with the same graph; (c) treat maps that contain only triangular and quadrilateral faces (except the pentagonal prism); (d) treat maps that contain something other than a triangular or quadrilateral face; and (e) treat pentagonal prisms.

In 1998, Hales announced that the program was now complete, with Samuel Ferguson (son of mathematician-sculptor Helaman Ferguson, who discovered the PSLQ algorithm) completing the crucial fifth step. This project involved extensive computation, using an interval arithmetic package, a graph generator, and *Mathematica*. The computer files containing the source code and computational results occupy more than three Gbytes of disk space. Additional details, including papers, are available at www.math.pitt.edu/~thales/kepler98. For a mixture of reasons – some more defensible than others – the *Annals of Mathematics* initially decided to publish Hales' paper with a cautionary note, but this disclaimer was deleted before final publication.

Hales subsequently embarked on a multi-year program to certify the proof by means of computer-based formal methods, a project he has named the "Flyspeck" project. This was completed in July 2014 [15]. As these computer-based formal proof techniques become better understood, we can envision a large number of mathematical results eventually being confirmed by computer, but this will take decades.

5 Why should we trust computations?

There are many possible sources of errors in large computations of the type discussed above:

- The underlying mathematical formulas and algorithms used might conceivably be in error, or the formulas may have been incorrectly transcribed.
- The computer programs implementing these algorithms, which typically employ sophisticated algorithms such as matrix operations or the fast Fourier transform (FFT), may contain bugs.
- Inadequate numeric precision may have been employed, invalidating some key steps of the algorithm.
- Erroneous programming constructs may have been employed to control parallel processing. Such errors are typically very hard to detect and rectify, since in many cases they cannot easily be replicated.
- Hardware errors may have occurred during the run, rendering all subsequent computation invalid. This was a factor in the 1986 computation of π, as noted above in Section 4.1.
- Quantum-mechanical errors, induced by stray subatomic particles, may have corrupted the results in storage registers.

So why should anyone believe any results of such calculations? The answer is that such calculations are double-checked, either by an independent calculation done using some other algorithm, or by rigorous internal checks. For instance, Kanada's 2002 computation of π to 1.3 trillion decimal digits involved first computing slightly over one trillion hexadecimal (base-16) digits, using formulas found by one of the present authors (Jonathan Borwein) and Peter Borwein. Kanada found that the 20 hex digits of π beginning at position $10^{12} + 1$ are B4466E8D21 5388C4E014.

Kanada then calculated these same 20 hexadecimal digits directly, using the "BBP" formula and algorithm, mentioned above in Section 4.1. The result of the BBP calculation was B4466E8D21 5388C4E014. Needless to say, in spite of the many potential sources of error in both computations, each of which required many hours of computation on a supercomputer, the final results dramatically agree, thus confirming in a convincing albeit heuristic sense that both results are almost certainly correct. Although one cannot rigorously assign a "probability" to this event, note that the probability that two 20-long random hexadecimal digit strings perfectly agree is one in $16^{20} \approx 1.2089 \times 10^{24}$.

This raises the following question: *Which is more securely established*, the assertion that the hexadecimal digits of π in positions $10^{12} + 1$ through $10^{12} + 20$ are B4466E8D21 5388C4E014, or the final result of some very difficult work of mathematics that required hundreds or thousands of pages, that relied on many results quoted from other sources, and that (as is frequently the case) has been read in detail by only only a relative handful of mathematicians besides the author? In our opinion, computation often trumps cerebration.

6 What will the future bring?

We have discussed numerous examples of the "experimental" paradigm of mathematics in action, both for "pure" and "applied" applications. It is clear that these experimental-computational methods are rapidly becoming central to all mathematical research, and also to mathematical education, where students can now see, hands-on, many of the principles that heretofore were only abstractions. What's more, the rapidly improving quality of mathematical software, when combined with the inexorable forward march of Moore's Law in hardware technology, means that ever-more-powerful software tools will be available in the future. Indeed, almost certainly one day we will look back on the present epoch with puzzlement, wondering how anyone ever got any serious done with such primitive tools as we use today!

It is not just *Maple* and *Mathematica* that are improving. In Section 2.2 we mentioned Neil Sloane's Online Encyclopedia of Integer Sequences (http://www.oeis.org) and the Inverse Symbolic Calculator (http://isc.carma.newcastle.edu.au. A few other valuable online resources include the Digital Library of Mathematical Functions (http://dlmf.nist.gov), a large compendium of formulas for special functions, LAPACK (http://www.netlib.org/lapack), a large library of highly optimized linear algebra programs, and SAGE (http://www.sagemath.org), an open-source mathematical computing package. We can certainly expect substantial improvements in these tools as well.

Of related interest is online collaborative efforts in mathematics, notably the PolyMath activity, which joins together a large team of mathematicians in online, computer-based collaborations to explore conjectures and prove theorems. One notable success was a dramatic lowering of the bound on prime gaps, completed in September 2014 [20].

But aside from steadily improving mathematical tools and online facilities, can we expect anything fundamentally different?

One possibility is a wide-ranging "intelligent assistant" for mathematics [10]. Some readers may recall the 2011 televised match, on the "Jeopardy!" quiz show in North America, between IBM's "Watson" computer and the two most accomplished champions of the show (Ken Jennings and Brad Rutter). This was the culmination of a multi-year project by researchers at the IBM Yorktown Heights research center, who employed state-of-the-art artificially intelligent and machine learning software that could first "understand" the clues (which are often quirky and involve puns) and then quickly produce the correct response. It has been reported that that the project cost IBM more than one billion U.S. dollars. While in their first match, the humans were competitive, in the second match Watson creamed its human opponents, and easily won the $1 million prize (which IBM donated to charity) [18]. IBM is now adapting the Watson technology for medical applications, among other things [13].

We can thus envision an enhanced Watson-class intelligent system for mathematics. Such a system would not only incorporate powerful software to "understand" and respond to natural-language queries, but it would also acquire and absorb the entire existing corpus of published mathematics for its database. One way or another, it is clear that future advances in mathematics will be intertwined with advances in artificial intelligence. Along this line, Eric Horvitz, Managing Director of the Microsoft Research Lab in Redmond, Washington, has launched a project to track the advance of artificial intelligence over the next 100 years, with progress reports every five years [22].

So will computers ever completely replace human mathematicians? Probably not, according to the 2014 Breakthrough Prize in Mathematics recipient Terence Tao – even 100 years from now, much of mathematics will continue to be done with humans working in partnership with computers [3].

But, as we mentioned above, more is at stake than merely accelerating the pace at which mathematical researchers, teachers and students can do their work. The very notion of what constitutes *secure mathematical knowledge* may be changing, before our eyes. It will be interesting in any event. We look forward to what the future will bring.

References

[1] F. J. ARAGON ARTACHO, D. H. BAILEY, J. M. BORWEIN and P. B. BORWEIN, Walking on real numbers, *Mathematical Intelligencer*, vol. 35 (2013), 42–60.

[2] D. H. BAILEY, R. BARRIO and J. M. BORWEIN, High-precision computation: Mathematical physics and dynamics, *Appl. Math. and Comp.*, vol. 218 (2012), 10106–10121.

[3] D. H. BAILEY and J. M. BORWEIN, Big bucks for big breakthroughs: Prize recipients give three million dollar maths talks, Nov. 11, 2014, http://experimentalmath.info/blog/2014/11/breakthrough-prize-recipients-give-math-seminar-talks.

[4] D. H. BAILEY and J. M. BORWEIN, Pi day is upon us again, and we still do not know if pi is normal, *American Mathematical Monthly* (March 2014), 191–206.

[5] D. H. BAILEY and J. M. BORWEIN, Nonnormality of Stoneham constants, *Ramanujan Journal*, vol. 29 (2012), 409–422; DOI 10.1007/s11139-012-9417-3.

[6] D. H. BAILEY, J. M. BORWEIN and R. E. CRANDALL, Integrals of the Ising class, *Journal of Physics A: Mathematical and General*, vol. 39 (2006), 12271–12302.

[7] D. H. BAILEY, J. M. BORWEIN, R. E. CRANDALL and J. ZUCKER, Lattice sums arising from the Poisson equation, *Journal of Physics A: Mathematical and Theoretical*, vol. 46 (2013), 115201.

[8] D. H. BAILEY, J. M. BORWEIN, A. MATTINGLY AND G. WIGHTWICK, The Computation of Previously Inaccessible Digits of π^2 and Catalan's Constant, *Notices of the American Mathematical Society*, vol. 60, no. 7 (2013), 844–854.

[9] D. H. BAILEY, P. B. BORWEIN and S. PLOUFFE, On the Rapid Computation of Various Polylogarithmic Constants, *Mathematics of Computation*, vol. 66, no. 218 (Apr 1997), 903–913.

[10] J. M. BORWEIN, The future of mathematics: 1965 to 2065, *MAA Centenary Volume*, S. Kennedy, et. al editors, Mathematical Association of America, 2015.

[11] J. M. BORWEIN and D. H. BAILEY, *Mathematics by Experiment: Plausible Reasoning in the 21st Century*, A K Peters, second edition, 2008.

[12] J. M. BORWEIN and M. K. TAM, Reflection methods for inverse problems with applications to protein conformation determination." Springer volume on *Generalized Nash Equilibrium Problems, Bilevel programming and MPEC* . In Press, August 2014.

[13] A. E. CHA, Watson's next feat? Taking on cancer, *Washington Post* (June 27, 2015), www.washingtonpost.com/sf/national/2015/06/27/watsons-next-feat-taking-on-cancer/.

[14] A. HADHAZY, A new way to reach Mars safely, anytime and on the cheap, *Scientific American* (Dec. 22, 2014), www.scientificamerican.com/article/a-new-way-to-reach-mars-safely-anytime-and-on-the-cheap.

[15] T. C. HALES, Announcement of completion, (Aug. 2014), https://code.google.com/p/flyspeck/wiki/AnnouncingCompletion.

[16] T. C. HALES, Formal proof, *Notices of the American Mathematical Society*, vol. 55, no. 11 (Dec. 2008), 1370–1380.

[17] G. HANNA and M. DE VILLIERS (Eds.), *On Proof and Proving in Mathematics Education*, the 19th ICMI Study, New ICMI Study Series, 15, Springer-Verlag, 2012.

[18] J. Markoff, Computer wins of 'Jeopardy!': Trivial it's not, *New York Times* (Feb. 16, 2011), www.nytimes.com/2011/02/17/science/17jeopardy-watson.html.

[19] M. MINOVITCH, A method for determining interplanetary free-fall reconnaissance trajectories, *JPL Tech. Memo TM-312-130*, (23 Aug 1961), 38–44.

[20] D. H. J. POLYMATH, The 'bounded gaps between primes' Polymath project – a retrospective. (Sep. 30, 2014), http://arxiv.org/abs/1409.8361.

[21] The Top 500 list, available at www.top500.org.

[22] JIA YOU, A 100-year study of artificial intelligence? Microsoft Research's Eric Horvitz explains, *Science* (Jan. 9, 2015), http://news.sciencemag.org/people-events/2015/01/100-year-study-artificial-intelligence-microsoft-research-s-eric-horvitz.

What is the impact of interactive mathematical experiments?

Albrecht Beutelspacher

In this article we look at mathematical experiments, in particular those shown in mathematical science centers. Although some mathematicians have the sneaking suspicion that such experiments are far too superficial and do not correspond to proper mathematics, we try to show that in fact mathematical experiments provide an ideal first step for the general public into mathematics.

Prehistory of mathematical experiments

Mathematical experiments are part of a long history. From the beginning of mathematics, an important aspect was to deal with the outside world. This covers cosmology (that is explaining and predicting the movements of the celestial bodies), use of mathematical instruments, applications of all kinds, and experiments.

Lots of stimuli for the development of mathematics came from real world problems. This goes back at least to mathematics in Mesopotamia and Egypt, where the rules of movement of the sun, the moon and the stars was of utmost importance.

A step towards experiments is mathematical instruments, for instance rulers, compasses, proportional rulers, calculating devices (such as Abacus, Napier sticks, and slide rules). These instruments are used to achieve an objective goal: construct a circle, solve an equation, and multiply two numbers. Although these instruments are based on structures and ideas (such as decimal system, intercept theorems, logarithms), the typical user is unaware of the mathematical background.

Experiments are man-made artefacts, which also offer real experience. Probably the first such experiments are due to the time of Galilei (for instance experiments with pendula). In mathematics, models and instruments became important in the 19th century. The book of Walter Dyck (1892) shows an impressive collection of mathematical models, apparatuses and instruments.

The idea of displaying experiments to the general public in order to get in touch with science goes back to the foundation of the Deutsches Museum in Munich in 1906. The inauguration of the Exploratorium in San Francisco in 1969 is usually considered as the birthday of modern science centers. The term "science center" was used since the content was science, mainly physics. Today science centers define themselves no longer by their contents, in fact there are science centers on nearly any subject. They see themselves as institutions, which bring science and humanities to a broad auditorium by displaying interactive exhibits, which enable the visitors to enter the subject by experience.

Mathematical experiments have been present in many science centers, mostly in a disguised way: For instance, the "Tower of Hanoi" was shown, symmetry experiments using mirrors always have been popular, and, of course, every center had its soap bubble experiments.

The first initiative to collect and develop mathematical experiments as such was undertaken by the Italian professors Franco Conti and Enrico Giusti, who very successfully developed and organized the exhibition "Oltre iI compasso – the mathematics of curves", which was first shown in 1992. Since 2004 it has been enlarged to form the "Giardino di Archimede" in Florence. Nearly at the same time, the first step towards the Mathematikum was taken: In 1994 the first German exhibition under the name "hands-on mathematics" ("Mathematik zum Anfassen") was shown in Giessen, Germany. This exhibition was a work of a group of students, who organized this exhibition as a follow-up of a mathematical seminar. Mathematikum, the world's first mathematical science center was opened in 2002. Since then quite a few institutions of different size followed these ideas, for instance "Adventure Land Mathematics" in Dresden, "Momath" in New York, and "Inspirata" in Leipzig.

The idea of all these institutions is primarily to use interactive exhibits – in contrast to texts, pictures or computer experiments – to popularize mathematics.

Mathematikum

The Mathematikum in Giessen, Germany (near Frankfurt) is a mathematical science centre founded in 2002. It aims to make mathematics accessible to as many people as possible, in particular to young people. On its 1200 square meters exhibition area it shows more than 170 interactive exhibits. From the very beginning, it was a great success. Between 120 000 and 150 000 people visit the Mathematikum each year. About 40 % are group visitors, mainly school classes, 60 % are private visitors, mainly families.

Visitors like the Mathematikum. In particular they like the way mathematics is presented there. They are entertained by performing the experiments and trying to understand what they experienced. The Mathematikum is a house full of communication. When one listens to what people are talking about, one notices that it is always about the exhibits.

The permanent exhibition of Mathematikum is complemented by several other formats, which address different target groups.

- *Temporary exhibitions* on special topics, such as randomness, calculating devices, mathematics in everyday life, mathematical games, etc.
- *Popular lectures* on special topics such as cryptography, astronomy, etc.
- *Lectures for children* on topics as, for instance, mathematics and – the bicycle, the bees, the heaven, the kitchen, the Christmas tree, and so on
- *Workshops*, which concentrate on special aspects of an exhibition. Etc.

What is a mathematical experiment?

An easier question is: What is a mathematical experiment not?

- An experiment has not the role of an example in a textbook, whose aim is to make an abstract definition or argument concrete.
- It also has not the role of examples in research where examples can lead to conjectures.
- Also, an experiment in mathematics is quite different to an experiment in science. In physics, chemistry, or biology, an experiment serves to confirm or verify a theory or, to be on the safe side, to falsify a wrong theory. In mathematics however, an experiment will never substitute a proof.

The basic property of a mathematical experiment is to stimulate thinking. Basically, a person working with a mathematical experiment is challenged by a mathematical problem. As in research, one has to get the right conception, the right idea of what's going on. And sometimes, after a while of thinking, and sometimes with luck, one finds the solution.

As an example we take the "Conway cube", an invention of John H. Conway. It is a puzzle consisting of three small cubes of side length 1 and six $2\times2\times1$-cuboids, which should be assembled to form a cube. One first calculates how big the cube will be. Even with this knowledge, most people struggle – until they get the idea where to locate the small cubes in the big cube.

A big advantage of such experiments is the fact that the solution is beyond any doubt, because it is materialized: the cube is there, the bridge is stable, the pattern is correct.

Developing a mathematical experiment is not an easy task. It is even more challenging, when we think of a museum or a science center, in which the experiments are openly displayed and normally no teacher or expert is present.

In Mathematikum, when a new experiment is developed and planned, we usually observe the following points.

Conway cube (Photo: Mathematikum Gießen)

Simplicity. All Mathematikum experiments look technically very simple. The reason is that a visitor should immediately see what she or he has to do. When an experiment needs too much technical background, the visitor's interest is more directed to the complicated background structure than to the core of the experiment.

Directness. The visitor should come as quickly as possible to the heart of the mathematical phenomenon. Therefore, the experiments in Mathematikum are not contextualized. (For instance, there is no kitchen, in which one could experience several mathematical forms, patterns and properties.) Also, there is no "motivation" such as fog, lights, strange colors and so on. If we would do so, we would make the visitor feel that these superficial attractors are the main objective and not the mathematics. On the contrary: We are convinced that people visit the Mathematikum because they want to do (experimental) mathematics.

Surprise. An experiment should have a very easy beginning. A visitor should be very confident that she or he will successfully perform the first step of the experiment. A good experiment, however, after a while, shows a surprise, a difficulty, or a stumbling block, in short: something which makes you think. If one encounters this difficulty, one tries once more, one starts to ask questions, one automatically starts a conversation with some other visitor – and finally one gets and sees the solution.

Success. Each experiment has a "positive result". In mathematics, there are also nonexistence theorems (for instance, "there exists no covering of the chess board, from which two opposite corners have been removed, by domino pieces"), but these kinds of theorems do not have a positive effect when dealing with experiments. At having successfully performed a "positive" experiment, the visitor can see that she or he has achieved the goal. Also, either the result or the process in getting the result provides insight in the mathematical problem. In fact, the visitor does mathematics, usually without being aware of it.

Interactivity. By an "interactive experiment" we mean a mostly physical experiment, the visitor has to work with. In other words, the experiment changes itself, at the end it looks different than at the beginning. A visitor is really working with the exhibit. She or he puts pieces together, tries several possibilities, experiences failures and so on.

Design. In Mathematikum one could say that the experiments have no design. Of course this is not true. There is a very clear concept how our exhibits should look like. In particular there is a clear color concept: Colors as red, blue, yellow are used only for the important pieces, mostly for pieces, the visitor has to work with. All other pieces are neutral (wood, black metal).

Labels. Each experiment in Mathematikum has a label. It shows (in German and English) the name of the experiment, a short description how to use it, sometimes hints for a solution, and credits, where the idea comes from. Intentionally, there is no mathematical background on the label. Again, this is due to the learning model. The visitors act as researchers: they do not know, how difficult the problem is, and they try to get the solutions by themselves.

Some experiments

Mathematikum shows more than 170 exhibits, which cover many mathematical disciplines, such as geometry (shapes and patterns), arithmetic (numbers and calculating), calculus (functions), probability (randomness and statistics), algorithms, and history of mathematics. Also, some exhibits open a window to other subjects, such as physics, biology, or music.

In the following we describe three exhibits. They do not belong to the "Wow exhibits", but visitors playing with them get the chance of good insight.

Tetrahedron in the cube.　This experiment deals with the simplest platonic solids and their relation. The task is to put a relatively large tetrahedron into a cube of glass with the top face removed. First tries fail. When one describes the failures, one could say "It doesn't work when a vertex is inserted first", also "it does not work when a face is inserted in the cube first". These words may suggest that one could try with an edge first – and it works like a miracle.

Satisfying as this experiment is, it has a great potential to go further on. We could start looking at numbers: The cube has 6 faces, the tetrahedron has 6 edges. The experiment shows that these numbers must be equal, since at each face of the cube there is an edge of the tetrahedron.

Secondly we look at the number of possibilities to insert the tetrahedron into the cube. Let us suppose that the cube is fixed and that the faces and edges of the tetrahedron are undistinguishable. Then there are exactly two ways to put the tetrahedron into the cube. If we imagine the two positions simultaneously, one could ask: What is the union of the two tetrahedral and what is their intersection. It turns out that the union is the stella octangula ("Kepler's star") and the intersection is an octahedron.

Finally, one can ask: Which ratio of the volume of the cube is covered by the tetrahedron? For this, it is useful to calculate the rest of the volume. This is composed of four pyramids, each of which has as its base one half of a face of the cube and as its height one edge of the cube. From this it follows easily that the volume of each corner pyramid is $1/6$ of the volume of the cube. Therefore, the tetrahedron has $1/3$ of the volume of the cube.

Tetrahedron in the cube (Photo: Mathematikum Gießen/Rolf K. Wegst)

Who sticks out the farthest? (Photo: Mathematikum Gießen/Rolf K. Wegst)

Who sticks out the farthest? On a table there are a number of wooden squares all of the same size and the same thickness. The task is to stack the pieces so that one of the pieces is completely beyond the edge of the table. In doing this all pieces should remain horizontal.

This is a particularly nice exhibit. Firstly there is an "obvious" mathematical solution: Stack all the pieces vertically to form a "tower" so that the lowest piece has an edge in common with the table. Now, push the top piece out as much as you can. Then push the second piece out as much as possible. Continuing this process, the top piece is clear of the table by a distance of $\frac{1}{2} + \frac{1}{4} + \frac{1}{6} + \frac{1}{8} + \ldots$, and with only four pieces you can achieve the goal. Here, one can recognize the harmonic series. The divergence of this series means that one can go as far as one wants off the table - provided that one has enough pieces.

Secondly there are many other, sometimes ingenious and good solutions. They lack the property that top piece sticks out the farthest.

Finally, as far as I know, the world record is unknown (how far can you make them stick out?) for, say, four quadratic pieces. (For a discussion of this problem see Paterson et al. 2009).

The queue of dice. This exhibit has a few rules, but at the end there is a real surprise – based on mathematical principles.

0. As a preparatory step throw about 40 dice and then arrange them in a "queue", so that the dice are in order.

1. Look at the first dice. It shows a number, for instance a "2". This means that you move two dice forward, arriving at the third dice. This dice shows a number, for instance a "4". You will move forward 4 and arrive at a new dice. Again, read the number and go forward the corresponding number of steps.

   Repeat this procedure to the end of the queue. Most likely you will not finish at the last dice. For instance, it could happen that you are at a dice showing a "4", but there are only two more dice left. Simply remove the remaining dice.

   In this way you ensure that the game ends at the last dice. So far there is no surprise.

The queue of dice (Photo: Mathematikum Gießen/Rolf K. Wegst)

2. For the second round, you throw the first dice once more. Now it shows, for instance, a "3".
   Follow the same rule as in the first round, which means: on arrival at a dice read its number
   and move forward the corresponding number of steps. Continue until the end of the queue
   – and now you will very likely arrive at the last dice.

You can repeat the experiment and you will notice that it is very "robust". Even if you miscount
or you don't start at the beginning of the queue – you will end at the last dice.

   In order to understand why this happens, look back at the first round. Mark the dice you
have stopped at. If, in the second round, you stop – by accident – at one of these dice, the
following steps are as in the first round. So, you will end at the last dice.

   This experiment shows two characteristic features of a random process: Firstly, the more,
the better. The probability of succeeding increases with the number of dice. Secondly there
are always exceptions: Imagine a queue, where all dice show a "6". This queue won't work,
regardless of length.

Effects and impact on the visitors

The main effect of all science centers is *experience.* Visitors experience real phenomena. This
is also what visitors like. It is not a virtual experience, which we have when we work with
computer programs. When we feel real physical objects and work with them, it is clear that we
cannot be cheated.

   Mathematical experiments stimulate *thinking.* One has to consider several possibilities, one
has to develop the right idea for a solution and one verifies whether a solution is correct.

   The unquestionable experience of many years of Mathematikum is that dealing with mathematical experiments makes the visitors happy. They become happy because they have *understood* something, which is very satisfying.

The fact that the people's brain is activated can be seen – or heard – by the noise in the exhibition. Sometimes it is really loud. But in fact it is *communication*. People talk to each other, ask questions, give advice – and enjoy the common solution.

Working with mathematical experiments goes far beyond "learning mathematics". It *empowers* people: When visitors see that they have achieved something by thinking by themselves, they become more self-confident.

To sum up, working with mathematical experiments is *a first step into mathematics.* This statement has two sides. Firstly, it is a step into mathematics. In fact, the problems posed by the experiments can only be solved by thorough thinking, by carefully observing, by looking for the right idea. On the other hand, dealing with experiments provides only a first step into mathematics. Many more steps could follow. In particular, in this context, there is no formal description of the mathematical phenomena.

In other words, mathematical experiments offer extremely good possibilities to "do" mathematics, but have also clear limitations: They stimulate enthusiasm and true motivation, but also they neither give formal arguments nor can replace a curse in a mathematical subject. *Course*

The picture of mathematics

Through experiments, mathematics shows a very *friendly face.* It is mathematics for everybody. The friendliness is also due to the fact that the visitors are considered as autonomous, self-responsible humans. They are not forced to visit Mathematikum or a similar institution. On the contrary, they come voluntarily and in a way they "do" mathematics. They may stop at every exhibit they want to. They may deal as deeply (or superficially) as they want with a particular exhibit. They are supported in understanding the phenomena primarily by carefully designed exhibits, and also by explainers and a catalogue (Beutelspacher 2015).

An exhibition with mathematical experiments shows the *power of the experiments.* People deal with them without a teacher, without being forced to. In school one often complains that an enthusiastic teacher is necessary and that there are so few of them. Mathematical experiments are a way to bring the fascination of mathematics to the people without the need of an enthusiastic teacher!

These two arguments make clear again that mathematical science centres aim at and succeed in changing the attitude towards mathematics of the general public. This is far beyond formal learning.

In Mathematikum, about half of the experiments have a close connection to school mathematics. This is useful and very attractive for school visits. The other half has a mathematical background which is nearly never taught at school. These experiments deal with minimal surfaces, optimization, prime numbers, codes, and so on. So, an exhibition like Mathematikum provides a much *richer picture of mathematics* than school. (Clearly, in school things are studied in more depth.)

A final point: If mathematics is interesting, then it is also interesting outside school. In mathematical science centers as the Mathematikum, mathematics is part of the visitor's leisure time. Adult people and whole families spend hours to experience the power of mathematics. Many, many times I have heard adult people sigh at me: "If I had a mathematical education with these experiments, I would have loved mathematics." Thus, mathematical experiments and mathematical science centers have a great impact on a *mathematical education of the general public.*

The future of math science centers and exhibitions

In recent years, the number of mathematical science centers seems to grow. Apart from Momath in News York, also the "Museu de Matèmatiques de Catalunya" in Cornellà, Spain, and the "Maison de Maths" in Mons, Belgium, opened. In addition to that, many museums and science centers established a complete section devoted to mathematics based on interactive experiments: Technorama, Swizzerland, Rami Koc-Museum Istanbul, Turkey, Phaenomenta Flensburg, Germany, and Deutsches Museum Munich, Germany.

It is worth mentioning that the German Goethe Institute has a math exhibition "Mathematik zum Anfassen" ("hands-on mathematics"), which travels since 2013 through the world's large cities and has everywhere a tremendous success.

In fact, if one visits any of these exhibitions and observes the visitors, one is convinced that "something happens in the brains of the visitors". One "feels" that the visitors leave the house happier than they entered it.

But this subjective impression is very hard to verify in an objective way. In fact, there is nearly no empirical study which looks at these effects on the visitors in a detailed way. Here, empirical research is very desirable. Key questions could be:

- How changes the attitude of the visitors during a visit? Does their knowledge increase?
- What about the communication between the visitors? Do they use a kind of mathematical arguments to solve the problems?
- Which properties of an exhibit stimulates mathematical thinking? Which properties impede it?

References

[1] A. BEUTELSPACHER, Lessons which can be learned from the Mathematikum. In: *Raising Public Awareness of Mathematics* (E. Behrends, Nuno Crato, José Francisco Rodrigues, Eds.) Springer-Verlag 2012, 101–108.

[2] A. BEUTELSPACHER, *Wie man in eine Seifenblase steigt. Die Welt der Mathematik in 100 Experimenten.* C.H. Beck 2015.

[3] W. DYCK, *Katalog mathematischer und mathematisch-physikalischer Modelle, Apparate und Instrumente.* Nachdruck 1994 Olms-Verlag.

[4] M. PATERSON, Y. PERES, M. THORUP, P. WINKLER and U. ZWICK, Maximum overhang. *Amer. Math. Monthly* 116 (2009), 763–787.

Mathematics and finance

Walter Schachermayer

This article consists of two parts. The first briefly discusses the history and the basic ideas of option pricing. Based on this background, in the second part we critically analyze the role of academic research in Mathematical Finance relating to the emergence of the 2007–2008 financial crisis.

1 Introduction

Mathematical Finance serves as a prime example of a flourishing application of mathematical theory. It became an important tool for several tasks in the financial industry and this "mathematization" of the financial business seems to be irreversible. Therefore in many curricula of mathematics departments, but also in business schools, mathematical finance is now regularly taught.

In this survey we want to summarize how these ideas developed, starting from the seminal work of Louis Bachelier [2] who defended his thesis "Théorie de la spéculation" in 1900 in Paris. Henri Poincaré was a member of the jury and wrote a very positive report. Bachelier used probabilistic arguments, thus introducing Brownian motion for the first time as a mathematical model, in order to develop a rational theory of option pricing.

This theme subsequently remained dormant for almost 70 years until it was taken up again by the eminent economist Paul Samuelson. In the sequel Fischer Black, Robert Merton, and Myron Scholes applied a slightly modified version of Bachelier's model and the resulting "Black-Scholes formula" for the price of a European option quickly became very influential in the world of finance. We shall sketch this development.

In the second part we want to give a critical view of the success or failure of these mathematical insights in the real world. We shall argue that the probabilistic approach turned out to be highly successful with respect to the original goal of pricing options on a liquidly traded risky asset, e.g. a share of a large company. On the other hand, the probabilistic approach was subsequently applied to many other tasks, such as credit risk, risk management, "real options" etc. We shall analyze to which extent mathematical models were involved in the financial crisis of 2007–2008. It is sometimes claimed in the public discourse that "nobody warned about the misuse of mathematical models". We shall see that such claims are not justified.

2 Louis Bachelier and Black–Scholes

We outline the remarkable work of L. Bachelier (1870-1946) by following the more extensive presentation [18] which I gave at the summer school 2000 in St. Flour.

It is important to note that the young Louis Bachelier did not attend any of the grandes écoles in Paris, apparently for economic reasons. In order to make a living he worked as a subordinate clerk at the Bourse de Paris where he was exposed on a daily basis to the erratic movements of prices of financial securities.

L. Bachelier was interested in designing a rational theory for the prices of term contracts. The two forms which were traded at the Bourse de Paris at that time also play a basic role today: forward contracts and options. We shall focus on the mathematically more interesting of these two derivatives, namely options.

DEFINITION 1. *A European call (resp. put) option on an underlying security S consists of the* right (but not the obligation) *to buy (resp. to sell) a fixed quantity of the underlying security S, at a fixed price K and a fixed time T in the future.*

The underlying security S, usually called the *stock,* can be a share of a company, a foreign currency, gold etc. In the case of Bachelier the underlying securities were *"rentes",* a form of perpetual bonds which were very common in France in the nineteenth century (compare [18]). The nominal value was 100 francs and it would pay 3 francs of interest every year. But the nominal capital was never paid. While the specifics of these assets are not relevant, it is worthwhile to note the following features (the terminology below will be explained later):

- the underlying asset S (the "rentes" in the concrete case of Bachelier) were liquidly traded.
- the value of the asset would typically not deviate too much from its nominal value of 100 francs.

In addition, they had the following properties:

- low volatility of the underlying asset;
- short term to maturity of the option (maximum: 2 months);
- approximately "at the money" options.

We mention these features explicitly as it is important in many applications to keep in mind for which purposes a mathematical model was originally intended, in particular, if it is later also applied to quite different situations.

Fixing the letter K for the strike price of the option, one arrives – after a moment's reflection – at the usual "hockey-stick" shape for the pay-off function of a call option at time T. We draw the value of the option as a function of the price S_T of the underlying asset S at time T.

Let $\hat{C}$ denote the upcounted (from time $t = 0$ to time $t = T$) price C of the option. We shall not elaborate on the rather boring aspects of upcounting and discounting and assume that the riskless rate of interest equals zero so that $C = \hat{C}$.

The graph displayed in Figure 1 appears explicitly in Bachelier's thesis. It gives the profit or loss of the option at time $t = T$ when we shall know the price S_T of the underlying asset S. But we have to determine the price C of the option which we have to pay at time $t = 0$. We note in passing that the special form of the above payoff function is not really relevant. Its only crucial feature is that it is not linear.

Louis Bachelier now passes to the central topic, *Probabilities in Operations on the Exchange.* Somewhat ironically, he had already addressed the basic difficulty of introducing probability

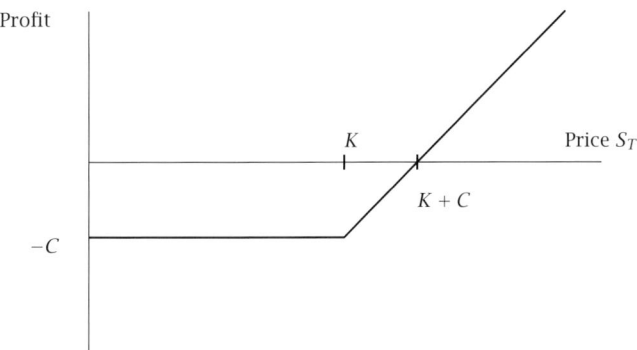

Figure 1. Pay-off function of a call option at time T

in the context of the stock exchange in the introduction to the thesis in a very sceptical way:

> The calculus of probabilities, doubtless, could never be applied to fluctuations in security quotations, and the dynamics of the Exchange will never be an exact science.

Nevertheless he now proceeds to model the price process of securities by a probability distribution distinguishing "two kinds of probabilities":

1. The probability which might be called "mathematical", which can be determined *a priori* and which is studied in games of chance.
2. The probability dependent on future events and, consequently impossible to predict in a mathematical manner.

> This last is the probability that the speculator tries to predict.

My personal interpretation of this – somewhat confusing – definition is the following: sitting daily at the stock exchange and watching the movement of prices, Bachelier got the same impression that we get today when observing price movements in financial markets, e.g., on the internet. The development of the charts of prices of stocks, indices etc. on the screen or on the blackboard resembles a "game of chance". On the other hand, the second kind of probability seems to refer to the expectations of a speculator who has a personal opinion on the development of prices. Bachelier continues:

> His (the speculator's) inductions are absolutely personal, since his counterpart in a transaction necessarily has the opposite opinion.

This insight leads Bachelier to the remarkable conclusion, which in today's terminology is called the "efficient market hypothesis":

> It seems that the market, the aggregate of speculators, at a given instant can believe in neither a market rise nor a market fall since, for each quoted price, there are as many buyers as sellers.

He then makes clear that this principle should be understood in terms of "true prices", i.e., discounted prices. Finally he ends up with his famous dictum:

> In sum, the consideration of true prices permits the statement *of this fundamental* principle: **The mathematical expectation of the speculator is zero.**

This is a truly fundamental principle and the reader's admiration for Bachelier's pathbreaking work will increase even more when continuing to the subsequent paragraph of Bachelier's thesis:

> It is necessary to evaluate the generality of this principle carefully: It means that the market, at a given instant, considers not only currently negotiable transactions, but even those which will be based on a subsequent fluctuation in prices as having a zero expectation.

> For example, I buy a bond with the intention of selling it when it will have appreciated by 50 centimes. The expectation of this complex transaction is zero exactly as if I intended to sell my bond on the liquiditation date, or at any time whatever.

In my opinion, in these two paragraphs, the basic ideas underlying the concepts of martingales, stopping times, trading strategies, and Doob's optional sampling theorem already appear in a very intuitive way. It also sets the basic theme of the modern approach to option pricing which is based on the notion of a martingale.

Let us look at the implications of the *fundamental principle*: In order to draw conclusions from it, Bachelier had to determine the probability distribution of the random variable S_T (the price of the underlying security at expiration time T) or, more generally, of the entire stochastic process $(S_t)_{0 \leq t \leq T}$. It is important to note that Bachelier had the approach of considering this object as a *process*, i.e., by thinking of the *pathwise behavior* of the random trajectories $(S_t(\omega))_{0 \leq t \leq T}$; this was very natural for him, as he was constantly exposed to observing the behavior of the prices, as t "varies in continuous time".

To determine the law of the process S, Bachelier assumes that, for $0 \leq t \leq T$, the probability $p_{x,t} dx$, that the price S of the underlying security, starting at time t_0 from S_{t_0}, lies at time $t_0 + t$ in the infinitesimal interval $(S_{t_0} + x, S_{t_0} + x + dx)$ is *symmetric around $x = 0$ and homogeneous in time t_0 as well as in space*.

Bachelier notices that this creates a problem, as it gives positive probabilities to negative values of the underlying security, which is absurd. But one should keep in mind the proportions mentioned above: a typical yearly standard deviation σ of the prices of the underlying stock S considered by L. Bachelier was of the order of 2.4%. Hence the region where the bond price becomes negative after one year is roughly 40 standard deviations away from the mean; anticipating that Bachelier uses the normal distribution, this effect is – in his words – "considered completely negligible". This was certainly justified as the horizons for the options were only fractions of a year. On the other hand, we should be warned when considering Bachelier's results asymptotically for $t \to \infty$ (or $\sigma \to \infty$ which roughly amounts to the same), as in these circumstances the effect of assigning positive probabilities to negative values of S_t is not "completely negligible" any more. But this was not Bachelier's concern. As J.M. Keynes phrased so nicely: in the long run we all are dead.

After these specifications, Bachelier argues that *"by the principle of joint probabilities"* (apparently he means the independence of the increments), we obtain

$$p_{z,t_1+t_2} = \int_{-\infty}^{+\infty} p_{x,t_1} p_{z-x,t_2} dx. \tag{1}$$

In other words, he obtains what we call today the Chapman-Kolmogoroff equation. Then he observes that "this equation is confirmed by the function"

$$p_{x,t} = \frac{1}{\sigma\sqrt{2\pi t}} \exp\left(-\frac{x^2}{2\sigma^2 t}\right), \tag{2}$$

concluding that "evidently the probability is governed by the Gaussian law already famous in the calculus of probabilities".

Summing up, Bachelier derived from some basic principles the transition law of Brownian motion and its relation to the Chapman-Kolmogoroff equation.

Bachelier then gives an *"Alternative Determination of the Law of Probability"*. He approximates the continuous time model $(S_t)_{t\geq 0}$ by a random walk, i.e., a process which during a time interval Δt moves up or down with probability $\frac{1}{2}$ by Δx. He clearly works out that Δx must be of the order $(\Delta t)^{\frac{1}{2}}$ and – using only Stirling's formula – he obtains the convergence of the one-dimensional marginal distributions of the random walk to those of Brownian motion.

Suming up, Bachelier arrives at the model for the stock price process

$$S_t = S_0 + \sigma W_t, \qquad 0 \leq t \leq T, \tag{3}$$

where, in modern terminology, $(W_t)_{0\leq t\leq T}$ denotes standard Brownian motion and the constant $\sigma > 0$ is the "volatility", which Bachelier has called the "coefficient de nervosité du marché".

Having fixed the model, Bachelier is now able to determine the price C of an option appearing in Figure 1. Indeed, the probability distribution in this picture is now given by a Gaussian distribution with mean S_0 (the current price of the underlying S) and variance $\sigma^2 T$. The "fundamental principle" (the mathematical expectation of the speculator is zero) states that the integral of the function depicted in Figure 1 with respect to this probability distribution equals zero. This yields the equation

$$-C + \int_{K-S_0}^{\infty} (x - (K - S_0)) f(x) dx = 0, \tag{4}$$

where

$$f(x) = \frac{1}{\sigma\sqrt{2\pi T}} e^{-\frac{x^2}{2\sigma^2 T}}, \tag{5}$$

which clearly determines the relation between the premium C of the option and the strike price K. In other words, equation (4) determines the price for the option and therefore solves the basic problem considered by Bachelier.

It is straightforward to derive from (4) an "option pricing formula" by calculating the integral in (4): denoting by $\phi(x)$ the standard normal density function, i.e., $\phi(x)$ equals (5) for $\sigma^2 T = 1$, by $\Phi(x)$ the corresponding distribution function, and using the relation $\phi'(x) = -x\phi(x)$, an elementary calculation reveals that

$$C = \int_{\frac{K-S_0}{\sigma\sqrt{T}}}^{\infty} \left(x\sigma\sqrt{T} - (S_0 - F)\right)\phi(x)dx \tag{6}$$

$$= (S_0 - K)\Phi\left(\frac{S_0 - K}{\sigma\sqrt{T}}\right) + \sigma\sqrt{T}\phi\left(\frac{S_0 - K}{\sigma\sqrt{T}}\right),$$

which is a very explicit and tractable formula. Note that the only delicate parameter is σ while all the other quantities are given.

Finally in Bachelier's thesis a section follows, which is not directly needed for the subsequent applications in finance, but which – retrospectively – is of utmost mathematical importance: "*Radiation of probability*". Consider the discrete random walk model and suppose that the grid in space is given by

$$\ldots, x_{n-2}, x_{n-1}, x_n, x_{n+1}, x_{n+2}, \ldots$$

having the same distance

$$\Delta x = x_n - x_{n-1},$$

for all n, and such that at time t these points have probabilities

$$\ldots, p_{n-2}^t, p_{n-1}^t, p_n^t, p_{n+1}^t, p_{n+2}^t, \ldots$$

for the random walk under consideration. What are the probabilities

$$\ldots, p_{n-2}^{t+\Delta t}, p_{n-1}^{t+\Delta t}, p_n^{t+\Delta t}, p_{n+1}^{t+\Delta t}, p_{n+2}^{t+\Delta t}, \ldots$$

of these points at time $t + \Delta t$? A moment's reflection reveals the rule which is so nicely described by Bachelier in the subsequent phrases:

> Each price x during an element of time radiates towards its neighboring price an amount of probability proportional to the difference of their probabilities.
>
> I say proportional because it is necessary to account for the relation of Δx to Δt.
>
> The above law can, by analogy with certain physical theories, be called the law of radiation or diffusion of probability.

Passing formally to the continuous limit and denoting by $P_{x,t}$ the distribution function associated to the density function (2)

$$P_{x,t} = \int_{-\infty}^{x} p_{z,t}\, dz \tag{7}$$

Bachelier deduces in an intuitive and purely formal way the relation

$$\frac{dP}{dt} = \frac{1}{c^2}\frac{dp}{dx} = \frac{1}{c^2}\frac{d^2 P}{dx^2} \tag{8}$$

where $c > 0$ is a constant. Of course, the heat equation was known to Bachelier: he claims that "*this is a Fourier equation*".

Hence Bachelier in 1900 very explicitly discovered the fundamental relation between Brownian motion and the heat equation; this fact was rediscovered five years later by A. Einstein [8] and resulted in a goldmine of mathematical investigation through the work of Kolmogoroff, Kakutani, Feynman, Kac, and many others up to recent research. It is worth noting that H. Poincaré in his (very positive) report on Bachelier's thesis saw the seminal importance of this idea when he wrote "On peut regretter que M. Bachelier n'ait pas developpé d'avantage cette partie de sa thèse" (One may regret that M. Bachelier did not further develop this part of his thesis.)

But unfortunately the thesis of Bachelier obtained only a "mention bien". Apparently the two other jury members did not have the same positive opinion as H. Poincaré towards this unusual student who was working at the stock exchange. But in order to make an academic career a "mention très bien" was an absolute must, just as it is today in France. Louis Bachelier subsequently had a rather difficult life and his work was not well received in France. On the other hand, A. Kolmogoroff or K. Itô did appreciate his writings.

We focused on the early work by L. Bachelier as his contribution is less known to a wider public than the "Black-Scholes option pricing formula". After Bachelier's pioneering work, it remained silent around the theme of option pricing for many decades. This is in sharp contrast to the progress made during this period in the mathematical theory of stochastic processes and their applications in physics and biology.

Eventually in 1965 the eminent economist P. Samuelson rediscovered Bachelier's thesis in the library of Harvard University, following a request of the statistician J. Savage. Samuelson was immediately fascinated by Bachelier's work and started a line of research on option pricing and related topics which at this time had much more repercussions than Bachelier's thesis. Samuelson [17] proposed a multiplicative version of Bachelier's model defined by the stochastic differential equation

$$\frac{dS_t}{S_t} = \sigma dW_t + \mu dt, \qquad 0 \le t \le T, \tag{9}$$

where $(W_t)_{0 \le t \le T}$ denotes a standard Brownian motion and $\sigma \in \mathbb{R}_+, \mu \in \mathbb{R}$ are constants.

Given the initial value S_0 of the stock, Itô's formula yields the solution

$$S_t = S_0 \exp\left(\sigma W_t + (\mu - \frac{\sigma^2}{2})t\right), \qquad 0 \le t \le T. \tag{10}$$

The SDE (9) states that the *relative* increments $\frac{dS_t}{S_t}$ of the price process are driven by a Brownian motion with drift. Today, the model (9) is usually called the *Black-Scholes model*.

In 1973, the papers by F. Black and M. Scholes [3] and R. Merton [15] appeared. Departing from the *no arbitrage principle* and using the concept of *dynamic trading* these authors derived the - by now famous - *Black-Scholes formula* for the price of a call option. Maintaining for convenience the above hypothesis that the riskless rate of interest equals zero (which presently happens to be close to the actual economic situation), one obtains the formula

$$C = S_0 \Phi(d_1) - K\Phi(d_2), \tag{11}$$

with

$$d_1 = \frac{\ln(\frac{S_0}{K}) + \frac{\sigma^2}{2}T}{\sigma\sqrt{T}}, \, d_2 = \frac{\ln(\frac{S_0}{K}) - \frac{\sigma^2}{2}T}{\sigma\sqrt{T}}. \tag{12}$$

This formula looks quite different from Bachelier's result (6) above. However, for moderate values of T and σ, as was the case in Bachelier's original application, the difference of the numerical values of (6) and (12) is remarkably small. In [19] the difference for typical data used by Bachelier has been estimated to be of the order of $10^{-8}S_0$. In a way, this is not too surprising as the difference between Bachelier's model (3) and the Black-Scholes model (9) is analoguous to the difference between linear growth and exponential growth. In the short run this difference is remarkably small.

We do not give the derivation of the Black–Scholes formula here as it may be found in many textbooks (e.g., [6]). It is remarkable that the solution (11) for the option pricing formula is eventually obtained by applying precisely Bachelier's *fundamental principle*, i.e., by choosing $\mu = 0$ in (9) and calculating the expectation of the payoff of the option under the law of S_T.

3 Mathematics and the financial crisis

The Black–Scholes formula and the related concepts of *hedging* and *replication* of derivate securities had enormous impact on the paradigms of financial markets. In particular, the use of *stochastic models* became ubiquitous in the financial industry. In this section we shall have a critical look at the effects of this probabilistic approach to the real world.

3.1 Value at risk

Let us start with the concept of *value at risk*. The CEO of J.P. Morgan, Dennis Weatherstone, asked the bank's quants in the wake of the 1987 crash to come up with a short daily summary of the market risk facing the bank. He wanted one single number every day at 4:15 pm which indicates the risk exposure of the entire bank. By that time the quants, i.e. the quantitative financial analysts, disposed of mathematical models for "market risk", such as the above considered price movements of stocks, options etc. Stochastic models were used to calculate the distribution of total profits or losses from these sources during a fixed period, e.g., the consecutive 10 business days. The *"value at risk"* was then defined as the 1% quantile of this distribution, i.e., the smallest number $M \in \mathbb{R}$ such that the probability of the total loss being bounded by M, is at least 99%. This was the famous "4:15 number".

It was quickly noticed by the quants that the above models of Bachelier and Black–Scholes are not well suited for the estimation of extreme events. After all, they are based on the Gaussian distribution which is derived from the central limit theorem.

As is very well known for almost 300 years, this theorem states that a random variable X, in our case the change of a stock price, which is the sum of "many" independent random variables X_n, where each of these random variables has little individual influence on the total effect $X = \sum X_n$, is approximately normally distributed. But in the financial world it happens quite often that a price movement is due to one big event (think, e.g., of 9/11) rather than due to the sum of many small events.

This is well illustrated by the following comment on the use of the Black–Scholes approach by a senior manager of an Austrian bank: "the Black–Scholes theory works very well, in fact surprisingly well, in 99% or even 99.5% of the days!" He then continued: "except for the one or two days per year which really matter."

But let us come back to the concept of *value at risk*. By choosing distributions with heavier tails it is not too difficult to correct for the above mentioned shortcomings of the Gaussian models. This was widely done, also by practitioners, in the context of risk management. As a general rule, when choosing a model it is always important to keep the applications in mind. If the purpose is to deal with the day-to-day business of pricing and hedging options, the Black–Scholes model, or even Bachelier's model, is a very efficient tool. However, when it comes to issues like risk management which deal with extreme events, the use of these models is highly misleading. After all, we have to keep in mind that these models were not invented for such purposes as risk management.

A similar fate of misuse happened to the "4:15 number" of Dennis Weatherstone which was originally designed as a very rough but focused information for the senior management of a bank. But this magic number quickly became very popular under the name of *value at risk* and used for other purposes, notably for the calculation of capital requirements. A risky portfolio of a bank requires sufficient underlying capital as a buffer against potential losses. According to the Basel II regulation this capital requirement is determined by calculating the value at risk of the portfolio and, in order to be on the safe side, eventually multiplying this number by three. Compare [14] for a more detailed discussion.

The use of *value at risk* for regulatory purposes is a prime example of what has become known as "Goodhart's law" which seems to hold true in many contexts: *when a measure becomes a target, it ceases to be a good measure.*

If banks (or traders) get the incentive to design their portfolios in such a way that the "value at risk" is kept low, this may lead to serious mis-allocations. To sketch the idea we give a somewhat artificial example. Suppose that a bank has a portfolio which causes a sure loss of one million Euros. The bank can decompose this portfolio into 101 sub-portfolios where each of these sub-portfolios makes a loss of one million with probability $\frac{1}{101}$, and zero loss otherwise. While the value at risk of the entire portfolio obviously is one million, each of the sub-portfolios has a value at risk of zero! Hence no capital requirement is necessary for these sub-portfolios. This effect is, of course, in sharp contrast to the basic idea of "diversification": by pooling sub-portfolios into one big portfolio, the capital requirement for the sum should be less than or equal to the sum of the capital requirements, and not vice versa.

Admittedly, the above example is too blunt to be realistic, but nevertheless it highlights what is happening in practice when *value at risk* is blindly used as a risk measure for risk management purposes. Mathematically speaking, the above effect is due to the fact that the *value at risk map*, which assigns to each random variable X the 1%-quantile of its distribution, fails to be sub-additive.

This shortcoming was soon and severely criticized in the academic literature. In 1999, Ph. Artzner, F. Delbaen, J.-M. Eber, and D. Heath [1] proposed a theory of "coherent risk measures" which do not suffer from this defect. According to *Google Scholar*, this paper has been cited more than 7000 times and there has been ample literature on this topic since.

Nevertheless, in practice the concept of *value at risk* still plays a central role for the determination of capital requirements.

3.2 The Gauss copula and CDO's

We now pass to a specific financial product which caused much harm during the financial crisis of 2007/2008, the so-called *collateralized debt obligations*, abbreviated CDOs. The basic idea looks quite appealing. In the banking and insurance business the notion of *risk sharing* plays a central role. If bank A is exposed to the risk of default of loan A and bank B to the risk of default of loan B, it is mutually beneficial if bank A passes over half of the risk of loan A to bank B and vice versa. This practice has existed for centuries and is the reason why, e.g., in the reinsurance business the financial damage of major catastrophes can be absorbed relatively smoothly by distributing the losses over several reinsurance companies.

Turning back to the example of bank A and B there is, however, a slight problem. As bank A negotiates with the obligor of loan A it disposes of better information on the status of this obligor than bank B. Of course, bank B is aware of this asymmetry of information which might work in favor of bank A, and therefore asks bank A for a higher recompensation when accepting half of the risk of loan A.

The original idea of a CDO is to find a mechanism which neutralizes this asymmetry of information. Suppose bank A has thousand loans $A_1, \ldots, A_{1000}$ in its portfolio and wants to pass over part of the involved risk to other financial institutions or investors. Bank A can pool these loans into one big *special vehicle* and then slice it into *tranches*, e.g. a senior, a mezzanine, and an equity tranche. The tranching might divide the collection of one thousand loans according to the proportion 70 : 20 : 10. When loans fail to perform, the equity tranche is hit first. Only when the losses exceed 10%, the mezzanine tranche is effected. When the losses exceed 30%, also the senior tranche has to start to absorb them. The basic idea is that the issuing bank A keeps the equity tranche – which is most effected by the asymmetry of information – in its own portfolio, while it tries to sell the senior tranche and, possibly, the mezzanine tranche to other financial institutions.

So far, so good. In fact, similar instruments exist for a long time, e.g., the good old German *"Pfandbriefe"* which were intoduced in Prussia under Frederick the Great. Their business model goes as follows: a bank gives loans to communities which are secured by mortgages on their property. To refinance, the bank then sells bonds (the *"Pfandbriefe"*), which are directly secured by the entity of the underlying mortgages, to private or institutional investors. It is worth noting that there is one essential difference to the concept of CDOs: the issuing bank remains fully liable to the owners of the *Pfandbriefe*. This seems to be an important reason why the *Pfandbriefe* have safely survived so many financial crises. During the past hundred years there was not a single failure of a Pfandbrief-bank.

Back to the CDOs: in order to determine e.g. the price of the senior tranche one tries to estimate the probability distribution of the losses of this tranche. Of course, if one assumes that the defaults of the one thousand loans $A_1, \ldots, A_{1000}$ are *independent*, the senior tranche has an extremely low probability of loss, even if each of the individual loans bears a relatively large default risk. But obviously nobody is so extremely naive to suppose independence in this context. Rather we expect some positive correlation of the failures of the individual loans. But how to model this dependence structure precisely?

D. Li [11] proposed in 1999 the *Gaussian copula* to handle this issue. For $0 \le \rho \le 1$, denote by $\mathbb{P}_\rho$ the centered Gaussian distribution on $\Omega = \mathbb{R}^{1000}$ defined in the following way. Denoting by $(X_i)_{i=1}^{1000}$ the coordinate projections, we prescribe $\mathbb{E}[X_i^2] = 1$, for each i, and $\mathbb{E}[X_i X_j] = \rho$, for each $i \ne j$. This covariance structure uniquely defines $\mathbb{P}_\rho$.

Now suppose that we know, for each $i = 1, \ldots, 1000$, the individual default probability p_i of loan A_i. This is not too problematic as banks have, of course, a long experience dealing with the frequency of defaults of individual loans. For simplicity we suppose that all loans have the same size and either fully pay the loan (with probability $1 - p_i$) or default totally (with probability p_i).

Denote by $x_i \in \mathbb{R}$ the $(1 - p_i)$-quantile of the standard Gaussian distribution so that $\mathbb{P}[X_i > x_i] = p_i$. We identify the event $\{loan\ A_i\ defaults\}$ with the event $\{X_i > x_i\}$. Having fixed $\rho \in [0, 1]$ as well as the p_i's we can now calculate all quantities of interest in an obvious and tractable way. For example, the probability that more than 30% of the loans default and therefore the senior tranche suffers a loss is given by

$$\mathbb{P}_\rho[\#\{i : X_i > x_i\} > 300].$$

The delicate task is to determine the correlation parameter ρ. As regards the senior tranche it is rather obvious that (for realistic choices of p_i) its expected loss is increasing in $\rho \in [0, 1]$.

Therefore it seems at first glance a reasonable approach to choose a realistic (i.e., sufficiently big) ρ by calibrating to observed prices on the market. This allows to calculate the price of the senior tranche as well as all the other quantities of relevance. In this way the rating agencies often granted a AAA to such senior tranches and other related products, obtained e.g. by pooling once again the mezzanine tranches of different CDOs into a new CDO (called "CDO-squared"). At least, they did so until 2007.

In 2007 it became very clear that the senior tranches of many CDOs were prone to suffer much bigger losses than predicted by "Li's formula". A Financial Times article in 2009 was entitled "The formula that felled Wall street" [12].

What had gone wrong? The sad fact is that David Li and other people applying the above method had not listened to people working in *extreme value theory*. In this theory it is well known that correlations of random variables tell only very little about the joint probabilities of extreme events. Only in the case of a (centered) *Gaussian* random variable X on $\mathbb{R}^{1000}$ the correlation matrix uniquely determines the law of X. As we shall presently see, it does so by giving rather small probabilities to *joint* extreme events, even if the correlation parameter ρ is close to 1.

This was made crystal clear in the paper [9] by P. Embrechts, A. McNeil, D. Straumann which has circulated since 1998. We give a short outline of the relevant concepts. Instead of thousand random variables $X_1, \ldots, X_{1000}$ we focus for simplicity on X_1, X_2.

DEFINITION 2 ([9]). *Let X_1, X_2 be random variables with distribution functions F_1, F_2. The* coefficient of tail dependence *is defined as*

$$\lambda := \lim_{\alpha \nearrow 1} \mathbb{P}[X_2 > F_2^{-1}(\alpha)|X_1 > F_1^{-1}(\alpha)],$$

provided the limit exists.

It is straightforward to calculate that for a *Gaussian* random variable (X_1, X_2) with $\rho(X_1, X_2) < 1$, we have $\lambda = 0$. This property is called *asymptotic independence* and has an obvious interpretation: whatever choice of $\rho \in [0, 1[$ in the Gaussian case is made, the probability of *joint extreme events* becomes small, as $\alpha \nearrow 1$, quicker than the probability of the corresponding individual extreme events. This is in sharp contrast to what happened in the real world of 2007 to the loans pooled in CDOs.

But, of course, the Gaussian copula is not the only way of modeling. There are plenty of other ways to model the joint probability of a vector (X_1, X_2) for given marginal distributions X_1, X_2 and correlation $\rho(X_1, X_2)$. As an example the *Gumbel copula*, for which we obtain a strictly positive value of λ, is thoroughly analyzed in [9]. We note in passing that the word "copula" refers to the rather obvious fact, observed by A. Sklar in 1959, that for the specification of the joint law of (X_1, X_2) for given marginals, there is no loss of generality to normalize the marginal distributions of X_1 and X_2 to be uniform on $[0, 1]$.

The subsequent highly instructive picture is taken from the paper [9]. For *identical marginal distributions and identical correlation coefficient ρ* the choice of the copula can make a dramatic change to the probability of joint extreme events (the upper right rectangle).

As P. Embrechts told me (and as is documented in [7]), he presented the paper [9] on March 27, 1999, at Columbia University. David Li was in the audience and introduced himself during the break. So much for the comments claiming that "nobody has warned."

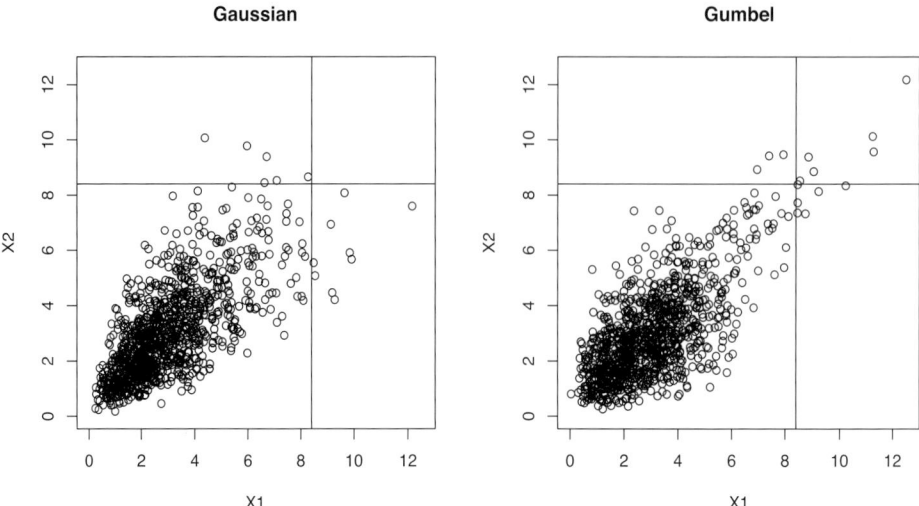

Figure 2. Figures showing 2000 sample points from the named copulas

To finish this section let me mention another highly respected mathematician in the field of Mathematical Finance, L. C. G. Rogers, who also warned very outspokenly and long before 2007 about the misuse of the Gaussian copula. The following quotation on the 2007/2008 crisis [16] dates from 2009:

> The problem is not that mathematics was *used* by the banking industry, the problem was that it was *abused* by the banking industry. Quants were instructed to build models which fitted the market prices. Now if the market prices were way out of line, the calibrated models would just faithfully reproduce those wacky values, and *the bad prices get reinforced by an overlay of scientific respectability*!

3.3   An academic response to Basel II
In the previous subsections we have looked at two concrete and important examples, value at risk and the Gauss copula. Academic criticism of their misuse arose early and was well argued, but failed to sufficiently influence the practitioners.

Actually, this did not only happen in these two specific examples, as the paper [5] shows very clearly. This paper dates from 2001 and bears the title of this section. Written by a number of highly renowned financial economists and mathematicians, among them the above mentioned Ch. Goodhart and P. Embrechts, it was an official response addressed to the Basel Committee for Banking Supervision. It was extremely visible, not only within academia.

Let me quote from the *Executive Summary* of [5]:

- The proposed regulations fail to consider the fact that risk is endogenous. Value-at-Risk can destabilize an economy and induce crashes when they would not otherwise occur.
- Statistical models used for forecasting risk have been proven to give inconsistent and biased forecasts, notably underestimating the joint downside risk of

different assets. The Basel Committee has chosen poor quality measures of risk when better risk measures are available.

- Heavy reliance on credit rating agencies for the standard approach to credit risk is misguided as they have been shown to provide conflicting and inconsistent forecasts of individual clients' creditworthiness. They are unregulated and the quality of their risk estimates is largely unobservable.
- Operational risk modeling is not possible given current databases and technology even if a meaningful definition of this risk were to be provided by Basel. No convincing argument for the need of regulation in this area has yet been made.
- Financial regulation is inherently procyclical. Our view is that this set of proposals will, overall, exacerbate this tendency significantly. In so far as the purpose of financial regulation is to reduce the likelihood of systemic crises, these proposals will actually tend to negate, not promote this useful purpose.

From today's perspective this reads like a clairvoyant description of the key issues of what went wrong in 2007–2008.

Let me try to make some personal comments on these five points.

Point 1: We have seen that for Bachelier as well as Black, Scholes, and Merton it was perfectly legitimate to model the risk involved in the price movements of a stock as *exogenous* and given by a stochastic model which is independent of the behavior of the agent. But the picture changes when all the agents believe in such a model or – making things worse – are forced by regulation to apply them. Value at risk plays an important negative role in this context.

Point 2: accurately summarizes what we have discussed in the above subsections 3.1 and 3.2.

Point 3: was strikingly confirmed by the crisis when the rating agencies, who did the above sketched ratings for the CDOs etc, turned out to have made very poor judgments of default probabilities. In addition, they may have been influenced by conflicts of interest.

Point 4: This is the only point which did not lead astray in 2007–2008. While the Basel II regulation of capital requirements for "operational risk", e.g., legal risks, IT failures etc, did not do much harm, it is important to note that is also did not do any good during the crisis.

Point 5: addresses the most fundamental issue, the procyclicality of financial regulation. While this is an inherent problem of regulation one should, of course, try to design the rules in a way to mitigate this effect. The prediction that, to the contrary, Basel II exacerbates the procyclicality has materialized in 2007–2008 in a dramatic way.

The final sentence of the introduction of the *Academic Response to Basel II* [5] could not have been more outspoken: "Reconsider before it is too late!" As we know today, the Basel Committee did not follow this urgent advice from academia.

The bottom line of these facts is that academia has not succeeded to influence financial practitioners sufficiently. Despite this rather sad story I do believe that academic research has to continue to try to thoroughly understand the problems at hand and to make itself understood in practice. To quote Sigmund Freud [10]:

> The voice of the intellect is soft. But it does not rest before it has made itself understood. (Die Stimme des Intellekts ist leise, aber sie ruht nicht, ehe sie sich Gehör verschafft hat.)

Acknowledgements. Partially supported by the Austrian Science Fund (FWF) under grant P25815, the Vienna Science and Technology Fund (WWTF) under grant MA14-008 and by Dr. Max Rössler, the Walter Haefner Foundation and the ETH Zurich Foundation.

References

[1] P. ARTZNER, F. DELBAEN, J.-M. EBER and D. HEATH, Coherent measures of risk, *Mathematical finance*, Vol. 9, No. 3 (1999), pp. 203–228.

[2] L. BACHELIER, Théorie de la Speculation. *Ann. Sci. Ecole Norm. Sup.*, Vol. 17 (1900) pp. 21–86. [english translation with excellent comments: M. Davis and A. Etheridge, *Louis Bachelier's Theory of Speculation: The Origins of Modern Finance*, Princeton University Press, 2006]

[3] F. BLACK and M. SCHOLES, *The pricing of options and corporate liabilities.* Journal of Political Economy, Vol. 81 (1973), pp. 637–659.

[4] J.-M. COURTAULT, Y. KABANOV, B. BRU, P. CRÉPEL, I. LEBON and A. LE MARCHAND, Louis Bachelier: On the Centenary of "Theorie de la Spéculation". *Mathematical Finance*, Vol. 10, No. 3 (2000), pp. 341–353.

[5] J. DANIELSSON, P. EMBRECHTS, C. GOODHART, C. KEATING, F. MUENNICH, O. RENAULT and H.S. SHIN, *An academic response to Basel II*, LSE Financial Markets Group. Special Paper Number 130 (2001).

[6] F. DELBAEN and W. SCHACHERMAYER, *The Mathematics of Arbitrage*, Springer Finance, 2006.

[7] C. DONNELLY and P. EMBRECHTS, The devil is in the tails: actuarial mathematics and the subprime mortgage crisis, *Astin Bulletin*, Vol. 40, No. 1 (2010), pp. 1–33.

[8] A. EINSTEIN, Über die von der molekularkinetischen Theorie der Wärme geforderte Bewegung von in ruhenden Flüssigkeiten suspendierten Teilchen, *Annalen der Physik*, Vol. 322, No. 8 (1905), pp. 549–560.

[9] P. EMBRECHTS, A. MCNEIL and D. STRAUMANN, *Correlation and dependence in risk management: properties and pitfalls*, Risk management: Value at risk and beyond, pp. 176–223, 2002.

[10] S. FREUD, *Totem und Tabu*, Wien: Internationaler Psychoanalytischer Verlag, 1913.

[11] D.X. LI, On default correlation: A copula function approach, *Journal of Fixed Income*, Vol. 9, No. 4 (1999), pp. 43–54.

[12] S. JONES, Of couples and copulas: The formula that felled Wall Street, *Financial Times*, April 24 (2009),

[13] P. JORION, (2007), *Value at Risk: the new Benchmark for Managing Financial Risk*, Vol. 3, New York: McGraw-Hill.

[14] A. MCNEIL, R. FREY and P. EMBRECHTS, (2015), *Quantitative Risk Management: Concepts, Techniques and Tools*, Princeton University Press.

[15] R. C. MERTON, Theory of rational option pricing, *Bell J. Econom. Manag. Sci.*, Vol. 4 (1973), pp. 141–183.

[16] L. C. G. ROGERS, *Financial Mathematics and the Credit Crisis.* Document in response to questions posed by Lord Drayson, UK Science and Innovation Minister, 2009.

[17] P. A. SAMUELSON, Proof that properly anticipated prices fluctuate randomly, *Industrial Management Review*, Vol. 6 (1965), pp. 41–50.

[18] W. SCHACHERMAYER, Introduction to the Mathematics of Financial Markets. In: S. Albeverio, W. Schachermayer & M. Talagrand: *Lectures on Probability Theory and Statistics, Saint-Flour summer school 2000*, Lecture Notes in Mathematics, 1816 (Pierre Bernard, editor), pp. 111–177. Heidelberg: Springer Verlag, 2003.

[19] W. SCHACHERMAYER and J. TEICHMANN, How close are the Option Pricing Formulas of Bachelier and Black-Merton-Scholes? *Mathematical Finance*, Vol. 18 (2008), No. 1, pp. 155–170.

Statistics in high dimensions

Aad van der Vaart and Wessel van Wieringen

High-dimensional data and models are central to modern statistics. We review some key concepts, some intriguing connections to ideas of the past, and some methods and theoretical results developed in the past decade. We illustrate these results in the context of genomic data from cancer research.

1 High-dimensional data and models

Although traditional statistics continues to play an undiminished role in many branches of science and society, present-day statistics deals increasingly often with high-dimensional problems. Since a few years *big data* is a collective term used for data of many different types, including records of internet clicks, measurements of far away galaxies, high-frequency time series of stock markets or other economic indicators, or historical documents and indicators.

In this article we shall review mostly big data questions that have arisen in the field of *genomics*, where high-dimensional data arose some 15 years ago, long before the hype word *big data* became fashionable. At the time medical and biological scientists invented new experimental procedures that allowed to measure thousands of characteristics simultaneously, whereas before measuring even a single characteristic had been a time-consuming procedure. A prime example were *gene expression data*, measured by *micro-arrays* (see Figure 1).

Genes are small segments in the DNA of living organisms (about 30 000 in humans). Each living cell contains a full (and identical) copy of the DNA (organised in pairs of chromosomes), but only a small number of the genes are active (one says *expressed*) in a given cellular environment. Unravelling which genes are expressed in which situation is not only a first step to revealing the "secrets of life", but also has great promise for developing medicines and treatments. For instance knowing which genes are active in tumor cells (or not) relative to healthy cells (such genes are called *differentially expressed*) may lead to cures for cancer. Already in the early days of genomics great promises were made, not only of curing diseases, but also of targeting treatments to specific patient profiles (*personalised medicine*). Certainly in the past 15 years much progress was made in this direction, albeit at a slower pace than the early pioneers (or scientists trying to acquire funding for their research) may have foreseen.

The initial and revolutionary experimental techniques of the early 2000s have been followed up by many other techniques, often more refined, but feasible at a fraction of the initial cost.

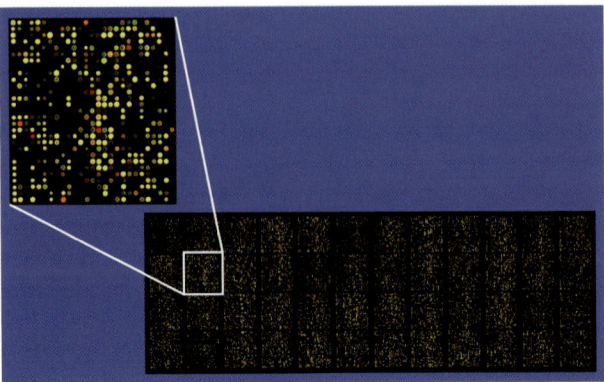

Figure 1. A micro-array is a small glass slide on which a (full) library of segments of DNA ("probes") are printed (about 50,000 in this case) in a matrix lay-out. Messenger RNA (mRNA) from cells under study is applied to the slide, and sticks to the corresponding (or, actually, complementary in the DNA sense) probes. By attaching a fluorescent to the mRNA the quantity of mRNA stuck to a given probe could be determined by measuring the luminosity at the probe site. This gives the characteristic pattern of dots in the picture. The geometric layout in a matrix has no significance (except possibly for statistical quality control).

Besides gene expression (now by *second generation sequencing* methods) one measures the full *DNA sequence* of an individual (e.g., all approximately 3 billion letters C, G, A, T, representing nucleotides, in a human), *genome-wide copy number variations* (a healthy cell contains two copies of the full DNA, arranged in pairs of chromosomes, but cancer cells may contain 0, 1, or many copies of given segments), *protein profiles* (proteins are another layer in the machinery of life; they are strings of amino acids that are coded by triples of three DNA letters), or the abundance of *metabolites* (the third layer of molecules in cellular processes). Instead of establishing simple differential expression between different types of cells (healthy versus cancer, or one part of the brain versus another, etc.) biological scientists now often try to understand a complete set of chemical reactions, involving a network (or *pathway*) of genes, proteins, and other molecules. The simple picture of DNA as a linear molecule (a string of letters), of which a small part (the genes, together making up 1 % of the letters) is translated into proteins, has been revised and augmented with the discovery of *alternative splicing* and *methylation* as important processes, also involving segments of DNA that ten years ago were considered to be *junk DNA*, and other stable environmental factors *epigenetics*. If 30 000 genes by itself is already an enormous number, the number of possible combinations and interactions, and variations in the down stream cellular processes are staggering. Even small segments of the machinery of life appear already very complex (see Figure 2). More and more of such information is accumulated in data-bases ready for combination with new data. The promise still is that (in say 15 years) medical scientists will have acquired so much insight that diseases as cancer, diabetes or schizophrenia will be curable ([10]).

In this article we connect this quest to some developments in statistics.

Summary. Changes in the size and complexity of data in the past decade call for reinventing statistical theory and methods, for instance such as used in cancer research.

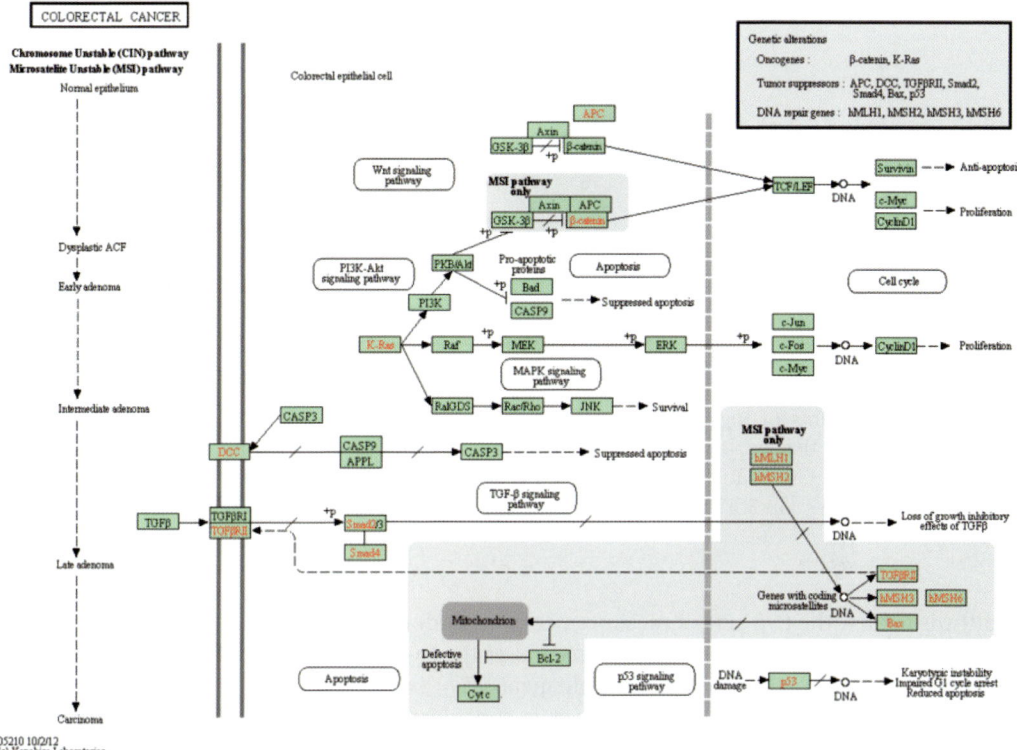

Figure 2. Detail from the KEGG pathway, thought to be involved in colorectal cancer, for the purpose of illustration arbitrarily selected from the KEGG pathway Database http://www.genome.jp/kegg/pathway.html. In general a biological pathway is a series of actions among molecules in a cell that can lead to the assembly of new molecules, turn genes on and off, spur a cell to move, or any other effect.

2 Multiple testing

An unfortunate side-effect of making many measurements is that one also measures more noise. Statisticians use the word *noise* in a general sense to refer to any variation in data that is not explainable by systematic factors. This may be variation among subjects or tissues, undesired variation over time, true measurement noise, or any other unavoidable variation. Statistical analysis is designed to extract the systematic signal from its noisy background, where success is measured by replicability of the conclusions.

As a simple case consider measurements of expression of a certain gene in a collection of m cancer tissues and a collection of n healthy tissues. Expression is a numerical value, and basic quality analysis has lead to standard normalisation procedures giving these values (after logarithmic transformation) on the scale of roughly a Gaussian distribution. Thus the m values $X_1, \ldots, X_m$ on the cancer tissues will be all different and follow a histogram as in Figure 3, as will the n values $Y_1, \ldots, Y_n$ on the healthy tissues. The standard statistical approach to answer the question whether there is a (systematic) difference between cancer and

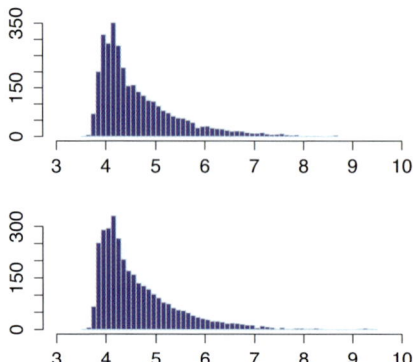

Figure 3. Histograms of log expression levels on 10 healthy (top) and 10 cervical squamous cell carcinoma (bottom) tissues, aggregated over 3372 micro-RNAs. Is there a difference? (Data collected at VU Medical Center, available from the Gene Expression Omnibus as Series GSE30656. See Wilting et al., Oncogene 2013 Jan 3;32(1): 106–116.)

healthy is to view the two sets of measurements as random samples from populations (possibly healthy) and then consider whether the means of the two populations differ. The randomness of the two samples makes that the sample means $\bar{X} = m^{-1} \sum_{i=1}^{m} X_i$ and $\bar{Y}$ may differ from the population means, with small probability even considerably so, and hence may send the wrong message. Standard *hypothesis testing theory* overcomes this problem by only concluding to a difference of the population means if the sample means differ "significantly". This concept is made precise by setting up the testing procedure so that the probability of wrongly making this decision will be smaller than a certain level, typically 5%, *in the case that in reality there is no difference between the population means.* For normally-distributed populations the well-known *t*-test is a concrete example of a procedure that fulfils this principle. In practice the procedure is typically carried out through the *p-value*, which is the probability that a more extreme value (in the sense of pointing to a difference) of the data would be found than actually observed, again under the assumption that in reality there is no difference. Such a *p-value* is constructed to be like a random draw from the numbers between 0 and 1 in the latter situation; a value less than 5% would typically be interpreted to mean that there is a real difference between cancer and healthy.

Hypothesis testing is a well established part of statistics. The challenge for modern statistics is that there are so many hypotheses. A medical scientist easily measures 30.000 genes simultaneously and will want to know about a possible difference between cancer and healthy for all of them. In a *genome wide association study* a geneticist may want to test all of 500.000 measured *single nucleotide polymorphisms (SNP)s.* This would not be a problem if the scientist measured a big multiple of 30.000 tissues of each type, but more realistic sample sizes are $m \approx n$ of the order 30–3000, even today. This is due to the cost of measuring, and also because there usually there are just not enough cancer tissues of a given type available. In the early days of genomics medical scientists with only little statistical training might proudly claim to have found as many as 50 significant (i.e. differentially expressed) genes out of 10.000 investigated, but at a 5% error probability per gene this is of course completely compatible with

Bayes's rule, discrete case

For a discrete random variable Δ and any event B the conditional probability that $\Delta = 0$ given B is given by

$$P(\Delta = 0 \mid B) = \frac{P(\Delta = 0)P(B \mid \Delta = 0)}{\sum_\delta P(\Delta = \delta)P(B \mid \Delta = \delta)},$$

where the sum is taken over all possible values of δ. The sum in the denominator gives the probability of the event B.

none of the 10.000 being differentially expressed. Statisticians know this as the *multiple testing problem*. This was addressed before, but at the scale of tens of thousands or more tests it was rather new in the early 2000s, and needed new approaches.

The classical solution is the *Bonferroni correction*, and consists simply of testing each of p hypotheses at the level $(5/p)\%$. This ensures that the probability of erroneously claiming one or more differences is smaller than 5%, but for large p requires the differences to be rather prominent in the data to find any significant one at all. Since many (expensive) experiments would end up inconclusive, this was considered too restrictive. The Bonferroni correction is based on the crude bound $P(A_1 \cup A_2 \cup \cdots) \leq P(A_1) + P(A_2) + \cdots$, which ignores intersections of the events A_j, here taken as the event that the jth gene is significant, and hence can be improved. However, if the events were stochastically independent, then the correct calculation would be

$$P(A_1 \cup A_2 \cup \cdots) = 1 - \prod_{j=1}^{p}(1 - P(A_j)) \approx \sum_{j=1}^{p} P(A_j),$$

where the last approximation is fairly accurate if all $P(A_j)$ are small. In that case the Bonferroni bound is not so bad after all, and no improvement will do much more than multiplying the 5% by a constant that is nearly 1. (E.g., $1 - (1 - 0.05/10000)^{10000} \approx 0.0488$.) Perhaps for this reason it was decided to change the notion of significance, from "making one or more mistakes" into controlling the "proportion of wrong findings". Finding a difference was called a "discovery" and the *false discovery rate* defined as the expected quotient (where the E is for "expectation")

$$\text{FDR} = E\frac{\#\text{ genes claimed different, but in reality not different}}{\#\text{ genes claimed different}}.$$

While allowing a greater number of significant genes, this quantity also makes sense, if the purpose is to find genes that are potentially involved and promising for further investigation, for instance by experiments such as "knockout" studies. It makes less sense if the consequences of making even a single wrong decision would be very harmful.

A testing procedure that controls the FDR had been introduced in 1995 by Benjamini and Hochberg in [1], probably with between 2 and 100 hypothesis tests in mind, rather than thousands. Since the genomic revolution in the early 2000s their proposal has become one of the most cited procedures in statistics. For given p-values $P^1, P^2, \ldots, P^p$ denote by $P^{(1)} \leq P^{(2)} \leq \cdots \leq P^{(p)}$ the same numbers, but ordered by size, and imagine the testing problems ordered in parallel so that each $P^{(j)}$ maintains its link to a test. Then the procedure declares significant (or "discoveries") all tests with $P^{(j)} \leq (j/p)\alpha$, and also any hypothesis with a p-value for which there is a bigger p-value that satisfies this inequality.

There is an interesting motivation for this procedure, as follows. Start by assuming that each gene is differentially expressed with a given probability $1 - \pi_0$. For notational convenience introduce an indicator variable Δ^j, taking the value 0 ("no difference") with probability π_0, and the value 1 otherwise. Notably there is no real experiment that produces these variables, or it should be "nature" constructing our reality in a grand, random scheme. However, it helps to *think* of gene j being not involved in cancer ($\Delta^j = 0$) or involved ($\Delta^j = 1$) as if this were the outcome of a chance experiment. Next assign the p-value P^j that summarizes our data concerning gene j one of two (conditional) probability distributions:

- if $\Delta^j = 0$ assign it the null distribution F_0, which is typically the distribution of a uniform draw from the interval $[0, 1]$. Specifically, set $F_0(q) = \mathrm{P}(P^j \le q | \Delta^j = 0) = q$, for $q \in [0, 1]$.
- if $\Delta^j = 1$ assign it a distribution F_1, the same for every gene j, but generally unknown. Specifically, set $F_1(q) = \mathrm{P}(P^j \le q | \Delta^j = 1)$, for $q \in [0, 1]$.

Together these specifications give a joint probability distribution of (Δ^j, P^j). By *Bayes's rule* (see box), it follows that

$$\mathrm{P}(\Delta^j = 0 | P^j \le q) = \frac{\pi_0 F_0(q)}{\pi_0 F_0(q) + (1 - \pi_0) F_1(q)}. \tag{1}$$

Statistical tests are constructed so that a small p-value is indicative of a difference. If we would decide to call gene j differentially expressed if $P^j \le q$, then the expression in (1) would be the probability, in our hypothetical chance setup, of making the wrong decision, and can be interpreted as the proportion of false discoveries, or a *Bayesian FDR*. For practical implementation one problem is that F_1 is not known. This can be solved by recognising that $F(q) := \pi_0 F_0(q) + (1 - \pi_0) F_1(q)$, the denominator in the quotient (1), is the distribution of an *arbitrary* gene having a P-value smaller than or equal to q. Having so many hypotheses now turns into an advantage, because the *empirical distribution*

$$\hat{F}(q) := \frac{1}{p} \sum_{j=1}^{p} 1\{P^j \le q\} = \frac{1}{p} \{\text{fraction of } j \text{ with } P^j \le q\}$$

is a good estimator of F if p is large. Then define $\widehat{\mathrm{FDR}}_{\mathrm{B}}(q) = \pi_0 F_0(q)/\hat{F}(q)$, and let us agree that we call gene j differentially expressed if this quantity is smaller than some prescribed α at $q = P^j$. Since $F_0(q) = q$ for every q and $\hat{F}(P^{(j)}) = j/p$, after ordering the tests by increasing p-value this becomes $\widehat{\mathrm{FDR}}_{\mathrm{B}}(P^{(j)}) \le \alpha$, or $\pi_0 P^{(j)} \le \alpha j/p$. This is the Benjamini–Hochberg rule, but with α replaced by α/π_0.

For Bayesian statisticians the preceding reasoning is compelling. (A true Bayesian would probably base a decision on the conditional probability $\mathrm{P}(\Delta^j = 0 | P^j)$, which conditions on the observed p-value, not on the set $(0, P^j]$ of more extreme p-values. See Section 5 for more discussion on the Bayesian perspective.) However, one may frown on the assumption of genes turned on at random, and demand a different justification. Fortunately Benjamini-Hochberg proved their procedure to be reasonable also without this assumption. Write $R^j = \#\{i: P^i \le P^j\}$ for the *rank* of P^j in the ordered set $P^{(1)} \le P^{(2)} \le \cdots \le P^{(p)}$: hence $P^{(j)} = P^{R^j}$.

THEOREM 1 (Benjamini-Hochberg). *If $P^1, \ldots, P^p$ are independent random variables with values in $[0, 1]$, with every P^j for $j \in I_0 \subset \{1, \ldots, p\}$ possessing the uniform distribution on $[0, 1]$, then*

$$\mathrm{FDR}_{\mathrm{BH}} := \mathrm{E} \frac{\#\{j: P^{(j)} \le j\alpha/p, R^j \in I_0\}}{\#\{j: P^{(j)} \le j\alpha/p\}} \le \frac{|I_0|}{p} \alpha.$$

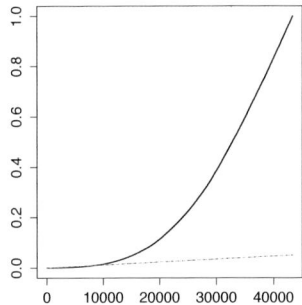 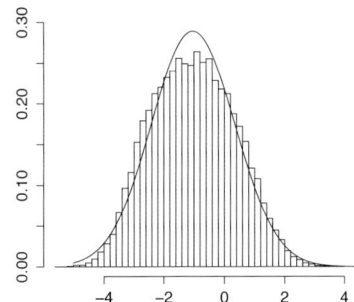

Figure 4. Histogram (left) and cumulative (right) of 43.376 p-values, transformed to a Gaussian scale, which could be used as estimates of the density $f = F'$ and cumulative distribution F of a random p-value. The straight line in the left picture gives the Benjamini–Hochberg procedure: all p-values left of the intersection point are declared significant. The curve overlaid on the histogram represents a Gaussian density of the same mean and variance and fits the data not very well. (Data collected at VU Medical Center, available from the Gene Expression Omnibus as Series GSE35477.)

The indices $j \in I_0$ correspond to the genes that are not differentially expressed, whence the fraction $|I_0|/p$ is similar to the probability π_0. If we make the identification $\pi_0 = |I_0|/p$, then the theorem can be interpreted as saying that the false discovery rate of the Benjamini-Hochberg procedure is bounded above by $\pi_0 \alpha \le \alpha$. It should be stressed that the expectation in the theorem is relative to *any* given marginal distributions of the P^j. In particular it does not refer to the Bayesian setup discussed previously.

The assumption that the p-values are independent can be relaxed. The theorem remains true if they are positively dependent, and can be adapted to general dependence. Also several methods for estimating the proportion π_0 of non-differentially expressed genes have been studied, allowing to control the FDR to a desired level α.

There are also many other innovations in multivariate testing, including the use of "permutations" or "closed testing". In applications such as *genome-wide association studies* one may attempt to model and estimate the dependence between the various tests, which should lead to better power to detect differences. Furthermore, an alternative empirical Bayesian method might estimate the probabilities $P(\Delta^j = 1 | P^j)$, using an estimate of the density of $P^1, \dots, P^p$ (see Figure 4).

Summary. When scientists investigate very many possible causes they always find some that seem active in their dataset. Statistics is needed to determine whether these are real causes or just pop up by chance. This is particularly true for the present day massive datasets. Some novel methods have become popular in the past decade.

3 Shrinkage and empirical Bayes

In the classical measurement error model of statistics one obtains n independent replicate observations $X_1, \dots, X_n$ of a given constant θ, each with a different random error from a centred univariate Gaussian distribution: $X_i = \theta + e_i$, where the probability $P(a \le e_i \le b)$ that the *error* e_i is between a and b is given by the area under a Gaussian curve (see Figure 5; we write

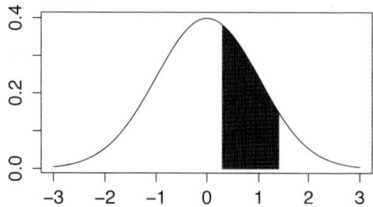

Figure 5. Gaussian density $x \mapsto \sigma^{-1}(2\pi)^{-1/2}e^{-\frac{1}{2}(x-\theta)^2/\sigma^2}$, shown with $\theta = 0$ and $\sigma = 1$

$X_i \sim N(\theta, \sigma^2)$, for σ the standard deviation of e_i). In this case computing the *sample mean* $\bar{X} = n^{-1}\sum_{i=1}^{n} X_i$ is the accepted method to recover θ from the data. It fulfils several desirable principles (maximum likelihood, objective Bayes, equivariance relative to the translation group) and enjoys several quality criteria (unbiased with minimum variance, best equivariant, minimax, admissible for symmetric loss). It is therefore a big surprise that when given simultaneously $p \geq 3$ samples of measurements on p constants $\theta^1, \ldots, \theta^p$, computing p sample means $\bar{X}^j$ may not be the optimal strategy. This was proved almost 60 years ago in [13].

THEOREM 2 (Stein, 1956). *The vector $\bar{X} = (\bar{X}^1, \ldots, \bar{X}^p)$ of sample means $\bar{X}^j$ of $p \geq 3$ independent random samples $X_1^j, \ldots, X_n^j$ of size n from $N(\theta^j, \sigma^2)$-distributions, for $j = 1, \ldots, p$, is inadmissible as an estimator of $\theta = (\theta^1, \ldots, \theta^p) \in \mathbb{R}^p$ relative to joint quadratic loss.*

The term *inadmissible* in this theorem has a precise definition: an estimator is inadmissible if there exists a uniformly (i.e. for any parameter value) better estimator. In the present situation the inadmissibility of $\bar{X}$ means that there exists a measurable function $T(\bar{X})$ of the vector of means such that

$$\mathrm{E}_\theta \|T(\bar{X}) - \theta\|_2^2 < \mathrm{E}_\theta \|\bar{X} - \theta\|_2^2, \qquad \forall \theta \in \mathbb{R}^p.$$

Here $\|\cdot\|_2$ is the Euclidean norm and E_θ denotes expectation under the Gaussian law as indicated in the theorem. The expectations in the display are known as the *mean square errors* of the respective estimators, and are a standard way of evaluating quality.

The square of the Euclidean norm is a sum over the coordinates, and taking a joint loss simultaneously in this way is crucial to the result. For instance, if one would measure the "loss of estimation" separately over the coordinates (but then one would have p measures of loss), then one would fall back on the optimality of the sample mean. It is a bit mysterious, that the improvement when measuring the sum of the losses is also impossible for $p = 2$, and the gain starts at $p = 3$. More interesting for our present story is that the potential gain increases with p. While Stein's result was always considered somewhat of an oddity, albeit a very intriguing one, it has become very significant in the era of high-dimensional statistics. For $p = 1000$ or $p = 100000$ the gains can be very significant indeed. Moreover, for data resulting from a single genomics experiment it makes perfect sense to measure the losses simultaneously.

There are many directions in which the improvement could go. The most interesting one from our current perspective is connected to *sparsity*, and reviewed in the next section. A general way of thinking of improvements is in terms of *empirical Bayes methods*, a second idea in theoretical statistics in the 1950/60s [12], which did not receive the attention that it deserves until recently.

The *Bayesian paradigm* in statistics is often contrasted with the *frequentist paradigm*. It is based on thinking of the parameter vector θ as a random object, generated from a *prior probability distribution*, which can next be updated into a *posterior probability distribution* after the data have been collected. Mathematically the latter is the conditional distribution of the parameter given the data, and Bayes's rule to obtain it is introductory probability (see box). The Bayesian approach is often criticised for the arbitrariness of a prior distribution (there exist so many). The *empirical Bayes method* alleviates this by tuning parameters of the prior distribution to the data.

In the situation of estimating $\theta \in \mathbb{R}^p$ based on Gaussian samples a simple, standard prior distribution is given by modelling $\theta^1, \ldots, \theta^p$ as independent variables with the univariate Gaussian distribution $N(0, A)$ (with mean 0 and variance A). Then the posterior distribution can be computed to be Gaussian with mean $A(A+\sigma^2/n)^{-1}\bar{X}$ (a special case of the *ridge estimator*) and variance $A\sigma^2(nA + \sigma^2)^{-1}$. If we would believe our prior guess to be realistic, with a particular value of A, then the latter posterior mean would be a natural candidate for estimating θ. However, there will rarely be a reason to believe the prior guess, where the value of A is to be most crucial part, as it determines the scale of the problem. Now the empirical Bayes argument runs as follows. If the prior guess were correct, then the means $\bar{X}^j$ would be distributed as independent variables from the $N(0, A + \sigma^2/n)$ distribution. In that case $\hat{A}: = \|\bar{X}\|_2^2/(p-2) - \sigma^2/n$ would be a reasonable estimator of A, and hence

$$T(\bar{X}):= \frac{\hat{A}}{\hat{A} + \sigma^2/n}\bar{X} = \left(1 - \frac{(p-2)\sigma^2}{n\|X\|_2^2}\right)\bar{X} \tag{2}$$

would be a reasonable estimator of θ. Now this intuitive argument does not prove anything, but it *can* be proved that the latter *James–Stein* estimator beats the estimator $\bar{X}$, significantly if p is large and θ not too far from the origin. (The James-Stein estimator is also not admissible, but further possible improvement is minor.) Bayesian reasoning aided by a frequentist touch thus solved a problem that could not be solved by principles such as maximum likelihood.

There are other possible Bayesian arguments. For instance, instead of assuming a Gaussian distribution with a single unknown A for the θ^j, we might adopt a completely general prior distribution G (cf. [8]). Then the $\bar{X}^j$ would have probability densities

$$x \mapsto \int (\sqrt{n}/\sigma)\, e^{-n(x-s)^2/(2\sigma^2)}\, dG(s),$$

and G could be estimated from the data by the maximum likelihood estimator

$$\operatorname*{argmax}_{G} \prod_{j=1}^{p} \int (\sqrt{n}/\sigma)\, e^{-n(\bar{X}^j-s)^2/(2\sigma^2)}\, dG(s).$$

Such schemes are promising also for more complicated models, and currently under investigation.

The possibility of combining, through prior probability distributions, various sources of information into an analysis is often quoted as an advantage of the Bayesian method. In particular, in the area of data science (including genomics), one may think of using the many existing data-bases to form priors. Although in 2013 we celebrated that Bayes's paper was published 350 years ago, the usefulness of Bayesian statistics still remains a point of debate [5,6].

Bayes's rule, continuous case

If a variable random θ follows a probability distribution Π (meaning that $P(\theta \in A) = \Pi(A)$ for every measurable set A) and given θ a random variable X follows a probability density $x \mapsto p(x \mid \theta)$ relative to a σ-finite dominating measure μ (meaning $P(X \in B \mid \theta) = \int_B p(x \mid \theta) \, d\mu(x)$ for every measurable set A), then θ given X follows the probability distribution given by

$$P(\theta \in A \mid X) = \frac{\int_A p(X \mid \theta) \, d\Pi(\theta)}{\int p(X \mid \theta) \, d\Pi(\theta)}.$$

In Bayesian statistics Π is called the *prior distribution* and $\Pi(\cdot \mid X) = P(\theta \in \cdot \mid X)$ the *posterior distribution*. The formula can be written in short in the form $d\Pi(\theta \mid X) \propto p(X \mid \theta) \, d\Pi(\theta)$, which shows that the prior weights $d\Pi(\theta)$ are multiplied with the likelihood to obtain the posterior weights $d\Pi(\theta \mid X)$.

Summary. Some mathematical ideas in statistics from the 1950s and 1960s have suddenly become relevant to tackle big data questions, for instance in medical research.

4 Sparsity and penalised estimation

The *linear regression model* is a very simple statistical model, which nevertheless has been a work horse in applied statistics for almost a century. It tries to explain an "outcome" or "response" y by a linear function of measured variables $x^1, x^2, \ldots, x^p$, and takes the form

$$Y_i = \beta^1 x_i^1 + \beta^2 x_i^2 + \cdots + \beta^p x_i^p + e_i, \qquad i = 1, \ldots, n,$$

where i labels measurements of different instances $(x^1, \ldots, x^p, y)$ or individuals, $\beta^1, \ldots, \beta^p$ are unknown numerical parameters, and an "error" e_i is added to the equation as in almost no practical situation the instances will belong to a single p-dimensional subspace of $\mathbb{R}^n$. (Superscripts refer to different input variables and parameters and do not denote powers.) The linearity of the model refers mostly to the dependence on the parameters β^j, as transformed values fo the inputs x^j, such as squares and higher powers, are routinely added as additional measured variables if that improves the fit of the model.

The model is better written in vector form as $Y = X\beta + e$, where X is an $(n \times p)$-matrix with entries $X_{i,j} = x_i^j$, and $\beta \in \mathbb{R}^p$ is the vector of parameters. The error vector e is a random vector that is assumed to be centred at 0; a convenient model is to let its coordinates be independent mean zero Gaussian variables with a common variance σ^2, i.e. $P(a \le e_i \le b)$ is the area under the Gaussian curve, shown in Figure 5, with $\theta = 0$.

Then the model has in total $(p + 1)$ unknowns and n observations $Y_1, \ldots, Y_n$. An old rule of thumb is that one needs at least five observations per unknown, i.e. $n \ge 5p + 5$, to obtain a reliable fit of the model. However if every einput x^j corresponds to a gene, then one may readily have more unknowns than measurements, as taking measurements is still laborious, expensive, or impossible, due to a lack of cancer patients. Genomics demands an analysis with numbers in the order $n \approx 30 - 300$ and $p \approx 500 - 10000$, which has lead to rethinking the

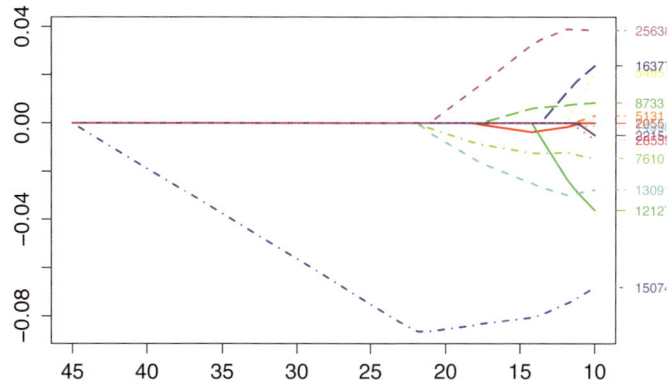

Figure 6. Solution path of the Lasso applied to genomics data: Y is an indicator for carcinoma (versus adenoma) and X gives expression levels on $p = 30000$ genes in tissue taken from the colon of $n = 68$ humans. Each piecewise linear curve gives the value (vertical axis) of a coordinate of the minimizer $\hat{\beta}_\lambda \in \mathbb{R}^p$ of the criterion (3) as a function of λ (horizontal axis, reversed). At most 13 of the 30.000 coordinates were ever nonzero. Collapsing of a piecewise linear curve to the horizontal axis for decreasing λ indicates a zero value of the corresponding coordinate. (Data collected at VU Medical Center, available from the Gene Expression Omnibus as Series GSE8067. See Muris et al, Br. J. Haematol., 2007 Jan; 136(1): 38-47.)

regression setup. In the last decade the phrase "$p \gg n$" has become to be understood by any statistician as referring to an analysis with (many) more unknowns than observations.

At first sight such an analysis is simply impossible. The old rule required multiple observations per parameter to allow to cut down the noise inherent in statistical observations. In the opposite setting $p \gg n$ the traditional analysis is impossible even with zero noise, for the parameters β are identifiable from measuring $Y = X\beta + e$ only if the matrix X is of rank p, which it fails to be by far if $p \gg n$.

There are two escapes. The first is that perhaps we are not interested in β, but rather in $X\beta$. This is true for certain applications and leads to a revaluation of bias and variance of estimators, which we discuss in the next section. The second is *sparsity*. This is the idea that most of the coordinates of β will be zero, or near zero. In many settings, including genomics, this is plausible. In a given biological process only a small number of genes will be active, and only those ones will have a nonzero coefficient β^j in the equation. If we knew beforehand which genes were active, then we would not include the other genes in the model and would have an old-fashioned $p \ll n$ problem. The novel aspect, which was studied vigorously in the past decade, is that we do *not*. Thus in the linear regression model there is a huge pool of potential explanations x^j, and the challenge is to filter out the relevant ones.

A simple method, which has become very popular, is the Lasso. This proposes to estimate β by minimising the criterion

$$\beta \mapsto \|Y - X\beta\|_2^2 + \lambda\|\beta\|_1. \tag{3}$$

Here $\|\cdot\|_r$ denotes the ℓ_r-norm (i.e. $\|u\|_r = (\sum_i^p |u_i|^r)^{1/r \vee 1}$), which in the display is used with $r = 2$ for vectors $u \in \mathbb{R}^n$ and with $r = 1$ for vectors in $\beta \in \mathbb{R}^p$. Furthermore, λ is a given number called the *smoothing parameter*, which must be set by the user. The Euclidean norm $\|Y - X\beta\|_2^2$ corresponds to the traditional least squares criterion, which is up to a constant equal to -2 times the logarithm of the likelihood of the observations, assuming i.i.d. centred Gaussian errors. For $\lambda = 0$ the criterion reduces to this criterion, but for increasing λ the term $\|\beta\|_1$, called a *penalty*, receives increasing weight, finally reducing the solution to 0 as $\lambda \uparrow \infty$.

The form of the ℓ_1-penalty can be motivated from various points of view. First one might guess that to model sparsity a penalty on the *number* of nonzero coefficients would be more natural, thus giving a criterion of the form

$$\beta \mapsto \|Y - X\beta\|_2^2 + \lambda \#\{j : \beta^j \neq 0\}. \tag{4}$$

The number $\#\{j : \beta^j \neq 0\}$ is often denoted by $\|\beta\|_0$, as it is the limit of the norms $\|\beta\|_r$, as $r \to 0$. The solution of minimisation problem (4) indeed possesses good statistical properties, but in high-dimensional situations this solution is difficult to compute, due to the fact that the criterion will have many local minima. In contrast, the Lasso problem (3) is convex in β, and a minimizer can be found by standard convex programming techniques, as well as specially dedicated algorithms. It is even easy to compute a complete "solution path" $\hat{\beta}_\lambda$, the minimum of (3) as a function of λ, as shown in Figure 6. A striking feature is that, in particular for large values of λ, many coordinates of $\hat{\beta}_\lambda$ are zero. Thus, although the ℓ_0-penalty seems better for the purpose, the Lasso also gives sparse estimators. A geometric explanation of this phenomenon is given in Figure 7.

Both algorithms give estimators with good statistical properties, provided the tuning parameter λ is set appropriately. In practice this is typically done by a *cross validation scheme* (fitting the model to half the data for a given λ; then estimating the error on the other half of the data; finally choosing the λ that minimises this error). A universally good theoretical choice is $\lambda \asymp \sigma \|X\| \sqrt{\log p}$, for $\|X\|^2$ the maximum diagonal element of $X^T X$, as shown by the following theorem (see e.g. [2], Theorem 6.1).

THEOREM 3. *If $Y \sim N(X\beta, \sigma^2 I)$ and $\lambda = 4\sigma \sqrt{2D + 2}\|X\| \sqrt{\log p}$, for a given positive constant D, then the Lasso estimator $\hat{\beta}_\lambda$ satisfies, with probability at least $1 - 2p^{-D}$,*

$$\|X\hat{\beta}_\lambda - X\beta\|_2^2 \leq \frac{4\sigma^2 \log p \, \|\beta\|_0}{\phi^2(S_0; X)},$$

$$\|\hat{\beta}_\lambda - \beta\|_1 \leq \frac{4\sigma \sqrt{\log p} \, \|\beta\|_0}{\|X\|\phi^2(S_0; X)}.$$

Here $S_0 = \{j : \beta^j \neq 0\}$, and $\phi(S_0; X)$ is the smallest constant such that

$$\|X\| \, \|b\|_1 \leq \frac{\|Xb\|_2 \sqrt{|S_0|}}{\phi(S_0; X)}$$

for all $b \in \mathbb{R}^p$ with $\sum_{j \notin S_0} |b^j| \leq 3 \sum_{j \in S_0} |b^j|$.

The set S_0 is the set of indices of nonzero coordinates β^j of the true vector β. If this set were known, then we would of course estimate the coordinates β^j with $j \notin S_0$ by 0, and we might apply ordinary least squares (the preceding criteria without the penalty terms) to find the nonzero coordinates. This would have resulted in an estimator $\hat{\beta}$ with errors $\|X\hat{\beta} - X\beta\|_2^2$ or $\|X\| \, \|\hat{\beta} - \beta\|_1$ of order the number $\|\beta\|_0$ of nonzero coefficients (a fixed cost per coefficient). The theorem shows that the estimation errors of the Lasso estimator, which does not use the same information, are bigger by factors $\log p / \phi^2(S_0; X)$ or $\sqrt{\log p} / \phi^2(S_0; X)$. If we think of the

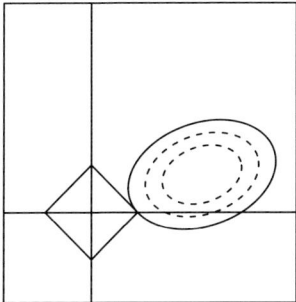

Figure 7. The level curves of the least squares criterion are ellipsoids around the least squares estimator, while the level curves of the ℓ_1-penalty are concentric to the ℓ_1-unit ball. The solutions of the Lasso problem tend to be taken at corners of an enlarged ℓ_1-ball, and therefore tend to possess zero coordinates.

logarithms as small factors (note, e.g., that $\log 10000 \approx 9$), then this is a modest price to pay, as long as the number $\phi(S_0; X)$ is not too small.

The latter number is known as a *compatibility number* of the design matrix X relative to S_0. If $p > n$, then a vector β cannot be recovered from $X\beta$ in general. The idea is that this may be different if it is known that β has few nonzero coordinates. The compatibility number measures this in a quantitative sense: if $b^j = 0$ for $j \notin S_0$, then $\|X\| \|b\|_1 \le \phi(S_0; X)^{-1} \|Xb\|_2 \sqrt{S_0}$, showing identifiability ($Xb = 0$ implies $b = 0$) and also invertibility (the norms $\|X\| \|b\|_1 / \sqrt{|S_0|}$ and $\|Xb\|_2$ are equivalent). Compatibility of a matrix X, and stronger but similar concepts with names as "restricted isometry property" or "mutual coherence property", are not easy to verify on a given matrix, but encouraging results show, for instance, that certain random matrices are with high probability compatible provided the cardinality of S_0 is not too big, i.e., the true state of nature is sufficiently sparse. Thus a typical design matrix in a genomics analysis, or other experiment, may well be compatible.

The Lasso estimator itself is always well defined, and is not limited by such theoretical questions on compatibility or recovery. In practice the usefulness of the Lasso estimator *is* limited by the reproducibility of the results. Application of the same estimation procedure to genomic data obtained under similar circumstances, or even subsampled from the original dataset, may yield drastically different estimators $\hat{\beta}$, although the estimators $X\hat{\beta}$ may be more stable. For this reason many other and new methods are explored. An older recipe originating from chemometrics, the *ridge estimator*, which replaces the ℓ_1-norm in (3) by the square ℓ_2-norm, is found more useful in some cases by some authors. Although it also shrinks the coefficients to zero relative to the least squares estimator, it does not induce sparsity, and for that reason is often combined with a selection and truncation procedure. With its smoothing parameter suitably estimated from the data the ridge estimator turns into the James-Stein estimator (see (2)) and has the good property that it uniformly beats the least squares estimator. In contrast the Lasso estimator, although good on sparse parameter vectors, does not have this property.

Summary. Even though there may be many candidate variables that potentially influence a process such as the genesis of cancer, only few may actually be relevant. New statistical methods are being developed that are dedicated to fit sparse models to measurements of such variables.

5 Bayes and uncertainty quantification

Lasso and ridge regression, and penalisation methods in general, have a *Bayesian interpretation*. The Bayesian approach in statistics consists of a description of a complete chance experiment that could have produced both the setting (i.e. model parameters) and the observed data, and then returns the conditional probability distribution of any characteristic of interest given the observed data. The assumption in Section 2 that genes are turned on or off at random is one example. In the case of the regression model of Section 4 the description must be more substantial, and could consist of an hierarchy of the form (the word "generate" may be interpreted as an "imperative" to use a random number generator or as "nature generates by some chance experiment"):

- generate a value for the error standard deviation σ from a probability density π_1 on $(0, \infty)$.
- generate values for the regression coefficients $\beta = (\beta^1, \dots \beta^p)$ from a probability density $\pi_2(\cdot | \sigma)$ on $\mathbb{R}^p$.
- generate values for the errors e_i from the $N(0, \sigma^2)$-distribution.
- compute $Y = X\beta + e$.

Different Bayesian methods are obtained by varying the *prior densities* π_1 and π_2 of the parameters. For given specifications of these densities, the vector (σ, β, Y) can then be considered generated from the joint probability density function (on the domain $(0, \infty) \times \mathbb{R}^p \times \mathbb{R}^n$)

$$(\sigma, \beta, y) \mapsto \pi_1(\sigma)\pi_2(\beta | \sigma) \frac{1}{\sigma^p (2\pi)^{p/2}} e^{-\|y - X\beta\|_2^2/(2\sigma^2)}.$$

The third factor in this product is the Gaussian density of Y given (σ, β), and referred to as the *likelihood* of the model. According to Bayes's rule given Y the probability density of (σ, β) should be updated to the *posterior density*

$$(\sigma, \beta) \mapsto \frac{\pi_1(\sigma)\pi_2(\beta | \sigma) \frac{1}{\sigma^p (2\pi)^{p/2}} e^{-\|Y - X\beta\|_2^2/(2\sigma^2)}}{\iint \pi_1(s\pi_2(b | s) \frac{1}{s^p (2\pi)^{p/2}} e^{-\|Y - Xb\|_2^2/(2s^2)} \, db \, ds}. \tag{5}$$

Depending on the choice of prior densities, computing this expression, in particular the integral in the denominator, may be a formidable task. To alleviate this the prior density π_1 is often chosen proportional to $\sigma^{-2a-1} e^{-b/\sigma^2}$, for given $a, b > 0$, a so-called Gamma prior on σ^{-2}, giving this factor the same functional form as the appearance of σ^2 in the likelihood.

The same trick can also be applied to the prior π_2 by choosing β to be a-priori Gaussian distributed with covariance matrix proportional to σ^2 (giving $\pi_2(\beta | \sigma) \propto \sigma^{-p} e^{-\beta^T \Lambda \beta / \sigma^2}$, for some matrix Λ). This leads to *ridge regression* and allows to derive the posterior density, which is also Gaussian, by simple analytic calculations.

However, for most choices of prior densities analytic calculations will not work. One may then utilise a simulation scheme (such as *Markov chain Monte Carlo* or *MCMC*) to generate pairs (σ, β) from the posterior density, and take empirical averages over these pairs to estimate any characteristic of interest. Alternatively one may analytically derive summary statistics from the analytic expression. The *posterior mode*, the point of highest density, is of prime interest, and for given σ is given by

$$\underset{\beta}{\operatorname{argmax}} \left[\log \pi_2(\beta \mid \sigma) - \frac{\|Y - X\beta\|_2^2}{2\sigma^2} \right].$$

For $\log \pi_2(\beta \mid \sigma) = -\lambda \|\beta\|_1 / (2\sigma^2)$ this precisely gives the Lasso criterion (3), which thus has a Bayesian interpretation. The corresponding prior density π_2 is the *Laplace density*, independently attributed to the coordinates β^j. In the 18th century Laplace used this density as a model for the errors e_i, before Gauss introduced the Gaussian form around 1810, which was later (and for good reasons) widely adopted for the purpose.

An important advantage of the Bayesian method is that it automatically comes with a method to quantify the remaining uncertainty after the analysis. The posterior density (5) gives the likelihood of the various values of the parameter (σ, β) given the data Y, and hence provides not only a "point estimate" (such as the mode), but gives a precise, quantitative indication of how far the truth could be from this estimate. In particular, a set of 95 % posterior probability around the mode might be used as an indication of the (im)precision of the mode. In finite-dimensional problems such a procedure is well documented, and is known to agree in many cases with "confidence sets", the standard non-Bayesian way of quantifying uncertainty. Unfortunately, in the high-dimensional situation this can be very different. For the Lasso the following result (see Theorem 7 in [4]) shows a complete failure of the Bayesian posterior.

For simplicity consider the regression model with design matrix X the identity (and hence $p = n$) and $\sigma = 1$. Write $\Pi_\lambda^{\text{Lasso}}(\cdot \mid Y)$, for the posterior distribution of β, the distribution on $\mathbb{R}^p$ with density given by (5) with σ removed and $\pi_2(\beta) \propto \lambda^p e^{-\lambda \|\beta\|_1 / 2}$ taken equal to the Laplace density with scale parameter $\lambda/2$.

LEMMA 4. *If $Y \sim N(\beta, \sigma^2 I)$, then for any $\lambda = \lambda_n \geq 1$ such that $\sqrt{n}/\lambda_n \to \infty$, there exists $\delta > 0$ such that, as $n \to \infty$,*

$$E_{\beta=0} \Pi_{\lambda_n}^{\text{Lasso}} \left(\beta : \|\beta\|_2 \leq \frac{\delta \sqrt{n}}{\lambda_n} \mid Y \right) \to 0.$$

As we noted previously the choice $\lambda \asymp \sqrt{2 \log n}$ endows the Lasso estimator (3) with its excellent properties. This easily satisfies the condition of the lemma, but for this choice the Lasso posterior distribution puts no mass on balls of radius of the order $\sqrt{n/\log n}$, which is substantially bigger than the concentration rate $(\log n \|\beta\|_0)^{1/2}$ of the Lasso estimator in the sparse situation (when $\|\beta\|_0 \ll n$). The Lasso posterior distribution may thus spread very widely, giving little information on the parameter β, even though its centre (mode) is a good estimator of β. Intuitively, this failure of the Bayesian method is explained by the fact that the Laplace prior density is used to model both the zero and nonzero coefficients, but has only a single "degree of freedom" λ. This parameter would have to be large (giving a Laplace density spiked near zero) to model the zero coefficients, but reasonable so that the prior can also model the nonzero coordinates. In other words, the Laplace prior is not a good prior to express sparsity of coefficients.

One better prior is the so-called *spike-and-slab* prior, which can be described as follows:

- generate a value for the error standard deviation σ from a probability density π_1.
- generate the number s of nonzero coefficients from a probability distribution π_2 on the set $\{0, 1, \ldots, p\}$.
- generate a random subset S of $\{1, 2, \ldots, p\}$ of size s.
- generate values for the regression coefficients $(\beta^j: j \in S)$ from a density $\pi_3(\cdot \mid \sigma, s)$ and set $\beta^j = 0$ for $j \notin S$.
- generate values for the errors e_i from the $N(0, \sigma^2)$-distribution.
- compute $Y = X\beta + e$.

The full posterior distribution relative to this prior has been shown to recover the parameter (in the sense of an analogue of Theorem 3) at the same rate as the Lasso (see [4]). So far its ability for uncertainty quantification has been shown only under restrictive conditions.

One of the challenges of current research is the general difficulty (and theoretical impossibility) of uncertainty quantification in high-dimensional models. Since not all dimensions can be seen in the data, high-dimensional uncertainty quantification appears to entail always some amount of extrapolation from the data. There is a need to develop prior models for the parameters that are reasonable in the application and that can guide the extrapolation. In the absence of such models one would fall back in worst case scenarios, which do not exploit the sparsity in the underlying phenomenon.

Summary. Big data usually also brings big noise. There is a need for methods that can quantify the confidence that one may have in a fitted mathematical model. How far is it from the truth? Bayesian methods are natural, but their validity is unclear for many of the new high-dimensional models.

6 Networks

Networks are ubiquitous in modern science and society. In genomics networks may describe how genes work together to produce proteins (genetic pathways), how copy number variations influence expression of RNA, how proteins interact with other proteins and RNA, and any other actions at the cellular level (see Figure 2).

Networks can mathematically be represented by directed or undirected graphs. A simple statistical model for network measurements in genomics is the *Gaussian graphical model*. Genes, or other actors, are represented as the nodes of a graph, labelled $1, 2, \ldots, p$, and on each such node j we obtain a measurement X^j, such that the vector $X = (X^1, X^2, \ldots, X^p)^T$ possesses a Gaussian distribution. The log expression levels of genes in a given environment may fit this scheme. Now two nodes k and l of the graph are connected by an edge if there is a certain dependence between the random variables X^k and X^l.

The assumption of Gaussianity entails that the probability density function of X takes the form

$$x \mapsto \frac{1}{\sqrt{\det \Sigma}(2\pi)^{p/2}} e^{-x^T \Sigma^{-1} x/2}, \qquad x \in \mathbb{R}^p. \tag{6}$$

Here we have assumed the vector X to be centred at zero, as we are interested in dependencies between the coordinates and not in sizes, leaving only the *covariance matrix* of X as an unknown. This is a $(p \times p)$ positive-definite, symmetric matrix with (k, l) element the covariance $\mathrm{cov}(X^k, X^l) = \mathrm{E}(X^k X^l)$ between X^k and X^l. The scaled version $\Sigma^{k,l}/\sqrt{\Sigma^{k,k}\Sigma^{l,l}}$ of this matrix has entries within the interval $[-1, 1]$ and is known as the *correlation matrix*. Gaussian vectors have the property that the coordinates X^k and X^l are stochastically independent if and only if their correlation is zero. This might motivate to form a network graph by connecting nodes k and l if and only if the (k, l)-element of Σ vanishes. However, it is considered more appropriate to use the inverse matrix $\Omega := \Sigma^{-1}$, which is called *precision matrix*, for this purpose. This matrix codes the so-called *partial correlations* of the X^k, through:

$$\mathrm{cor}(X^k, X^l \mid X^j : j \notin \{k, l\}) = -\frac{\Omega^{k,l}}{\sqrt{\Omega^{k,k}\Omega^{l,l}}}.$$

Under the Gaussian assumption the variables X^k and X^l are stochastically independent *given* the other coordinates if and only if the partial correlation is zero. Thus the partial correlation "corrects" the dependence between X^k and X^l for dependence due to other variables and is a measure of "direct dependence". This is usually viewed as a more informative expression of interaction in a network. Thus we form a *dependence graph* by inserting an edge between nodes k and l if and only if $\Omega^{k,l} \neq 0$.

In practice we need to infer Ω from data. Suppose we have expression measures $X_1, \ldots, X_n$ on a sample of n tissues, each a vector $X_i = (X_i^1, \ldots, X_i^p)^T$ of length p considered sampled from the Gaussian distribution with density (6). In the classical case that $n \gg p^2$ estimation is easy: we simply estimate each covariance $\Sigma^{k,l}$ by the *sample covariance* $S^{k,l} = n^{-1} \sum_{i=1}^{n} X_i^k X_i^l$ and obtain an estimator $\hat{\Omega}$ of the precision matrix as the inverse S^{-1} of the sample covariation matrix S. Unfortunately, although each sample covariance $S^{k,l}$ is a good estimator of $\Sigma^{k,l}$ for each (k, l) as soon as n is not too small in an absolute sense ($n \geq 30$ often suffices), if $p^2 \gg n$ taken together in the $(p \times p)$ matrix S the result is too noisy to reflect the matrix properties of Σ.

Therefore we again need dedicated methods to deal with high-dimensional data. One possibility is the *graphical Lasso* ([7])

$$\hat{\Omega} = \underset{\Omega}{\mathrm{argmax}}\Big[\ell(\Omega) - \lambda \underbrace{\sum\sum_{k \neq l} |\Omega^{k,l}|}_{\text{penalty}}\Big],$$

where $\ell(\Omega)$ is the log likelihood of the model:

$$\ell(\Omega) = \log \prod_{i=1}^{n}\Big[\sqrt{\det \Omega}\; e^{-\frac{1}{2}X_i^T \Omega X_i}\Big] = \tfrac{1}{2}\log \det \Omega - \tfrac{1}{2}\mathrm{trace}(S\Omega).$$

As for the Lasso, the ℓ_1-penalty is a compromise between computability and shrinking coefficients $\Omega^{k,l}$ to zero. Using the square ℓ_2-norm gives again a *ridge estimator*, which is even easier to compute and has some good properties. There are several other popular recipes, in some cases together with theoretical results analogous to Theorem 3 justifying their use.

Given p genomic entities, the number of parameters is presently of the order p^2, which makes network estimation challenging in practice, even in sparse situations. For obtaining stable estimates it is generally a good idea to incorporate prior knowledge in the analysis. As illustration we discuss one Bayesian approach, which incorporates direct prior knowledge of the edges in the network [9]. Such knowledge may come from a database, or perhaps from analysing healthy tissues as a preparation to studying the same network in cancer tissues.

Assume that the prior information consists of guesses on all of the possible edges, coded in a $(p \times p)$ adjacency matrix P:

$$P^{j,k} = \begin{cases} 0, & \text{if } j \text{ and } k \text{ are thought not to be related,} \\ 1, & \text{if } j \text{ and } k \text{ are thought to be related.} \end{cases}$$

Next assume the following generative model for the data $X_1, \ldots, X_n$, in the form of a hierarchical Bayesian model, where each following step is conditional on the previous one. For $j = 1, \ldots, p$:

- generate $\tau^{j,0}$ from a root inverse Gamma distribution with parameters (a_0, b_0), i.e., from the density proportional to $\tau \mapsto \tau^{-2a_0-1} e^{-b_0/\tau^2}$.
- generate $\tau^{j,1}$ from a root inverse Gamma distribution with parameters (a_1, b_1).
- generate σ^j from a root inverse Gamma distribution with small parameters (e.g., $(0.001, 0.001)$), making it widely spread on $(0, \infty)$.
- generate $\beta^{j,k}$ from a centred Gaussian distribution with standard deviation $\sigma^j \tau^{j,0}$ if $P^{j,k} = 0$ and $\sigma^j \tau^{j,1}$ if $P^{j,k} = 1$.
- generate e_i^j from a centred Gaussian distribution with standard deviation σ^j, for $i = 1, \ldots, n$.
- set $X_i^j = \sum_{k \neq j} \beta^{j,k} X_i^k + e_i^j$, for $i = 1, \ldots, n$.

In the last step we take X^k, for $k \neq j$, as given, just as in the regression model of Section 4. By known properties of Gaussian distributions the coefficient $\beta^{j,k}$ (in the "linear regression of X^j onto the other variables X^k" given in the last line) will be proportional to the partial correlation between X^j and X^k. Hence a zero value for $\beta^{j,k}$ will indicate the absence of an edge in the network. The idea of the Bayesian model is now that the coefficients $\beta^{j,k}$ for which $P^{j,k} = 0$ are generated from a prior distribution that is concentrated near zero. (A Dirac point mass at zero would have been more logical, but increases the computational burden.) This prior guess is included in the specification of $\beta^{j,k}$ at the third last line by allowing its standard deviation $\sigma^j \tau^{j,0}$ to be small if $P_{j,k} = 0$ (giving a Gaussian distribution that is peaked at zero), smaller than the standard deviation $\sigma^j \tau^{j,1}$ in case of the opposite guess $P_{j,k} = 1$. The exact sizes of these standard deviations are determined, on the average, by the parameters (a_0, b_0) and (a_1, b_1). These would typically be determined from the data by the empirical Bayes method. Besides setting the general scale of the problem this would also account for the fact that we cannot be absolutely sure that our prior guesses are correct.

Approximations to the posterior distributions of the parameters $\beta^{j,k}$ are computed in [9] using a "variational method". As the model does not explicitly include sparsity in its specification, a coefficient $\beta^{j,k}$ is declared nonzero if its posterior mean is sufficiently different from zero (relative to its posterior standard deviation). For an illustration of the results, see Figure 8. Although the procedure works well in computer simulations, as yet mathematical justifications are lacking. This is true of many other recent proposals for network analysis as well.

Summary.   Statistical models defined in terms of graphs find increasing application to analyse phenomena such as biological reaction networks and pathways, or social networks.

7 Causality

Every scientist knows that "correlation is not causation". However, causation is usually what we are interested in, not correlation. Drawing causal conclusions is best served through the

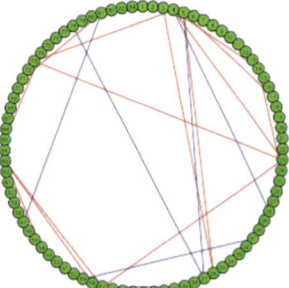 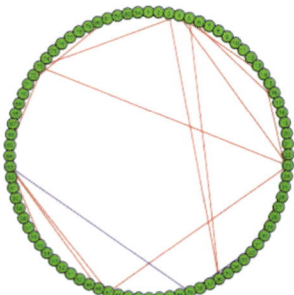

Figure 8. Right: Network of 84 genes in the Apoptosis pathway estimated on log expression in 58 cancer tissues from the lung (see [9]). Nodes, arranged in a circle for clarity of representation, represent genes. The left picture served as prior guess for computing the right picture, and was obtained by fitting a network to a sample of 49 healthy tissues. The smaller number of connections in the right picture may be the effect of a loss of functionality due to cancer, or to heterogeneity of cancer tissues. (Data publicly available as DataSet Record GDS3257 in the Gene Expression Omnibus. See Landi, M. T. et al. Gene expression signature of cigarette smoking and its role in lung adenocarcinoma development and survival. PLoS ONE, 3:e1651, 02 2008.)

design of an experiment or investigation. If the conditions of an experiment are controlled, and other factors or individuals are distributed over the conditions according to a random design, then observed differences in outcome must be due to the conditions, or to chance. The role of a statistical analysis is next to rule out (or make unbelievable) the possibility of a chance finding, leaving the conditions as a *causal* explanation.

A clinical trial for a cancer medicine or treatment provides a good example of a designed experiment. Every individual in a random sample of patients from a population is either administered the cancer medicine or not, determined by the flip of a coin. Patients who are not assigned to the cancer medicine receive a placebo, in order to ensure that the two groups of patients differ only in the treatment. At the end of the trial the two groups are compared; a statistically significant difference is attributed to be caused by the medicine.

However, much of the big data of today is not the outcome of a controlled experiment; it is *observational data*, in some cases even registered for some other purpose than the research question. Can one use databases of medical records of patients who themselves chose to come to the hospital, or were driven there for unknown reasons, to draw causal conclusions? The answer is that one should always be extremely cautious, but high-dimensional data opens up new possibilities.

Drawing causal conclusions from observational data is not new. In economics or social science most data is observational, and epidemiologists too are used to dealing with much more complicated data than provided by clinical trials. The past decades have seen much progress in developing methods for causal inference based on observational data, with *causal inference* sometimes portrayed as a new branch of mathematical science [11].

The main idea is to control for other factors that could also explain the observed outcomes, so-called *confounding variables*. Here the question which set of possible confounders should be included in the analysis seems unsolvable by purely mathematical means, and will always remain an important limitation of causal inference. The question *how* to control for the confounders may be more open to analysis. A classical way for controlling is to insert the confounders as explanatory variables in a linear regression model, of the form discussed in Section 4. One is interested in the causal effect of an explanatory variable x^1 on an outcome Y, but believes that variables $x^2, \ldots, x^p$ may also affect the outcome, and at the same time affect x^1. A regression of Y only on x^1 would also reflect indirect effects of $x^2, \ldots, x^p$ on Y via their correlation with x^1. A regression of Y onto $x^1, \ldots, x^p$ jointly may take such indirect effects away, since it allows $x^2, \ldots, x^p$ to "talk" directly. However, it is not clear that effects of the confounders is linear and additive, as the linear regression model assumes.

Now the good news is that in modern times this may not be necessary, as there may be sufficient data to fit also nonlinear models and non-additive models. There may also be sufficient data to fit the networks of confounders and direct effects.

Summary. For true progress it is usually necessary to understand a phenomenon in terms of causes and effects. Present day databases hold great promise, but new mathematical models and methods are necessary to find causes rather than mere correlations.

References

[1] Yoav Benjamini and Yosef Hochberg, Controlling the false discovery rate: a practical and powerful approach to multiple testing. *J. Roy. Statist. Soc. Ser. B*, 57(1):289–300, 1995.

[2] Peter Bühlmann and Sara van de Geer, *Statistics for high-dimensional data.* Springer Series in Statistics. Springer, Heidelberg, 2011. Methods, theory and applications.

[3] Emmanuel Candes and Terence Tao, The Dantzig selector: statistical estimation when p is much larger than n. *Ann. Statist.*, 35(6):2313–2351, 2007.

[4] Ismaël Castillo, Johannes Schmidt-Hieber and Aad van der Vaart, Bayesian linear regression with sparse priors. *Ann. Statist.*, 43(5):1986–2018, 2015.

[5] Bradley Efron, *Large-scale inference*, volume 1 of *Institute of Mathematical Statistics (IMS) Monographs.* Cambridge University Press, Cambridge, 2010. Empirical Bayes methods for estimation, testing, and prediction.

[6] Bradley Efron, Bayes' theorem in the 21st century. *Science*, 340(6137):1177–1178, 2013.

[7] Jerome Friedman, Trevor Hastie and Robert Tibshirani, Sparse inverse covariance estimation with the graphical lasso. *Biostatistics*, 9(3):432–441, 2008.

[8] Wenhua Jiang and Cun-Hui Zhang, General maximum likelihood empirical Bayes estimation of normal means. *Ann. Statist.*, 37(4):1647–1684, 2009.

[9] Gino B. Kpogbezan, Aad W. van der Vaart, Wessel N. van Wieringen, Gwenael G.R. Leday and Mark A. van der Wiel, An empirical bayes approach to network recovery using external knowledge. *preprint*, 2016.

[10] E. Landers, *De anatomische les.* VUMC en Volkskrant, Amsterdam. 2015.

[11] Judea Pearl, *Causality.* Cambridge University Press, Cambridge, second edition, 2009. Models, reasoning, and inference.

[12] Herbert Robbins, The empirical Bayes approach to statistical decision problems. *Ann. Math. Statist.*, 35:1–20, 1964.

[13] Charles Stein, Inadmissibility of the usual estimator for the mean of a multivariate normal distribution. In *Proceedings of the Third Berkeley Symposium on Mathematical Statistics and Probability, 1954-1955, vol. I*, pages 197–206. University of California Press, Berkeley and Los Angeles, 1956.

Filtering theory:
Mathematics in engineering, from Gauss to particle filters

Ofer Zeitouni

The evolution of engineering needs, especially in the areas of estimation and system theory, has triggered the development of mathematical tools which, in turns, have had a profound influence on engineering practice. We describe this interaction through one example, the evolution of filtering theory.

1 Introduction: Least square and MMSE estimators

In the late 18th century, because of advances in astronomy, the problem of fitting data to a model became important, both as practical problem (in particular, with implications for ocean navigation) and as a tool in validation of scientific models. Some of the leading scientists of the time participated in the effort to find a good procedure for such fitting. A famous dispute over priority between Legendre [14] and Gauss [7] relates to the invention of the least square method; for more on this dispute, see [15,16]. Gauss' claim of priority is based on a paper of his which is devoted to a very concrete problem: computing the meridian quadrant, the distance from the north pole to the equator along a parallel of latitude passing through Paris. Indeed, as part of revolutionary fervor, the French decided to base the (then new) metric system on $1/10^7$ of that length. Equipped with data from a 1795 geodetic survey, which measured distances along the arc Dunkirk to Barcelona, Gauss in 1799, using what he describes as "my method" and has been interpreted as a variant of the least-squares method, computed the ellipticity of the earth and the length of the meridian quadrant (in the then prevailing units, modules). See [16] for a modern analysis of Gauss' computations.

In its simplest version, which is adequate for the purpose of this paper, the least square method solves the problem of finding an unknown *parameter* θ (real or, more generally, taking values in the d-dimensional Euclidean space) given a known *model* $y = f(y,\theta)$, independent variables $\{y_i\}_{i=1}^n$ and observations $y_i = f(y_i,\theta), i = 1,\ldots,n$ supposedly fitting the model. The method of least squares consists of finding an estimator $\hat{\theta}$ of θ by minimizing the cost function $\sum_{i=1}^n (y_i - f(y_i,\theta))^2$ over θ. In the particular case of a *linear* model $f(y,\theta) = h(y) \cdot \theta$, the solution turns out to be linear in the observations $Y = (y_1,\ldots,y_n)$: with A denoting the matrix with rows $h(y_i)$,

$$\hat{\theta} = (A^T A)^{-1} A^T Y. \tag{1}$$

Regardless of the dispute on priority mentioned above, there is general agreement that Gauss was a first to give a probabilistic interpretation to the least square criterion; in modern language, Gauss assumed that $y_i = f(y_i, \theta) + v_i$ where the *errors* v_i are independent, zero mean identically distributed random variables. In fact, Gauss gave two interpretations: first, in [7], he showed that the least square method yields the maximum likelihood estimator if the distribution of the v_is takes a special form: the Gaussian distribution was born. (Shortly afterward, Laplace [13] used his recently proved Central Limit Theorem (CLT) to show that if the y_is were themselves large sample averages, the assumptions in Gauss' approach held.) A few years later, Gauss [8] showed that for linear models, the optimal unbiased linear estimator $\hat{\theta}$, in the sense of minimizing the variance $E(\hat{\theta} - \theta)^2$ (expectation over the errors), is the least square estimator. Note that the variance of the errors plays no role in the least square method.

Laplace's work [13] also serves as a prelude to the Bayesian point of view: in modern language, if θ is actually a random variable, distributed according to the Gaussian distribution, then under a Gaussian assumption on the errors and a linear model, the MMSE estimator (defined below) coincides with the least square estimator, in the limit of vanishing prior information.

Because of the fundamental importance of these notions to the continuation of the story, we digress and give it a modern mathematical formulation, somewhat simplified by ignoring the extra independent variables $\{y_i\}$ and making them part of the observation function g. Let $X \in \mathbb{R}^d$ denote a random vector, let $g : \mathbb{R}^d \to \mathbb{R}^m$ be a known function, and let $V \in \mathbb{R}^m$ denote a random vector whose entries are independent, identically distributed (i.i.d.) zero mean random variables (the *noise*), independent of X. The *observation* is the random vector

$$Y = g(X) + V. \tag{2}$$

An estimator is a (measurable) map $Z : \mathbb{R}^m \to \mathbb{R}^d$. We denote by $\mathcal{Z}$ the set of all estimators, and by $\mathcal{Z}^L$ the set of linear estimators, i.e. those of the form $Z = BY + C$ for appropriate deterministic B, C. The *Minimum Mean Square Error* (MMSE) estimator is the estimator $\hat{X} \in \mathcal{Z}$ satisfying

$$E(\|X - \hat{X}\|_2^2) = \min_{Z \in \mathcal{Z}} E(\|X - Z\|_2^2), \tag{3}$$

where $\| \cdot \|_2$ denotes the Euclidean norm. One similarly defines the optimal *linear* MMSE estimator (with minimization over $\mathcal{Z}^L$), which we denote by $\hat{X}^L$.

It is straight forward to check that $\hat{X} = E(X | \mathcal{F}_Y)$, where $\mathcal{F}_Y$ is the sigma-algebra generated by the observations; in what follows, we will write $E(X|Y)$ for the latter conditional expectation. With these definitions, the last sentence in the previous paragraph is a consequence of the following.

THEOREM 1. *Consider the observation model* (2). *Assume* $g(X) = AX$ *for some (known) deterministic matrix* A. *Assume that the entries of* V *are standard Gaussian. Assume* X *is Gaussian with zero mean and covariance matrix* R_X. *Then*

$$\hat{X} = \hat{X}^L = (A^T A + R_X^{-1})^{-1} A^T Y. \tag{4}$$

Further, if one drops the Gaussian assumption on X, V *but keeps the covariance assumptions, then the second equality in* (4) *remains true.*

The identification of the MMSE and the least square estimator occur in the limit $R_X^{-1} \to 0$.

2 Filtering of stationary processes and time series

The notions discussed in the introduction form the basis of the science of statistics, with its crucial role in modern science. However, we now shift attention to a different type of statistical analysis, this time motivated by the needs of a younger field, that of engineering and in particular mathematical engineering. (I learnt the last term from the essay [9], who attributes it to Wiener.)

In the early 20th century, the role of stochastic processes in science began to emerge somewhat sporadically. Already in 1827, the biologist Brown observed what will become Brownian motion in the movement of pollen suspended in water. Bachelier observed such motion in the prices of stocks, and began to lay the mathematical foundation for a theory of stochastic processes in his thesis [1]. Einstein [6] showed how stochastic processes (and, in fact, Brownian motion, although he was apparently unaware of Brown's observations) are relevant to the kinetic theory of gases. But it was not until Wiener [17] that a rigorous foundation for the definition of Brownian motion was established. Wiener did that by introducing the measure that bears his name on the space of continuous functions.

Wiener's original interest was purely mathematical. But then came the source of great progress in engineering and science: war. As part of the World War II United States war effort, Wiener was presented with the problem of predicting trajectories of airplanes (with the goal of shooting them down ...): he proposed a project (which was subsequently funded) on the "design of a lead or prediction apparatus in which, when one member follows the track of an airplane, another anticipates where the airplane is to be after a final lapse of time". Originally, Wiener thought that available methods would suffice for designing what, in modern language, would be called a tracking device. Soon however, Wiener discovered that noise played a major role in limiting the performance of such designs, and he came to realize that the theory of stochastic processes he was familiar with is crucial in both modeling the problem and in formulating a solution. As it turned out, in formulating his solution Wiener used a theory of factorization for integral equations that he had earlier developed with Hopf, and whose motivation, somewhat like Gauss', came from astronomy. As a practical design, Wiener's design never made it beyond a crude prototype, and in particular eventually did not contribute to the war effort. However, it had a dramatic influence on post-war engineering. See [9] for a lively account of this part of Wiener's work, and in particular his role in introducing into engineering practice the notion of an optimal solution.

We now describe the mathematical story behind these developments. There are two basic approaches to constructing stochastic processes, which can be shown to be equivalent under appropriate assumptions that we do not discuss. Wiener's approach was to construct a measure P_W on $C(\mathbb{R}_+; \mathbb{R}^d)$, the space of continuous functions taking values in $\mathbb{R}^d$, so that the increments of the (random) path are statistically independent, stationary and Gaussian; these properties turn out to uniquely determine P_W up to scaling. Later, another more general approach emerged, following the foundational work of Kolmogorov and later Doob: given a probability space $(\Omega, \mathcal{F}, P)$, a stochastic process is a measurable map from $\mathbb{R} \times \Omega$ to $\mathbb{R}$. (There are subtle measurability issues that we do not discuss in the above definition, in particular as stated it is not clear that one can speak of "paths" or "continuous paths". That this is possible is one of the great achievements of the modern foundation of probability theory.) A process $\{X(t)\}$ is *stationary* if for fixed s, the processes $\{X(t + s)\}$ has the same distribution as that of $\{X(t)\}$. Wiener modeled the prediction problem he was faced with as follows[1]. Let $X(t)$ be the signal process, scalar for simplicity, with known distribution, and suppose one observes

$$Y(t) = X(t) + N(t) \tag{5}$$

where $N(t)$ is *noise*, a zero mean process independent of the process $\{X(t)\}$, and with known distribution. How can one recover say $X(1)$ from the observation $\{Y(s), 0 \le s \le 1\}$, or even predict $X(2)$ from the same observation? specifically, how to construct an estimator $\hat{X}(t)$, function of the observations, so that $E((X(t) - \hat{X}(t))^2)$ is minimal? Note the similarity with the MMSE setup!

As we will see, in this generality the problem is hard. Motivated by what could actually be built (recall that this is the pre-computer era), Wiener added the restriction that $\hat{X}(t)$ must be a *linear* transformation of the observation, viz., for some function $h(\cdot)$,

$$\hat{X}(t) = \int_{-\infty}^{\infty} h(t-s)Y(s)ds. \tag{6}$$

The resulting estimator is a (linear) filter of the observation $\{Y(t)\}$, and the function $h(\cdot)$ is called the *impulse response* of the filter. The simplest solution is when both $X(t)$ and $N(t)$ are stationary processes, and no further restrictions are put on h. In that case, simple calculations show that

$$R_X(t) = \int_{-\infty}^{\infty} h(t-s)(R_X(s) + R_N(s))ds, \tag{7}$$

where for a zero mean stationary process $Z(t)$,

$$R_Z(t) = E(Z(0)Z(t)) = E(Z(s)Z(s+t)).$$

The Fourier transform of R_Z is called the *spectral density* of the process $Z(t)$, and denoted $S_Z(f)$. Taking the Fourier transform of (7), one obtains

$$H(f) = \frac{S_X(f)}{S_X(f) + S_N(f)}, \tag{8}$$

where H, the *transfer function* of the filter, is the Fourier transform of h. In case $N(t)$ is a multiple of white noise, formally the derivative of Brownian motion, $S_N(f) = \sigma^2$. Compare this with (4).

The solution described in (7) suffers from a serious inconvenience: it assumes knowledge of the whole path $\{Y(t)\}$, and in particular the resulting filter $h(t)$ is not *causal*: observations at time t influence the output at time smaller than t. For some applications, this can rectified by introducing a delay, that is by considering the filter $h(\cdot - \tau)$ (and truncating h at small negative time). However, in the fire-control problem that Wiener was considering, one cannot afford a delay, neither for practical reasons (after all, one is trying to shoot down an airplane before it can cause damage) not for theoretical reasons (this is a general feature of feedback systems, where delay may introduce instability). To remedy this, Wiener appealed to his earlier work with Hopf, namely the Wiener–Hopf factorization (itself motivated by physical applications,

---

1. Wiener's formulation is more general in that it allows for correlation of the noise and the signal; we simplified the setup for the sake of exposition.

most notably in astrophysics). Indeed, under the constrain of causality, instead of (7), the optimal filter satisfies

$$R_X(t) = \int_0^\infty h(t-s)(R_X(s) + R_N(s))ds, \quad t \geq 0. \tag{9}$$

One cannot apply the Fourier transform directly to solve for h; instead, define the function

$$g(t) = R_X(t) - \int_0^\infty h(t-s)(R_X(s) + R_N(s))ds,$$

and note that $g(t) = 0$ for $t \geq 0$. Taking now Fourier transforms and extending to the complex plane, one gets that

$$G(z) = S_X(z) - H(z)[S_X(z) + S_N(z)],$$

where G is analytic on the left hand plane. Since h is causal, its Fourier transform is analytic on the right hand plane. Wiener and Hopf observed that any non-negative even function $\Psi(f)$ on the real line can be factored uniquely as $\Psi = \Psi_+ \cdot \Psi_-$ with Ψ_+ analytic and bounded on the RHP, $1/\Psi_+$ analytic in the RHP, and $\Psi_-(z) = \Psi_+^*(-z^*)$. Applying this to $S_X(z) + S_N(z) = \Psi_+(z)\Psi_-(z)$ one obtains

$$\frac{G(z)}{\Psi_-(z)} = \frac{S_X(z)}{\Psi_-(z)} - H(z)\Psi_+(z). \tag{10}$$

Note that the left side is analytic on the LHP while $H(z)\Psi_+(z)$ is analytic on the RHP, with inverse Fourier transforms causal for the latter and causal in $-t$ for the former. It follows by taking inverse Fourier transforms that the inverse Fourier transform of $H(z)\Psi_+(z)$ must be equal to the causal part of the inverse Fourier transform of S_X/Ψ_-. Since taking the causal part of a function is a projection in Fourier space, denoted $\mathcal{P}_c$, we conclude after Wiener [18] that

$$H(f) = \frac{\mathcal{P}_c\left(\frac{S_X(f)}{\Psi_-(f)}\right)}{\Psi_+(f)}. \tag{11}$$

(Wiener's work appeared in 1942 as a classified report.) It is worthwhile noting that, contemporary with Wiener, Kolmogorov [12] described a general prediction procedure for stationary time series, based on the notion of innovation. That the two notions are closely related was later demonstrated by Bode and Shannon, and by Zadeh and Ragazzini.

Wiener's approach had an immense impact on the nascent field of communication and systems theory; in particular, it heavily influenced Shannon in his invention of Information Theory, as well as laid the foundation for what will become the study of systems and control. However, as a practical solution to the filtering problem, the filter in (11) has several significant drawbacks: first, the Wiener–Hopf factorization does not generalize well to the vector case (i.e., to matrix spectral densities); almost any real-life problem involves vector data, and even Wiener's original motivation, the anti-aircraft fire control problem, involved three coordinates for position and potentially three more for velocity data. More important, Wiener's approach does not deal well with non-stationary signals, which however come from structured sources. Finally, Wiener's theory computes the optimal *linear* filter, which is optimal in case the signal process is Gaussian; with the dramatic increase in computing power one is naturally led to seeking the optimal estimator, with no restriction on linearity. The next chapters in the filtering story address these issues.

3 Markov processes and state space methods

A large part of mathematics deals with dynamics, and in particular the evolution in time of certain quantities, as described by difference or differential equations. This point of view has an analogue in the theory of stochastic processes through the notion of *Markov processes.* Without being pedantic with the definition, a Markov process is a process in which, given the current value (or *state*) of the process, the future of the process becomes independent of its past. Such processes are determined by their initial distribution $\pi_0(dx) = P(X_0 \in dx)$ and the transition probabilities $P_t(x, dy) = P(X_t \in dy | X_0 = x)$ (with certain continuity assumptions on the x-dependence). In the continuous time case, an important role is played by the generator L of the process, defined as

$$Lf(x) = \lim_{t \to 0} \frac{1}{t} (P_t f(x) - f(x)),$$

where $P_t f(x) = \int f(y) P_t(x, dy)$. Under appropriate regularity conditions, the generator is a linear, elliptic second order differential operator.

We are interested in stationary stochastic processes that, at time t, take values in Euclidean space. If we insist that the process is Gaussian, and we work in discrete time, then such process can be written in general as

$$X(n + 1) = AX(n) + BW(n) \tag{12}$$

where A, B are deterministic matrices and $\{W(n)\}$ is a sequence of i.i.d. zero mean Gaussian vectors with identity covariance. A similar description holds in the continuous setup, with

$$X(t) = X(0) + \int_0^t AX(s)ds + BV(t), \tag{13}$$

where $V(t)$ is (multidimensional) Brownian motion. (We note in passing that scalar stationary Gaussian processes with rational spectral densities can always be written as components of processes of the form (12) and (13) [5]. In particular, for such processes the Wiener-Hopf factorization can be performed and Wiener's filter computed.) With such a description, we have a pathwise description of the process, and given an observation

$$Y(n) = CX(n) + W(n) \quad \text{or} \quad Y(t) = C \int_0^t X(s)ds + W(t), \tag{14}$$

where $W(t)$ is multidimensional Brownian motion and $\{W(n)\}$ are i.i.d. zero mean Gaussian vectors of identity covariance, one may return to the filtering problem discussed before: find the MMSE optimal estimator of $X(t)$ give $\{Y(s), 0 \le s \le t\}$.

One could hope that given a Markovian description, it would be possible to find a simple (read: Markovian) description of the filter. This became important during the 50s and 60s, the years of great engineering advance (again, somewhat driven by military needs and by space exploration), because of implementation issues. Indeed, often one does not know the exact statistics of the signal in advance, and one has to estimate it also from the data (in the Markov setup, this means one wants to estimate the matrices A, B, C). The spectral factorization inherent in the Wiener filter made it impossible to compute, in real time and with limited computing power, the optimal filter. Kálmán's observation [10] (see also [11] for the continuous time setup) is that spectral factorization is not needed: by taking (13) and (14) as starting points, one can write down the optimal filter itself as the solution of a difference (or differential equation). In continuous time one obtains

$$d\hat{X}(t) = A\hat{X}(t)dt + P_t C^T (dY(t) - C\hat{X}(t)dt)$$

$$\frac{dP_t}{dt} = AP_t + P_t A^T + BB^T - P_t C^T C P_t. \tag{15}$$

(In the Bayesian framework that we work in, $\hat{X}(t)$ is the conditional mean of the process given the observations, while P_t is the covariance matrix of the estimation error. The differential equation for P_t is the well studied *Riccati* equation.) We note in passing that this solution works also if one takes A, B, C time dependent, thus departing further from the stationary setup of Wiener's solution, and allowing for online adaptation. Further, it turns out that the filtering (or prediction) step is a crucial element in control systems which seek to use the observation to perform an action that in turns influences the observed process. (The theory of such control systems developed in parallel with our filtering story, with some of the main contributors to the filtering story, such as Wiener and Kálmán, also contributing to the development of control theory, and in particular stochastic control theory.) Thus, Kálmán's filter found wide range application not just in the task of cleaning observations from noise, but also in virtually every control system built, from fire control and guidance of airplanes and missiles to moon landing, robots, and innumerable other engineering tasks.

4 Nonlinear filtering and stochastic PDEs

Recall that the filtering problem, now stated in the Markovian setup, can be formulated as follows: given a Markov process $X(t)$ (usually, taking values in $\mathbb{R}^d$) with generator L and observation

$$Y(t) = \int_0^t h(X(s))ds + \sigma W(t) \tag{16}$$

where h is some known function and $W(t)$ is Brownian motion, find the optimal (in the sense of MMSE) estimator of $X(t)$ given the observations $\{Y(s), 0 \le s \le t\}$. Kálmán's theory of filtering (and some of its ad-hoc variants) is based on a Gaussian assumption, which translates to $X(t)$ being a Gauss-Markov process and h being linear. In practice, it works well in situations where the observation of the signal is rather complete, and the observation noise small. (This was observed by practitioners already in the 1960s, and much later, starting from the 1980s, given a full theoretical explanation.) However, often the measurements are incomplete, or far from being linear in the signal. Two particular examples are worth mentioning: first, in a radar measurement of a target, one measures, besides the distance to the target, also the angle. The latter is a periodic function and not well approximated by a linear model. Another related example comes from communication theory: in FM radio one modulates the frequency of a carrier by the information transmitted; modeling the latter as a stochastic process $X(t)$, the observations now uses $h(x) = \cos(x)$, which again is not a linear model. In situations of moderately strong observation noise, that is of σ moderately large, the Gaussian approximation breaks down, and a linear filter is simply inadequate.

   Thus one is faced with the task of computing the optimal filter without imposing a linearity assumption. Eventually, one would need to take implementation constraints into account, but it is remarkable that in the pre-computer area, mathematically inclined engineers set out to find the optimal filter, with no additional constraints. Following initial results of Wonham, who solved the problem completely in case $X(t)$ is a finite state space Markov process, a complete solution was obtained for $\mathbb{R}^d$ valued Markov processes in the mid 1960s by Kushner, Shiryayev and Stratonovich. A particularly elegant formulation of the solution was provided by Zakai [19]: given the a-priori density of the signal $p_0(x)$, and with L^* denoting the formal adjoint operator to L, solve the partial differential equation

$$\rho(t,x) = p_0(x) + \int_0^t L^* \rho(s,x)ds + \sigma^{-2}h(x) \int_0^t \rho(s,x) \circ dY(s). \tag{17}$$

(The right most integral in the last equation is not an ordinary integral but rather the Stratonovich form of Ito's stochastic integral.) Then,

$$\hat{X}(t) = \frac{\int x\rho(t,x)dx}{\int \rho(t,x)dx}. \tag{18}$$

The function $\rho(t,x)/\int \rho(t,x)dx$ is the conditional density of $X(t)$ given the observations up to time t, and $\hat{X}(t)$ is the conditional mean. At least in principle, the filtering problem is now completely solved. Of course, this solution requires solving, in real time, a partial differential equation. In the Gaussian case, one can check that the (un-normalized) Gaussian distribution with appropriate time-dependent mean and covariances solves Zakai's equation, and that the equations for the mean and covariance are precisely the Kálmán filter given by (15). Mathematically, this situation is the exception, not the rule, and the class of filtering problems for which solving the Zakai equation can be reduced to solving a finite number of (stochastic) *ordinary* differential equations has been classified and essentially contains exactly the linear case and some minor variations [3]. This classification was achieved using geometric ideas and in particular a Lie-algebraic framework proposed by Brockett and by Mitter. In general, however, solving the equation directly was in 1969 a computationally formidable task, and in spite of considerable improvements in computing power (and an interesting attempt to implement the equation on a chip [2]), solving directly the Zakai PDE remains so today.

5 Particle filters

A particularly nice interpretation of the solution of Zakai's equation (17) (and, in fact, a crucial element in the proof) is as a conditional expectation: one has, for any smooth bounded test function f,

$$\int \rho(t,x)f(x)dx = E\left[f(Z(t)) \exp \frac{1}{\sigma^2} \left(\int_0^t h(Z(s))dY(s) - \frac{1}{2}\int_0^t h^2(Z(s))ds \right) \right], \tag{19}$$

where $\{Z(t)\}$ is an independent copy of the Markov process $\{X(t)\}$. Such a representation is called a *Feynman–Kac* path integral, and appeared in physics in the early 1950s as Feynman's path interpretation of Schroedinger's equation. The equality in (19) suggests that simulating the expectation could yield an efficient algorithm for the evaluation of the conditional expectation. With the dramatic increase in available computing power, such a procedure becomes reasonable, if carried out effectively. It is interesting to note that such algorithms were already proposed in the early '50s as a tool for performing Feynman–Kac integration, but in the context of filtering (and hence, of simulating a family of observation-driven path integrals), this was only seriously studied starting from the 1990's.

In practice, such simulations are performed by using a variant of *genetic algorithms*. That is, a large number of individual particles are simulated, each evolving as a Markov process according to the original generator L, and in addition the population of particles is subjected to mutations and selections that replicate the weighting in the Feynman–Kac representation. On the mathematical side, proving that adding such mutations and selections effectively allow one to evaluate the expectation in (19) is not a trivial task, and requires input from several areas of modern probability, such as interacting particle systems and their mixing properties, the theory of measure valued diffusion, the stability of solutions of SPDEs to perturbations in their initial conditions and coefficients, etc; we refer to the book [4] for mathematical story behing the analysis of particle filters.

Particle filters have found a wealth of applications, from traditional applications of filtering to guidance and control to the forcasting of financial data, computational chemistry and more. As is appropriate for post 1990s development, we refer to the web site http://web.maths. unsw.edu.au/~peterdel-moral/simulinks.html for a panorama of different applications of such filters.

6 Some concluding remarks

In this note I have tried to present the relation between engineering needs and the development of mathematical theories to address them, through the evolution of the theory (and practice) of filtering theory. In the sake of brevity, I have omitted many contributions; in particular, I have not discussed the detailed developments of stochastic analysis and of the theory of stochastic partial differential equations that were necessary in order to put filtering theory on firm ground. I have not discussed aspects that led to important practical and theoretical developments, such as the study of performance bounds and of the stability of the optimal filter. It should be perhaps emphasized that the interplay between engineering applications and the mathematical theory of filtering does not end with the introduction of particle filters. In particular, the latter work well for systems of moderate dimension. In certain modern applications, and particularly in those in which biologocal or financial data is involved, the dimensionality of the system is such that particle filters cannot work well in practice. These challenges have to be met by new tools (one possibility is to harness the theory of machine learning, which may allow for an effective reduction in the size of the system). These challenges ensure that the interplay between the engineering needs and mathematical tools will continue to advance both fields.

References

[1] L. BACHELIER, Théorie de la speculation, *Ann. Sci. École Norm. Sup* **17** (1900), 21–86.

[2] J. S. BARAS, Real time architectures for the Zakai equation and applications. In *Stochastic analysis – Liber amicorum for Moshe Zakai (E. Mayer-Wolf, E. Merzbach and A. Shwartz, Eds.)*, Academic Press, Boston, MA (1991), 15–38.

[3] M. CHALEYAT-MAUREL and D. MICHEL, Des résultats de non existence de filtre de dimension finie, *Stochastics* **13** (1984), 83–102.

[4] P. Del Moral, Feynman-Kac formulae. Genealogical and interacting particle approximations. Springer, Berlin, New York (2004).

[5] J. L. Doob, The elementary Gaussian process, *Annals Math. Stat.* **15** (1944), 229-282.

[6] A. Einstein, Über die von der molekularkinetischen Theorie der Wärme geforderte Bewegung von in ruhenden Flüssigkeiten suspendierten Teilchen, *Annalen der Physik* **322** (1905), 549-560.

[7] C. F. Gauss, *Theory of the Motion of the Heavenly Bodies Moving About the Sun in Conic Sections.* Perthes and Besser, Hamburg (1809). Translation by Charles Henry Davis, reprinted by Dover, New York (1963).

[8] C. F. Gauss, *Theory of the Combination of Observations Least Subject to Errors,* translated by G. W. Stewart. SIAM, Philadelphia (1965).

[9] T. Kailath, Norbet Wiener and the development of mathematical engineering. In: The Legacy of Norbert Wiener: A Centennial Symposium (D. Jerison, D. W. Stroock, eds.), *Proc. Symp. Pure Math.* **60** (1997). AMS, Providence.

[10] R. E. Kálmán, A new approach to linear filtering and prediction problems. *Trans. ASME (Ser D) - J. Basic Eng.* **82** (1960), 34-45.

[11] R. E. Kálmán and R. S. Bucy, New results in linear filtering and prediction theory. *Trans. ASME (Ser D) - J. Basic Eng.* **83** (1960), 95-107.

[12] A. N. Kolmogorov, Interpolation and extrapolation of stationary sequences. *Izvestia Akad. Nauk. SSSR* **5** (1941), 3-14.

[13] P. S. Laplace, *Théorie analytique des probabilités,* 3rd edition, Paris (1820). Reprinted by Editions Jacques Gabay, Paris (1995).

[14] A. M. Legendre, *Nouvelles méthodes pour la determination des orbites des comètes.* F. Didot (1805), Paris.

[15] R. L. Plackett, The discovery of the method of least squares. *Biometrika* **59** (1972), 239-251.

[16] S. M. Stigler, Gauss and the invention of least squares. *Annals Stat.* **9** (1981), 465-474.

[17] N. Wiener, The average value of a functional, *London Math. Soc.* **S2-22** (1922), 454-467.

[18] N. Wiener, *Extrapolation, Interpolation and Smoothing of Stationary Time Series. With Engineering Applications.* Wiley, New York (1949). Reprinted by MIT Press.

[19] M. Zakai, On the optimal filtering of diffusion processes. *Z. Wahr. Verw. Geb.* **11** (1969), 230-243.

Mathematical models for population dynamics: Randomness versus determinism

Jean Bertoin

Mathematical models are used more and more frequently in Life Sciences. These may be deterministic, or stochastic. We present some classical models for population dynamics and discuss in particular the averaging effect in the setting of large populations, to point at circumstances where randomness prevails nonetheless.

1 Introduction

Without any doubt, Biology is amongst the sciences in which advances accomplished during the last century have been the most spectacular. For mathematicians, it is both a source of inspiration (for instance, genetic algorithms mimic natural selection to solve optimization problems), and raises formidable challenges, notably in the field of modeling. Actually, most Life Sciences require pertinent mathematical models, which should not only fit experimental data, but more importantly, should enable practitioners to make reliable analysis and predictions (for instance, one wishes to predict the outbreak of an epidemic and prevent its occurence by an appropriate vaccination program). These mathematical models may be *deterministic*, in the sense that the outputs are entirely determined by the values of the parameters of the model, or *stochastic*, when the model incorporates inherent randomness and outputs then depend not only on the parameters but also on some additional stochastic elements.

The study of the dynamics of populations is a tool of fundamental importance in this area, notably in Genetics, Ecology, and Epidemiology, to name just a few. In general, deterministic models in this field concern global or averaged features of the population, typically the size of certain sub-populations, or the proportion of individuals sharing certain characteristics. That is, the features of the population are averaged and the model aims at depicting the evolution of those averaged quantities as time passes. They are based on the implicit assumption that, roughly speaking, all individuals in a given sub-population behave essentially the same. Dynamics are usually modeled in discrete times through some difference equations, and through differential equations in continuous times. The reader is referred to the textbooks by Allman and Rhodes [1], Edelstein-Keshet [10] or Hofbauer and Sigmund [15] for some basic models in this framework and their biological motivations, and to the monograph by Bürger [5] for a comprehensive overview.

In turn, stochastic models are built either by adding noise terms to deterministic evolution equations, in order to take random fluctuations into account, or more interestingly, by considering individual behaviors which are then viewed as stochastic processes. Individual-based models permit in particular to consider how individuals collaborate or compete with each other for resources, or interact with their environment. Stochastic models of population dynamics rely essentially on Markov chains in discrete times, Markov jump processes and stochastic differential equations in continuous times, including notably branching processes and coalescent processes. We refer in particular to the monographs by Durrett [9], Haccou, Jagers and Vatutin [13], and Hein, Schierup, and Wiuf [14], and the lecture notes by Dawson [8] and Etheridge [11].

"All models are wrong but some are useful", George Box used to say. The mechanisms driving dynamics of populations in nature are extremely intricate and involve a number of diverse features, whereas mathematical models must remain tractable and thus can only incorporate a few of them. In general, mathematical models for highly complex phenomena focus on a few key variables, and view the effects of the remaining ones as small (possibly random) perturbations of the simpler model. In this respect, deciding whether to opt for a deterministic versus a stochastic model may be a delicate issue. Deterministic models are simpler to solve analytically or numerically; random models can be considerably more complicated, in particular in the individual-based case, but it is generally admitted that they may be also more realist. One may wonder whether it is useful to handle more complex random models when a deterministic answer is expected anyway, and at the opposite, one may be concerned with the risk of missing some important consequences of randomness by making an oversimplified deterministic analysis.

Of course, a first key question is whether a given mathematical model accurately describes a phenomenon of interest, which is usually answered by checking the agreement with experimental measurements. Once the scope of a model has been validated and the model is applied in concrete situations, another fundamental problem for practitioners is the comparison with available data in order to estimate its parameters, and then to be able to make reliable predictions about the future (or inferences about the past) of the population. Thus statistical analysis plays a crucial role in this area; see for instance the books by Allman and Rhodes [1] and Turchin [20]. However here we shall not discuss applied statistical aspects and rather focus on more theoretical issues.

In this text, we shall first briefly present some of the simplest, best known and widely used models for population dynamics, both in the determinist and the random frameworks. Starting from the most elementary models, which were introduced in the 18th and 19th centuries, and in which the reproduction mechanism is assumed to independent of the characteristics of the population, we shall then discuss how models have been thereafter modified and complexified in order to incorporate some more realistic features. We further address the question of possible averaging effects, which may suggest that for large populations, determinist models should prevail, and then point at simple situations for which naive intuitions may fail.

2 Some classical population models

In this section, we briefly review some classical population models, deterministic or random, both for discrete and continuous times. We shall mainly focus the simple case where each individual has a single parent, which corresponds to haploid populations (i.e. individuals have only one set of chromosomes). However this can be also relevant for sexual reproduction, either by considering the subpopulations of individuals of the same sex (in most diploid populations, mitochondria DNA are inherited exclusively from the mother), or by viewing a diploid population with N individuals each carrying a pair of chromosomes as a haploid population with $2N$ individuals each having a single chromosome.

2.1 Exponential growth model and branching processes

The simplest of all population models was considered by T.M. Malthus at the very end of the 18th century. If the size of a population at date t is measured by $P(t)$, where $t \geq 0$ is either an integer or a real number, then one assumes that $P(t)$ grows at constant rate $r \in \mathbb{R}$ in time. That is, in discrete times, the increment of the population has the form

$$\Delta P(t) := P(t + 1) - P(t) = rP(t), \tag{1}$$

whereas in continuous times,

$$\frac{\mathrm{d}P(t)}{\mathrm{d}t} = rP(t).$$

In terms of the initial population size $P(0)$, one thus gets

$$P(t) = (1 + r)^t P(0) \quad \text{for the discrete time version,}$$

and

$$P(t) = \mathrm{e}^{rt} P(0) \quad \text{for the continuous time version.}$$

An individual-based stochastic counterpart of the Malthus growth model in discrete time was introduced first in the middle of 19th century by I. J. Bienaymé, and then re-discovered nearly 20 years later by F. Galton and H. W. Watson. Originally, F. Galton was motivated by the study of the extinction of family names; the model can also be useful for describing, for instance, the initiation of a nuclear chain reaction, or the early stages of the spread of contagious diseases.

The building block is an integer valued random variable ξ, which represents the number of children of a typical individual. The probability distribution of ξ is called the reproduction law. One imagines that at each generation, each individual i is replaced by a random number ξ_i of individuals (the children of i), where ξ_i has the same law as ξ, and for different individuals, the ξ_i's are given by independent random variables. If $Z(n)$ denotes the number of individuals at the n-th generation, the chain $(Z(n))_{n \in \mathbb{N}}$ is known as a *Bienaymé–Galton–Watson process* (thereafter in short, BGW process), and its dynamics are depicted at each generation by the identity

$$Z(n + 1) = \xi_1^N + \cdots + \xi_{Z(n)}^N \tag{2}$$

where $\xi_1^N, \ldots$ denote independent copies of the variable ξ.

BGW processes are merely elementary prototypes of more sophisticated branching processes, which can be used to model more accurately a variety of dynamics. In short, multi-type branching processes cover the situation where individuals in the population may have different types, and the reproduction law of each individual depends on its type. Branching processes can also be defined in continuous time, possibly with values in $[0, \infty)$ rather than merely in $\mathbb{N}$. They can incorporate phenomena such as immigration, spatial displacements, migration, mutations, random environments, etc. The so-called Crump–Mode–Jagers processes deal with situations where the rate of reproduction of an individual may depend on a number of characteristics of that individual, including for instance, its age, and in particular the reproduction rate varies as time passes. This gives access to probabilistic models of age-structured populations. We refer to the book by Haccou, Jagers and Vatutin [13] for much more on this topic.

The connection with the Malthus growth model in discrete times is easy to explain. Assume that the variable ξ has a finite mathematical expectation $\mathbb{E}(\xi)$. Then it follows from (2) that there is the identity
$$\mathbb{E}(Z(n+1)) = \mathbb{E}(\xi) \times \mathbb{E}(Z(n)),$$
so that, if we set $P(n) = \mathbb{E}(Z(n))$, then we get (1) with $r = \mathbb{E}(\xi) - 1$.

BGW processes fulfill the fundamental *branching property*, which can be viewed as the stochastic analogue of the elementary additivity property of the Malthus growth model. If $(Z(n))_{n\in\mathbb{N}}$ and $(Z'(n))_{n\in\mathbb{N}}$ are two independent BGW processes with the same reproduction law, then their sum, $S(n) := Z(n) + Z'(n)$ is again a BGW process, of course with the same reproduction law. Combining the branching property with the Law of Large Numbers, one sees that the Malthus growth model can be viewed as the limit of BGW process started with a large initial population. Indeed, if $(Z_1(n))_{n\in\mathbb{N}}, (Z_2(n))_{n\in\mathbb{N}}, \ldots$ denote independent BGW processes with the same reproduction law, each started with a single ancestor, then $S_k(n) := Z_1(n) + \cdots + Z_k(n)$ is a BGW process started with k ancestors, and the Law of Large Numbers entails that, provided that $\mathbb{E}(\xi) < \infty$,

$$\lim_{k\to\infty} \frac{1}{k} S_k(n) = \mathbb{E}(Z(n)) = P(n).$$

This suggests that, on average, when the number of ancestors is large, the population should increase exponentially fast when $\mathbb{E}(\xi) > 1$ (the super-critical case), should decay exponentially fast when $\mathbb{E}(\xi) < 1$ (the sub-critical case), and should remain roughly stable when $\mathbb{E}(\xi) = 1$ (the critical case). However, this is not entirely correct; in fact a critical BGW process always become eventually extinct, despite of the fact that the mathematical expectation of $Z(n)$ remains constant, and no matter how large the initial population is. The analysis of the extinction is a cornerstone of the theory of branching processes; see in particular Haccou, Jagers and Vatutin [13].

The fact that in the super-critical case, the size of the population increases indefinitely exponentially fast is clearly unrealistic. Thus modeling the dynamics of a population by a BGW process can only be pertinent at early stages of its development, and one needs different models to describe the long term behaviors.

2.2   Models with regulated growth

The fact that exponential growth of populations is unrealistic for a large time horizon yield P.-F. Verhulst to introduce in 1838 the so-called *logistic equation*[1] to describe the dynamics of populations with self-limiting growth. Roughly speaking, the underlying idea is that the rate of growth should be proportional to both the existing population and the amount of available

---

1.   In turn, the equation was re-derived several times in the sequel, notably by R. Pearl, and is sometimes also known as the Verhulst-Pearl equation.

resources, and informally, the effect of the latter is to slow down the growth of the population when it is already large. The equation, in continuous time, has the form

$$P'(t) = \frac{dP(t)}{dt} = rP(t)(1 - P(t)/K)$$

where $r \geq 0$ should be thought of as the growth rate in absence of self-regulation (typically when the population is small), and $K > 0$ is known as the carrying capacity.

The logistic equation can also be written in the form

$$P'(t)/P(t) = r(1 - P(t)/K);$$

observe that the rate of growth $P'(t)/P(t)$ is positive when $P(t) < K$, negative when $P(t) > K$, and approaches 0 when $P(t)$ is close to K. The carrying capacity K corresponds to the limiting size of the population when times goes to infinity. Indeed, the logistic equation can be solved,

$$P(t) = \frac{KP(0)e^{rt}}{K + P(0)(e^{rt} - 1)},$$

so in particular $\lim_{t \to \infty} P(t) = K$.

During the early 20th century, A.J. Lotka proposed a related equation for predator-prey systems, which was then later on re-derived independently by V. Volterra. Typically, consider a population of prey, whose size at time t is denoted by $N(t)$, and a population of predators, whose size at time t is denoted by $P(t)$. Imagine that the population of prey grows naturally at constant rate and that the presence of predators induces an additional rate of decay proportional to the size of the population of predators. That is,

$$\frac{dN(t)}{dt} = (a - bP(t))N(t) \tag{3}$$

where a and b are two positive constants. In turn, the population of predators declines naturally at a constant rate, and only increases in the presence of preys with rate proportional to the size of the population of prey:

$$\frac{dP(t)}{dt} = (-c + dN(t))P(t), \tag{4}$$

where c and d are also two positive constants. Informally, the growth of the prey population is regulated by the predator population, and vice-versa. The system formed by (3) and (4) is known as the *Lotka-Volterra equations*; although it has no simple explicit solution, it can be proved that solutions are always periodic.

The Lotka–Volterra equations can be modified and incorporate a further logistic growth element reflecting the competition between individuals of the same species (these are known as the competitive Lotka-Volterra equations), or to any number of species competing against each other. We refer the interested reader to Chapter 2 in Hofbauer and Sigmund [15]. We also mention that there exist other differential equations for modeling regulated growth, for instance those named after, among others, Gompertz, von Bertalanffy and Weibull.

In turn, there exist stochastic versions of the logistic growth equation and of the Lotka-Volterra equations, which mainly rely on stochastic calculus. In particular, Lambert [19] introduced branching processes with logistic growth, which, in the simpler case of continuous processes, are viewed as solutions to the stochastic differential equation

$$dZ(t) = aZ(t)dt - bZ^2(t)dt + c\sqrt{Z(t)}dW(t)$$

with $(W(t))_{t\geq 0}$ a Brownian motion. Providing a detailed account of the meaning of such stochastic differential equation would drift us too far away from our purpose, let us simply mention that, in comparison with the deterministic logistic equation, the additional stochastic term $c\sqrt{Z(t)}\mathrm{d}W(t)$ is meant to take into account random fluctuations of the model.

Somewhat similarly, the stochastic version of the Lotka-Volterra equations takes the form

$$\begin{cases} \mathrm{d}X(t) = (aX(t) - bY(t)X(t))\mathrm{d}t + \sigma_1\sqrt{X(t)}\mathrm{d}W_1(t) \\ \mathrm{d}Y(t) = (cY(t) - dX(t)Y(t))\mathrm{d}t + \sigma_2\sqrt{Y(t)}\mathrm{d}W_2(t) \end{cases}$$

where $(W_1(t))_{t\geq 0}$ and $(W_2(t))_{t\geq 0}$ are two (possibly correlated) Brownian motions. Let us simply observe that, since stochastic integrals have (usually) zero mathematical expectation, $\mathbb{E}(Z(t))$ (respectively, $\mathbb{E}(X(t))$ and $\mathbb{E}(Y(t))$) solve the deterministic logistic (respectively, Lotka-Volterra) equation.

In Section 4, we shall further see that the logistic growth model is also related to other random individual-based models with large constant size population, which describe the evolution of the size of a sub-population carrying a favorable allele in the regime of strong selection.

2.3   Constant size populations

We shall now present some basic (stochastic) population models with constant total size $N \geq 2$, where subpopulations of different types can be distinguished. One is then interested in the evolution of these subpopulations as time passes. First, the *Wright–Fisher model* was introduced around 1930 for studying the transmission of genes for diploid populations, but for the sake of simplicity, we shall consider here a haploid version. Imagine thus a population with fixed size N and non-overlapping generations, such that for each individual i at generation $n+1$, the parent of i is an individual chosen uniformly at random amongst the N individuals at generation n, independently of the other individuals. Because for a given individual j at generation n, the event that j is the parent of i has probability $1/N$, and is independent of the event that j is the parent of another individual i' at generation $n + 1$, the number of children v_j of j is given by the sum of N independent Bernoulli variables with parameter $1/N$, and has therefore the binomial distribution with parameter $(N, 1/N)$, that is

$$\mathbb{P}(v_j = k) = \binom{N}{k}N^{-k}(1 - 1/N)^{N-k} \quad \text{for } k = 0, 1, \ldots, N.$$

Note that the sequence of the numbers of children for individuals at the same generation, $(v_j : j = 1,\ldots,N)$, must fulfill the identity $v_1 + \cdots + v_N = N$, and thus does not consist of independent variables.

In this setting, one often supposes that individuals have types, or that their chromosomes carry certain alleles, which are transmitted to their descend; and one is interested in the propagation of those types or alleles. Assume for simplicity, that one gene has just two alleles, a and A, which are neutral for the reproduction, in the sense that the reproduction laws are the same for all individuals, no matter which allele they carry. If $P_N(n)$ denotes the number of individuals carrying allele a at generation n, then $(P_N(n) : n \geq 0)$ is a Markov chain with values in $\{0, 1,\ldots,N\}$. That is to say, roughly speaking, that conditionally on $P_N(n)$, the statistics of $P_N(n + 1)$ are independent of the values of the chain P_N for the preceding generations.

Specifically, conditionally on $P_N(n) = j$, as each individual at generation $n + 1$ has probability j/N of having a parent that carries allele a, independently of the other individuals at the same generation, one has thus

$$\mathbb{P}(P_N(n+1) = k \mid P_N(n) = j) = \binom{N}{k}\left(\frac{j}{N}\right)^k\left(1 - \frac{j}{N}\right)^{N-k}$$

for $k = 0, 1, \ldots, N$. The states $j = 0$ and $j = N$ are called absorbing, as once the chain reaches state 0 (respectively, N), there are no more individuals carrying allele a (respectively, A), and allele a (respectively, A) has therefore disappeared forever in the population. One then says that fixation occurred for allele A (respectively, a). It is easy to check, that the probability that allele a eventually fixates is simply given by the proportion of individuals carrying allele a in the initial population.

In 1958, P. Moran introduced a somewhat even simpler model, now in continuous time, that can be described as follows. Each individual lives for an exponential time, say with fixed rate $r > 0$ (that is the probability that the lifetime is larger than t equals e^{-rt}), and then dies and is instantaneously replaced by a clone of an individual sampled uniformly at random in the remaining population. Using elementary properties of independent exponential variables, we can also reformulate the evolution as follows. Starting from a population with N individuals, after an exponential time with parameter $2rN$, we select a pair of individuals uniformly at random and then replace one of them by a copy of the other.

If one assumes, just as in the Wright-Fisher model, that the population has a fixed size N and that individuals carry a neutral allele a or A, and if $P_N(t)$ denotes the number of individuals carrying allele a at time t, then the process in continuous time $(P_N(t))_{t\geq 0}$ is Markovian, and more precisely is a so-called *birth and death process*. Its dynamics are determined by the transition semigroup

$$T_{N,t}f(j) = \mathbb{E}_j\big(f(P_N(t))\big), \quad j = 0, 1, \ldots, N,$$

where $f : \{0, 1, \ldots, N\} \to \mathbb{R}$ denotes a generic function on the state space and $\mathbb{E}_j$ the mathematical expectation given $P_N(0) = j$ (that is, there are j individuals carrying allele a at the initial time $t = 0$). However, the transition semigroup is not so simple to express explicitly, and one rather works with its time differential. Specifically, the infinitesimal generator of the process is defined as

$$\mathcal{A}_N f(j) = \lim_{t \to 0+} \frac{T_{N,t}f(j) - f(j)}{t},$$

and fulfills the Kolmogorov's forward and backward equations

$$\frac{dT_{N,t}f(j)}{dt} = T_{N,t}(\mathcal{A}_N f)(j) = \mathcal{A}_N(T_{N,t}f)(j).$$

This forms a systems of linear differential equations, and enables to recover the transition semigroup as the exponential of a matrix:

$$T_{N,t} = \exp(t\mathcal{A}_N).$$

Roughly speaking, the infinitesimal generator provides an analytic description for the evolution of a Markov process in terms of its rates of jumps. In the setting of the Moran process, one has for every $j = 1, \ldots, N - 1$, that

$$\mathcal{A}_N f(j) = r(f(j+1) + f(j-1) - 2f(j))\frac{j(N-j)}{N-1}. \tag{5}$$

Indeed, $j(N - j)/(N - 1)$ is both the rate at which an individual is picked amongst the j individuals carrying the allele a and is replaced by a clone of one of the $N - j$ individuals carrying allele A, resulting in a decay of one unit for P_N, and the rate at which an individual is picked amongst the $N - j$ individuals carrying the allele A and is replaced by a clone of one of the j individuals carrying allele a, resulting in an increase of one unit for P_N. Finally, the states 0 and N are absorbing, therefore $\mathcal{A}_N f(0) = \mathcal{A}_N f(N) = 0$. It is easy to check, that, just as for the Wright-Fisher model, the probability that allele a eventually fixates when j individuals carry allele a in the initial population equals j/N.

Both the Wright-Fisher model and the Moran model can be considerably generalized by considering several alleles, or by changing the reproduction laws (which yields the Cannings model [6]), or by incorporating further phenomena such as mutations, selection, competition, etc. In this direction, we shall see in the forthcoming Section 4.2 that introducing a strong selection mechanism in Moran's model yields the deterministic logistic growth model of Section 2.2 in the limit when the size of the population goes to infinity.

2.4 Genealogies

In the early 1980's, J. F. C. Kingman [17,18] formalized the idea of building the genealogical tree of a population by tracing backwards the ancestral lineages, and since then, his new approach has had a considerable impact on the way genealogies are viewed and studied. Roughly speaking, the object of interest is the process obtained by letting time run backward and observing the partition of the present population into the sub-populations, called blocks thereafter, which have the same ancestors at time $-t < 0$ as t increases.

To describe both its mechanism and its purposes as simply as possible, we start by considering the Moran population model which was presented in the preceding section. Recall that for a population with size N, a pair of individuals is picked uniformly at random after an exponential time with parameter rN, one of them is replaced by a copy of the other, and thereafter the process continues to evolve according to the same dynamics, independently of its past. We now assume that $r = (N - 1)/2$, which induces no loss of generality since we can simply rescale time as a function of the size of the population. Loosely speaking, this means that one unit of time corresponds in average to $(N - 1)/2$ generations. Observe that then $rN = \binom{N}{2}$ is the number of pairs of individuals at any given time. Now imagine that the present time is used as the origin of times, and that we follow the ancestral lineages of individuals backward in time. Specifically, we label the individuals in the present population by $\{1, \ldots, N\}$ uniformly at random, and for every $t \geq 0$, we obtain a partition $\Pi_N(t)$ of $\{1, \ldots, N\}$ into subpopulations which stem from the same ancestor at time $-t$; see Figure 1 below. The process $(\Pi_N(t))_{t \geq 0}$ is called the N-coalescent. Plainly, as t increases, these partitions get coarser, and $(\Pi_N(t))_{t \geq 0}$ evolves by coalescent events which are related to certain reproduction events in the past of the Moran process. More precisely, $\Pi_N(0)$ is the partition into singletons, and the first instant $t > 0$ at which $\Pi_N(t)$ does not only consists of singletons, corresponds to the last reproduction event before the present time. It has an exponential distribution with parameter $\binom{N}{2}$. At this instant, the ancestral lineages of two individuals chosen uniformly at random coalesce (i.e. merge).

By iteration, one can check that the process $(\Pi_N(t))_{t \geq 0}$ is Markov and its rates of jumps are given as follows. When the state of the process is given by some partition of $\{1, \ldots, N\}$, say with blocks $B_1, \ldots, B_k$, where $k \geq 2$, then it stays at this state during an exponential time with

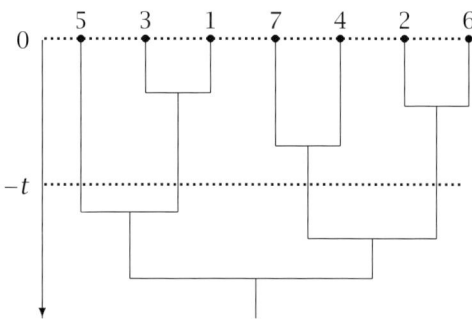

Figure 1. N-coalescent for $N = 7$. $\Pi_N(t) = (\{1,3\}, \{2,6\}, \{4,7\}, \{5\})$

parameter $\binom{k}{2}$, and then two blocks chosen uniformly at random amongst $B_1, \ldots, B_k$ (and independently of the waiting time) merge into a single block of the new partition. This corresponds to focussing on the k ancestors in the population at time $-t$ who generate the k subpopulations that form the entire present population, and letting time run further backward until a pair of their ancestral lineages meets. Equivalently, we may equip each pair of blocks $\{B_i, B_j\}$ with an independent standard (i.e., with mean 1) exponential variable, say $\mathbf{e}_{i,j}$, so that the minimum of those exponential variables over all possible pairs, $\mathbf{e} = \min_{1 \le i < j \le k} \mathbf{e}_{i,j}$, has the exponential distribution with parameter $\binom{k}{2} = k(k-1)/2$. If we denote by $\{i_0, j_0\}$ the pair of indices for which the minimum is achieved, then the two blocks B_{i_0} and B_{j_0} are merged at the instant $\mathbf{e}$, that is, they are replaced by $B_{i_0} \cup B_{j_0}$. The process eventually reaches the trivial partition with a single block, which is the absorbing state. The time ζ_N to absorption is the age of the most recent common ancestor to all individuals in the present population. Note that it can be expressed as

$$\zeta_N = \sum_{k=2}^{N} \frac{2}{k(k-1)} \epsilon_k$$

where the ϵ_k are independent standard exponential variables, since then $\epsilon_k / \binom{k}{2}$ has the exponential law with parameter $\binom{k}{2}$. In particular there is the bound

$$\mathbb{E}(\zeta_N) \le \sum_{k=2}^{\infty} \frac{2}{k(k-1)} = 2.$$

Kingman pointed at the remarkable property of sampling consistency of N-coalescents. For every integer $N' < N$, if we write $\Pi'_N(t)$ for the restriction of the random partition $\Pi_N(t)$ to $\{1, \ldots, N'\}$, then the process $(\Pi'_N(t))_{t \ge 0}$ is an N'-coalescent. By taking projective limits, this enables the construction of a version for infinite populations. Namely, there is a process with values in the space of partitions of $\mathbb{N} = \{1, 2, \ldots\}$, which we denote by $(\Pi(t))_{t \ge 0}$, such that for every integer N, the restriction of $\Pi(t)$ to $\{1, \ldots, N\}$ is an N-coalescent. One calls $(\Pi(t))_{t \ge 0}$ *Kingman's coalescent*.

The calculations for the expected absorption time in the paragraph above show, that even though $\Pi(0)$ is the partition into singletons and has thus infinitely many blocks, for every

$t > 0$, $\Pi(t)$ has almost surely finitely many blocks. One says that the coalescent comes down from infinity. This property has notably a key role in the problem of estimating the age of "mitochondrial Eve", i.e. the most-recent common female ancestor of all present-date humans; see Chang [7] and the discussion thereafter. This is an interesting instance of a concrete problem where a mathematical model is crucially needed in order to infer estimations of quantities to which it impossible to have a direct access.

During the last 15 years or so, various extensions of Kingman's coalescent have been considered. The so-called Λ-coalescents, which were introduced independently by Pitman and Sagitov, cover situations where multiple mergers may occur, whereas in Kingman's coalescent, each coalescent event involves exactly 2 blocks. In particular, this may be relevant to describe genealogies for species with extreme reproductive behavior (occasionally, a single parent may have a huge offspring of the same order as the whole population), such as certain marine species. In a different direction, the analysis of the genealogy of spatially structured populations motivated the introduction of spatial versions of coalescent processes. Overall, coalescent theory has been mainly developed in the neutral case (absence of selection), with the notable exception of recent developments initiated by Brunet and Derrida. These authors considered population of branching type in which only the best fitted children are selected for the next generation. This changes considerably the genealogy. Roughly speaking it is no longer described by Kingman's coalescent as one might have expected, but rather by the Bolthausen-Sznitman coalescent, an important special case of a Λ-coalescent which has first appeared in connection with spin-glass models in Statistical Physics. See Berestycki, Berestycki and Schweinsberg [3] and references therein for much more on this topic.

Kingman's analysis of genealogies relies crucially on the hypothesis that populations have a constant size, which is of course not very realistic when applied to real life models (think for instant of the growth of the human population in history), and a difficult question is to develop useful models which cover the case when the size of the population is time-varying. Further Kingman's coalescent only applies to haploid populations, and genealogies for in the diploid situation is far more complex. In particular, key biological phenomena such as recombination have to be taken into account, see notably the work of Baake [2] and collaborators in this area.

3　When should one expect deterministic averages?

Roughly speaking, the pertinence of deterministic models is often justified by the assertion that, due to the Law of Large Numbers, a quantity evaluated for each individual and averaged over a large population shall be nearly deterministic. A possible objection when applying bluntly this simple rule of thumb, is that the implicit hypothesis that the population can be modeled as a family of independent individuals with the same distribution may be unrealistic in practice, putting in doubt the legitimacy of the conclusions. Actually, much less restrictive requirements than independence are needed for the validity of the conclusions of Law of Large Numbers; and the issue of whether an average over a large population is essentially deterministic or random, can be clarified by a simple covariance analysis. Recall that, roughly speaking, a square integrable random variable is close to a constant (which then coincides with its mathematical expectation) if and only if its variance is small. We shall now present this covariance analysis tailored for our purposes.

Consider for every integer $n \geq 1$, a random population of size $P(n) \geq 1$, say $\mathcal{P}_n = \{x_i : i = 1, \ldots, P(n)\}$, where the labelling of the individuals is irrelevant for our purposes, and let $f_n : \mathcal{P}_n \to \mathbb{R}$ denote some real-valued function evaluated for each individual of the population. We may think of $f_n(x)$ as some real trait, that is, a real number measuring some characteristic of the individual x; for instance, $f_n(x)$ may denote the adult size of x, or the weight of x at birth, etc. We are interested in the average of f_n over the population,

$$\bar{f}_n = \frac{1}{P(n)} \sum_{i=1}^{P(n)} f_n(x_i),$$

which is a random quantity since the population is random. Roughly speaking, the next result provides an elementary answer – actually, almost a tautology – to the question of whether $\bar{f}_n$ is nearly deterministic when n is large. Recall that when ξ and ξ' are two (real) random variables, their covariance is denoted by

$$\mathrm{Cov}(\xi, \xi') = \mathbb{E}(\xi\xi') - \mathbb{E}(\xi)\mathbb{E}(\xi'),$$

whenever this quantity is well-defined.

LEMMA 1. *For each $n \geq 1$, sample two individuals X_n and X'_n in the population $\mathcal{P}_n$ uniformly at random with replacement, and set $\xi_n = f_n(X_n)$ and $\xi'_n = f_n(X'_n)$. Assume further that*

$$\sup_{n \geq 1} \mathbb{E}(\xi_n^2) < \infty.$$

Then for every $\ell \in \mathbb{R}$, the following two assertions are equivalent:

(i) $\lim_{n \to \infty} \bar{f}_n = \ell$ in $L^2(\mathbb{P})$.
(ii) $\lim_{n \to \infty} \mathbb{E}(\xi_n) = \ell$ and $\lim_{n \to \infty} \mathrm{Cov}(\xi_n, \xi'_n) = 0$.

Proof. Because ξ_n and ξ'_n are two traits sampled uniformly at random with replacement, we have

$$\mathbb{E}(g(\xi_n, \xi'_n)) = \mathbb{E}\left(\frac{1}{P(n)^2} \sum_{i=1}^{P(n)} \sum_{j=1}^{P(n)} g(f_n(x_i), f_n(x_j))\right)$$

for every measurable function $g : \mathbb{R}^2 \to \mathbb{R}_+$. We deduce that

$$\mathbb{E}(\bar{f}_n) = \mathbb{E}(\xi_n) = \mathbb{E}(\xi'_n) \quad \text{and} \quad \mathbb{E}(\bar{f}_n^2) = \mathbb{E}(\xi_n \xi'_n).$$

As a consequence, the variance of $\bar{f}_n$ is simply given by

$$\mathrm{Var}(\bar{f}_n) = \mathrm{Cov}(\xi_n, \xi'_n),$$

and our claim then follows easily.   □

REMARK 2. *This elementary second moment analysis lies at the heart of the notion of propagation of chaos, developed by Marc Kac [16], which is also relevant for asymptotic study of large populations.*

Roughly speaking, Lemma 1 shows that for the empirical average of a real trait over a large random population to be close to a constant, the correlation between the traits of two randomly sampled individuals has to be small. Let us now conclude this section by illustrating Lemma 1 with the following simple example. Consider a population model with non-overlapping generations, and some real trait t for that population. Imagine now that traits are transmitted from parents to children up-to an independent perturbation η which has a fixed distribution. That is, if y is a child of x, then $t(y) = t(x) + \eta_y$, where η_y is a random variable distributed as η. Assume also that the η_y are further independent for different individuals, and that η is centered [$\mathbb{E}(\eta) = 0$], and has finite variance $\sigma^2 = \mathbb{E}(\eta^2) < \infty$.

We write $\mathcal{P}_n$ for the population at the n-th generation, which has size $\text{Card}(\mathcal{P}_n) = P(n)$. For every $r \in \mathbb{R}$, we are interested in the proportion of individuals at the n-th generation having a trait smaller than $r\sigma^2\sqrt{n}$, that is

$$\bar{f}_n(r) = \frac{\text{Card}\{x \in \mathcal{P}_n : t(x) \le r\sigma^2\sqrt{n}\}}{\text{Card}(\mathcal{P}_n)}.$$

Let us assume that if X_n and X'_n denote two individuals picked uniformly at random in the population $\mathcal{P}_n$, and if $y_n \le n$ denotes the generation of the most recent common ancestor of X_n and X'_n, then

$$\lim_{n \to \infty} \mathbb{E}\left(\frac{y_n}{\sqrt{n}}\right) = 0. \tag{6}$$

This is a very mild assumption which is fulfilled by many natural models; note that it also implies that $\lim_{n \to \infty} P(n) = \infty$, as otherwise the probability that $X_n = X'_n$ would not tend to 0 as $n \to \infty$ and (6) would fail.

We assert that then, the repartition of traits in the population is asymptotically Gaussian, viz.

$$\lim_{n \to \infty} \bar{f}_n(r) = \frac{1}{\sqrt{2\pi}} \int_\infty^r \exp(-x^2/2)dx. \tag{7}$$

Let us now briefly show how this follows from Lemma 1. Note that the traits of X_n and X'_n can be expressed in terms of the trait, say τ_n, of their most recent common ancestor, in the form

$$t(X_n) = \tau_n + \beta_n, \quad t(X_n) = \tau_n + \beta'_n$$

with

$$\beta_n = \eta_{y_n+1} + \cdots + \eta_n, \quad \beta'_n = \eta'_{y_n+1} + \cdots + \eta'_n,$$

where y_n stands for the generation of the most recent common ancestor of the two individuals, and $\eta_1, \ldots, \eta_n$ and $\eta'_1, \ldots, \eta'_n$ are independent copies of η.

On the one hand, the trait of the most recent common ancestor of X_n and X'_n has the same law as $t_0 + \eta_1 + \cdots + \eta_{y_n}$, where t_0 denotes the trait of the ancestor of the entire population and may be assumed deterministic for the sake of simplicity, and as a consequence

$$\mathbb{E}(\tau_n) = t_0 \quad \text{and} \quad \text{Var}(\tau_n) = \mathbb{E}(\eta) \times \mathbb{E}(y_n).$$

We then see from (6) that $\mathbb{E}(\tau_n^2/n)$ converges to 0 as $n \to \infty$.

On the other hand, an easy application of the Central Limit Theorem combined with the assumption (6) shows that $\beta_n/\sigma\sqrt{n}$ and $\beta'_n/\sigma\sqrt{n}$ converge in distribution as n tend to ∞ towards a pair of independent standard Gaussian variables. We conclude from above that the same holds for the rescaled traits $t(X_n)/\sigma\sqrt{n}$ and $t(X'_n)/\sigma\sqrt{n}$. In particular, if f_n denotes the indicator function of the interval $(-\infty, r\sigma^2\sqrt{n}]$, then, as $n \to \infty$

$$\mathbb{E}(f_n(X_n)) = \mathbb{P}(t(X_n) \le r\sigma^2\sqrt{n}) \longrightarrow \frac{1}{\sqrt{2\pi}}\int_\infty^r \exp(-x^2/2)\mathrm{d}x,$$

and $\mathrm{Cov}(f_n(X_n), f_n(X'_n)) \to 0$. Since $\bar{f}_n(r)$ can be expressed in the form

$$\bar{f}_n(r) = P(n)^{-1}\sum_{x\in\mathcal{P}_n} f_n((x)),$$

the conclusion (7) follows from Lemma 1.

REMARK 3. *The example presented above can be considerably generalized in the setting of branching random walk, where (7) can be viewed as an elementary version of much deeper limit theorems due mainly to Biggins; see e.g. [4].*

4 Sources of random averages in large populations

We shall now discuss situations where the empirical average of a trait remains intrinsically random, even for large populations. Keeping in mind the elementary Lemma 1, we shall describe circumstances where, even though the size of the population tends to infinity, the traits of two randomly sampled individuals remain correlated. We stress that these simple examples aim at illustrating typical phenomena rather than at describing realistic population models.

4.1 The effect of small ancestral populations
The first example illustrates the classical effect of small ancestral populations; the informal idea is that random events should have in general a larger stochastic impact on small populations than on large ones, due to averaging effects for large populations, and that this randomness can then propagates to the next generations.

We consider a population modeled by a simple *Pólya urn*. That is, imagine that at initial time, we have two individuals, one carrying allele A and the other one carrying allele a, which are viewed as two balls, one labelled A and one labelled a, placed in a urn with infinite capacity. At each step, we pick a ball in the urn uniformly at random, note its label, and then replace it into the urn together with an additional ball having the same label. The interpretation in terms of population dynamics is that at each step, an individual is picked uniformly at random in the current population, and gives birth to a clone, that is that types are transmitted without change from parents to children.

At first sight, it may seem awkward to model a population as an urn[2], however, this is precisely what happens in the following situation. A *Yule process* $(Y(t))_{t\ge 0}$ is one of the simplest

---

2. Actually, urn models can be quite useful for population modeling. A further important example is Hoppe's urn which can be depicted as follows. We start with an urn with two balls, one with a label, say A, and one unlabeled ball and proceed as in Pólya's model, except that when the unlabelled ball is picked, then it is replaced in the urn together with a ball having a new label which was not present in the urn before. This can be used as a simple model for neutral mutations, and explain the celebrated Ewens sampling formula (which is further discussed in the next section) and its connections to Poisson-Dirichlet random partition for the repartition of alleles.

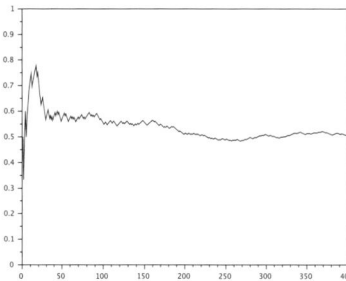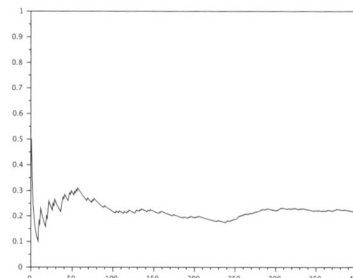

Figure 2. Two simulations of a Pólya urn illustrating convergence to a random limit

random population process in continuous time. It is a pure birth process, which describes the evolution of the size of a population in which each individual gives birth to a clone at rate 1, independently of the other individuals. Imagine now that a two-type Yule process starts from an individual with allele A and an individual with allele a (alleles are implicitly assumed to be neutral for the reproduction). When the population reaches size n, it remains unchanged for an exponentially distributed time with parameter n, and then an individual is selected uniformly at random in the current population and duplicated. So if we observe a two-type Yule process at the sequence of times when individuals duplicate, we get precisely the dynamics of a Pólya urn.

We are interested in the repartition of alleles in the population, and write $R(n)$ for the proportion of individuals carrying allele A when the total population reaches size n, that is after $n - 2$ steps. It is well-known and easy to prove that as $n \to \infty$, $R(n)$ converges to a random variable $R(\infty)$ which has the uniform distribution on $[0, 1]$. For instance, one can use the easy fact that a Yule process started from a single individual grows exponentially fast as a function of time; more precisely $\lim_{t \to \infty} e^{-t} Y(t) = W$ where W is a random variable with the standard exponential distribution, viz. $\mathbb{P}(W > x) = e^{-x}$ for all $x \geq 0$. In our situation, we have two independent Yule processes, say $(Y_A(t))_{t \geq 0}$ and $(Y_a(t))_{t \geq 0}$ in the obvious notation, and the proportion of individuals carrying allele A thus fulfills

$$\lim_{t \to \infty} \frac{Y_A(t)}{Y_A(t) + Y_a(t)} = \lim_{t \to \infty} \frac{e^{-t} Y_A(t)}{e^{-t} Y_A(t) + e^{-t} Y_a(t)} = \frac{W_A}{W_A + W_a}$$

where W_A and W_a are two independent standard exponential random variables, so that the ratio $W_A/(W_A + W_a)$ has the uniform distribution on $[0, 1]$.

So even though the size of the population tends to infinity, the repartition of alleles remains intrinsically random. Actually, the randomness of the repartition is essentially built at the early stages of the process when the population is still small. This phenomenon is of course much more general than discussed in this elementary example.

Last, it may be also interesting to compare with Lemma 1. What goes wrong when we try to apply Lemma 1, is that when we sample at random two individuals in the population when it reaches size n, and for $i = 1, 2$, write $\xi_i(n) = 1$ if the i-th sampled individual carries allele A

and $\xi_i(n) = 0$ otherwise, then when n is large

$$\mathbb{P}(\xi_1(n) = \xi_2(n) = 1) = \mathbb{P}(\xi_1(n) = \xi_2(n) = 0) \sim 1/3$$

and

$$\mathbb{P}(\xi_1(n) = 1, \xi_2(n) = 0) = \mathbb{P}(\xi_1(n) = 0, \xi_2(n) = 1) \sim 1/6.$$

That is, the correlation between two randomly sampled individuals persists even when the population is large.

4.2   Role of the regimes

Lemma 1 shows that non-deterministic averages over large populations should be expected whenever two randomly picked individuals remain asymptotically correlated when the size of the population tends to infinity. Intuitively, the correlation between individuals stems from their common history, whereas the genetic material transmitted from an ancestor tends to fade away generation after generation, for instance due to new mutations. Roughly speaking, we may expect that the older the common ancestors are and the faster traits evolve, the smaller the correlation between individuals is.

We shall illustrate this idea with two fairly different examples which are both based on Moran's model for a population with large size. In the first example, neutral mutations are superposed to the model, whereas in the second example, we shall be interested in the invasion of an advantageous allele.

Rare mutations.   So let us consider Moran's population model, for a population with a large size N. Imagine that some gene has different alleles, which are all neutral for the reproduction. Let $\mathcal{A} = \{a_1, \ldots, a_j\}$ denote the set of alleles, and for the purpose of modeling, simply assume that a mutation process is superposed (independently) to the evolution of the Moran process. Specifically, consider a transition kernel q on $\mathcal{A}$, that is, for each $a \in \mathcal{A}$, $q(a, \cdot)$ is a probability measure on $\mathcal{A}$ with $q(a, a) = 0$ for all $a \in \mathcal{A}$. So for two different alleles, a and a', $q(a, a')$ is the probability that, provided that a mutation occurs while an individual carrying allele a duplicates, this produces an individual with allele a'. In other words, during a reproduction event when an individual, say x, is replaced by a copy y' of an individual, say y, if y carries allele $a \in \mathcal{A}$, then y' also carries allele a with probability $1 - p(N)$, and carries a different allele $a' \in \mathcal{A}$ with probability $p(N)q(a, a')$, where $p(N)$ is a small parameter which depends only on N and represents the rate of mutations.

We are interested in the repartition of alleles in the current population, and more precisely, in the proportion $\bar{f}_N$ of individuals carrying a given allele a. That is, for every individual x, we write $f_N(x) = 1$ if x carries allele a and $f_N(x) = 0$ otherwise, and $\bar{f}_N = \frac{1}{N} \sum f_N(x)$, where the sum is taken over the N individuals of the present population. For this, we pick two individuals X_N and X'_N uniformly at random in the present population, and write $\xi_N = f_N(X_N)$ and $\xi'_N = f_N(X'_N)$. With no loss of generality, we assume that the lifetime of each individual in the Moran model has the exponential distribution with parameter $r = (N-1)/2$, so that the genealogy is described by an N-coalescent; see Section 2.4. In particular, the age of the most recent common ancestor, say Y_N, of X_N and X'_N has a standard exponential distribution, say $\mathbf{e}$, and therefore there are about $\mathbf{e} \times N/2$ individuals along the ancestral lineage from X_N (respectively, X'_N) to Y_N.

We now see that the asymptotic correlation between ξ_N and ξ'_N depends on whether the rate of mutations $p(N)$ is much smaller than $1/N$, or much larger that $1/N$, or of order $1/N$. Specifically:

- if $p(N) \ll 1/N$, then the probability that a mutation occurred on the ancestral lineages from X_N and from X'_N to their common ancestor is small when N is large, and therefore $\xi_N = \xi'_N$ with high probability. In this situation, the population at the present time is essentially monomorphic, that is nearly all individuals carry the same allele (which is the allele carried by their most recent common ancestor Y_N). The precise value of that allele is random, its distribution being given by the invariant law π of the Markov chain on $\mathcal{A}$ with transition kernel q. Then $\bar{f}_N$ is statistically close to a Bernoulli random variable β, with $\mathbb{P}(\beta = 1) = \pi(a)$ and $\mathbb{P}(\beta = 0) = 1 - \pi(a)$.

- if $p(N) \gg 1/N$, then with high probability, a large number of random mutations have occurred on the ancestral lineages from X_N and from X'_N to their common ancestor, and ξ_N and ξ'_N are asymptotically uncorrelated. The proportion of individuals carrying allele a is nearly deterministic, with $\bar{f}_N \sim \pi(a)$.

- In the critical case when $p(N) \sim \theta/N$ for some constant $\theta > 0$, then one can prove that the pair (ξ_N, ξ'_N) converges in distribution to a pair of random variables which are neither identical, nor independent. It follows that $\bar{f}_N$ converges in distribution to a random variable, which has not a Bernoulli law.

We further mention that in the same circle of ideas, but now in the setting of the infinite allele model of Kimura and Crow, one of the most remarkable applications of Kingman's coalescent is an illuminating explanation of the celebrated sampling formula due to Warren Ewens. Roughly speaking, imagine that each individual in Moran's model may mutate at (critical) rate θ/N, and then bears a new (neutral) allele which was never observed before. Each allelic population eventually becomes extinct, and one can check that the partition of the population at time t into subfamilies sharing the same allele converges in distribution to a certain statistical equilibrium as t goes to infinity. Ewens obtained an explicit formula for the equilibrium distribution, which arises not only in the context of population models, but much more generally in a variety of combinatorial structures. Kingman has shown that Ewens sampling formula can be recovered by superposing random marks on the branches of the genealogical tree describing an N-coalescent, and then analyzing the clusters of individuals which are connected by branches having no mark.

Strong or weak selection.   Here, just as in Section 2.4, we work with Moran's model for a population of size $N \gg 1$, and suppose for simplicity that individuals carry either allele a or allele A. Whereas we assumed in Section 2.4 that these two alleles are neutral for reproduction, we suppose here that allele a is advantageous with selection coefficient $s \in (0, 1)$, which may depend on the size of the population. This means that for each reproduction event, when a pair of individuals of different types (a, A) is picked, then it is replaced by a pair (a, a) with probability $(1 + s)/2$, and by a pair (A, A) with probability $(1 - s)/2$ (so the neutral case would correspond to setting the selection coefficient $s = 0$). Writing $P_N^s(t)$ for the number of individuals carrying allele a at time t, just as in the case without selection discussed in Section 2.4, one easily checks that the process $(P_N^s(t) : t \geq 0)$ is Markovian, now with infinitesimal generator

$$\mathcal{A}_N^s f(j) = r\left((1 + s)f(j + 1) + (1 - s)f(j - 1) - 2f(j)\right) \frac{j(N - j)}{N - 1}$$

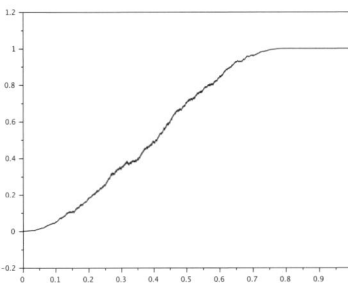 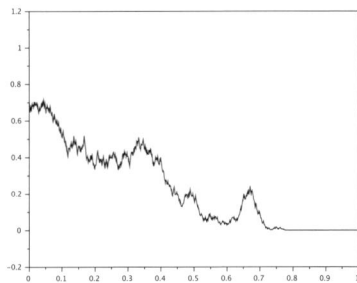

Figure 3. Ratio process for a Moran process with selection: Strong selection (left) and weak selection (right)

Note that this can also be expressed in the form

$$\mathcal{A}_N^s f(j) = \mathcal{A}_N f(j) + rs(f(j+1) - f(j-1))\frac{j(N-j)}{N-1},$$

where $\mathcal{A}_N f(j) = \mathcal{A}_N^0 f(j)$ is given in (5).

The upshot of computing explicitly infinitesimal generators is that this enables the use of powerful techniques known as *diffusion-approximation* to establish limit theorems (in a strong or in a weak sense) for Markov processes from the asymptotic behaviour of their infinitesimal generators. We refer to Ethier and Kurtz [12] or Etheridge [11] for a detailed account of this concept, which is especially useful for the study of large population models. In the present setting, this enables to prove that as the size N of the total population goes to infinity, in the regime of called *strong selection* when the parameter s does not depend on N, the ratio process $R_N^s(t) = P_N^s(t)/N$ converges, $\lim_{N \to \infty} R_N^s(t) = R^s(t)$, where $R^s : [0, \infty) \to [0, 1]$ is a deterministic function which solves the logistic growth equation

$$\frac{dR^s(t)}{dt} = rsR^s(t)(1 - R^s(t)).$$

On the other hand, when the selection parameter s depends on N such that $s \sim c/N$ with $c > 0$ constant, which is known as a regime of *weak selection*, the ratio process properly time-rescaled, $R_N^s(tN) = P_N^s(tN)/N$, converges, now merely in distribution, to a Wright-Fisher diffusion process with selection; see for instance Etheridge [11]. Figure 3 above illustrates these two cases; note that in the strong selection case, the ratio process is close to the curve of a deterministic logistic growth function, and that in the weak selection case, the favorable allele may nonetheless disappear (which would be much more unlikely in the strong selection regime).

5 Conclusions

Modeling biological phenomena has been an important source of studies in applied mathematics for many years. The main issues are to come up with models which are both realistic enough, and thus capture the essence of the phenomena of interest, and nonetheless simple enough, and thus remain tractable for statistical analysis. Finding the right trade-off between realism and simplicity is often a difficult problem, and depends on the nature of the questions to be answered. The same applies when deciding whether to opt for a deterministic or a stochastic model.

In this text, we have merely discussed a few important population models, even though of course, mathematical modeling in Biology, and more generally in Life Sciences, concerns a great variety of aspects aside from population dynamics. Further, many more population models can be found in the literature, e.g. for propagation of infectious diseases, evolutionary invasion or adaptive dynamics, parasites, etc. For individual-based models, traits of individuals are correlated through common ancestors, and in general cannot be viewed as independent variables. This may play an important role for determining whether average features over large populations are nearly deterministic or intrinsically random quantities. We have illustrated the possible impact of the small size of ancestral populations (which is often referred to as bottleneck), of the rates of mutations, and of the strength of selection. Other phenomena may of course also have a crucial role, for instance rare extreme events (think of the impact of the collision of a large asteroid with Earth).

Without any doubt, Life Sciences will continue to motivate frontline researches in mathematical modeling for many more years. Let us merely point at one challenging problem amongst others in this area. Most models of evolution focus on the haploid case and on already-existing genes, and describe how natural selection affects their frequencies depending on their relative fitnesses. The problem of modeling sexual reproduction, including genetic recombination and generation of new alleles that may appear through mutations, and of describing their dynamics and their genealogies, has only been partly addressed so far and should be the subject of deeper investigations in the future.

References

[1] ALLMAN, ELIZABETH S. and RHODES, JOHN A., *Mathematical Models in Biology: An Introduction*, Cambridge University Press, Cambridge, 2004.

[2] BAAKE, ELLEN, Deterministic and stochastic aspects of single-crossover recombination. In: *Proceedings of the International Congress of Mathematicians* **IV**, 3037–3053, Hindustan Book Agency, New Delhi, 2010.

[3] BERESTYCKI, JULIEN, BERESTYCKI, NATHANAËL and SCHWEINSBERG, JASON, The genealogy of branching Brownian motion with absorption. *Ann. Probab.* **41** (2013), 527–618.

[4] BIGGINS, JOHN D., Uniform convergence of martingales in the branching random walk, *Ann. Probab.* **20** (1992), 137–151.

[5] BÜRGER, REINHARD, *The Mathematical Theory of Selection, Recombination, and Mutation*, Wiley Series in Mathematical and Computational Biology, John Wiley & Sons, Ltd., Chichester, 2000.

[6] CANNINGS, CHRIS, The latent roots of certain Markov chains arising in genetics: a new approach. I. Haploid models. *Advances in Appl. Probability* **6** (1974), 260–290.

[7] CHANG, JOSEPH T., Recent common ancestors of all present-day individuals (with discussion and reply by the author), *Adv. in Appl. Probab.* **31** (1999), 1002–1038.

[8] DAWSON, DONALD A., *Introductory lectures on stochastic population systems.* (2010) Available at: www.researchgate.net/profile/Donald_Dawson

[9] DURRETT, RICK, *Probability Models for DNA Sequence Evolution*, Probability and its Applications, Springer-Verlag, New York, 2002.

[10] EDELSTEIN-KESHET, LEAH, *Mathematical Models in Biology*, SIAM Classics in Applied Mathematics, 2005.

[11] ETHERIDGE, ALISON, Some Mathematical Models from Population Genetics. In *École d'été de Probabilités de Saint-Flour*, Lecture Notes in Mathematics vol. 2012. Springer, 2011.

[12] ETHIER, STEWART N. and KURTZ, THOMAS G., *Markov Processes: Characterization and Convergence.* Wiley Series in Probability and Mathematical Statistics, John Wiley & Sons, Inc., New York, 1986.

[13] HACCOU, PATSY, JAGERS, PETER and VATUTIN, VLADIMIR A., *Branching Processes: Variation, Growth, and Extinction of Populations.* Cambridge Studies in Adaptive Dynamics, Cambridge University Press, Cambridge, 2007.

[14] HEIN, JOTUN, SCHIERUP, MIKKEL H. and WIUF, CARSTEN, *Gene Genealogies, Variation and Evolution, A Primer in Coalescent Theory.* Oxford University Press, Oxford, 2005.

[15] HOFBAUER, JOSEF and SIGMUND, KARL, *The Theory of Evolution and Dynamical Systems.* London Mathematical Society Student Texts, Cambridge University Press, Cambridge, 1988.

[16] KAC, M., Foundations of kinetic theory, in: *Proceedings of the Third Berkeley Symposium on Mathematical Statistics and Probability, 1954-1955, vol. III*, 171-197, University of California Press, Berkeley and Los Angeles, 1956.

[17] KINGMAN, JOHN F.C., On the genealogy of large populations, *J. Appl. Probab.* **19A** (1982), 27-43.

[18] KINGMAN, JOHN F.C., The coalescent, *Stochastic Process. Appl.* **13** (1982), 235-248.

[19] Lambert, Amaury, The branching process with logistic growth, *Ann. Appl. Probab.* **15** (2005), 1506-1535.

[20] TURCHIN, PETER, *Complex Population Dynamics: A Theoretical/Empirical Synthesis*, Monographs in Population Biology **35**, Princeton University Press, Princeton, NJ, 2003.

The quest for laws and structure

Jürg Fröhlich

The purpose of this paper is to illustrate, on some concrete examples, the quest of theoretical physicists for new laws of Nature and for appropriate mathematical structures that enables them to formulate and analyze new laws in as simple terms as possible and to derive consequences therefrom. The examples are taken from thermodynamics, atomism and quantum theory.[1]

1 Introduction: Laws of nature and mathematical structure

> *Truth is ever to be found in the simplicity, and not
> in the multiplicity and confusion of things.* - Isaac Newton

The editor of this book has asked us to contribute texts that can be understood by readers without much formal training in mathematics and the natural sciences. Somewhat against my natural inclinations I have therefore attempted to write an essay that does not contain very many heavy formulae or mathematical derivations that are essential for an understanding of the main message I would like to convey. Actually, the reader can understand essential elements of that message without studying the formulae. I hope that Newton was right and that this little essay is worth my efforts.

Ever since the times of Leucippus (of Miletus, 5th Century BC) and Democritus (of Abdera, Thrace, born around 460 BC) - if not already before - human beings have strived for the discovery of universal laws according to which simple natural processes evolve. Leucippus and Democritus are the originators of the following remarkable ideas about how Nature might work:

1. Atomism (matter consists of various species of smallest, indivisible building blocks);
2. Nature evolves according to eternal laws (processes in Nature can be described mathematically, their description being derived from laws of Nature);
3. The law of causation (every event is the consequence of some cause).

Atomism is an idea that has only been fully confirmed, empirically, early in the 20th Century. Atomism and Quantum Theory turn out to be Siamese twins, as I will indicate in a little more

---

1. One might want to add to the title: "…and for Unification" - but that would oblige us to look farther afield than we can in this essay.

detail later on. The idea that the evolution of Nature can be described by precise mathematical laws is central to all of modern science. It has been reiterated by different people in different epochs – well known are the sayings of Galileo Galilei[2] and Eugene P. Wigner[3]. The overwhelming success of this idea is quite miraculous; it will be the main theme of this essay. The law of causation has been a fundamental building block of classical physics [4]; but after the advent of quantum theory it is no longer thought to apply to the microcosm.

In modern times, the idea of *universal natural laws* appears in Newton's Law of Universal Gravitation, which says that the trajectories of soccer balls and gun bullets and the motion of the moon around Earth and of the planets around the sun all have the same cause, namely the gravitational force, that is thought to be universal and to be described in the form of a precise mathematical law, Newton's celebrated "$1/r^2$-law". The gravitational force is believed to satisfy the "equivalence principle", which says that, locally, gravitational forces can be removed by passing to an accelerated frame, (i.e., locally one cannot distinguish between the action of a gravitational force and acceleration). This principle played an important role in Einstein's intellectual journey to the General Theory of Relativity, whose 100th anniversary physicists are celebrating this year. Incidentally, the $1/r^2$- law of gravitation explains why a mechanics of point particles, which represents a concrete implementation of the idea of "atoms", is so successful in describing the motion of extended bodies, such as the planets orbiting the sun. The point is that the gravitational attraction emanating from a *spherically symmetric* distribution of matter is *identical* to the force emanating from a point source with the same total mass located at the center of gravity of that distribution. This fact is called *"Newton's theorem"*. It is reported that it took Newton a rather long time to understand and prove it. (I recommend the proof of this beautiful theorem as an exercise to the reader.) Newton's theorem also applies to systems of particles with electrostatic Coulomb interactions. In this context it has played an important role in the construction of thermodynamics for systems of nuclei and electrons presented in [1].

But rather than meditating Newton's law of universal gravitation, I propose to consider the Theory of Heat and meditate the Second Law of Thermodynamics; (see Section 2, and [2]). This will serve to illustrate the assertion that discovering a Law of Nature is a miracle far deeper and more exciting than cooking up some shaky model that depends on numerous parameters and can be put on a computer, with the purpose to fit an elephant; (a rather dubious activity that has become much too fashionable). Our presentation will also illustrate the claim that discovering and formulating a law of Nature and deriving consequences therefrom can only be achieved once the right mathematical structure has been found within which that law can be formulated precisely and further analyzed. This will also be a key point of our discussion in Section 4, which, however, is considerably more abstract and demanding than the one in Section 2.

---

2. "Philosophy is written in that great book which ever lies before our eyes – I mean the Universe – but we cannot understand it if we do not first learn the language and grasp the symbols, in which it is written. This book is written in the mathematical language."

3. "The Unreasonable Effectiveness of Mathematics in the Natural Sciences," in: Communications in Pure and Applied Mathematics, vol. 13, No. I (1960).

4. "Alle Naturwissenschaft ist auf die Voraussetzung der vollständigen kausalen Verknüpfung jeglichen Geschehens begründet." – Albert Einstein, (talk at Physical Society in Zurich, 1910).

New theories or frameworks in physics can often be viewed as "deformations" of precursor theories/frameworks. This point of view has been proposed and developed in [3] and references given there. As an example, the framework of quantum mechanics can be understood as a deformation of the framework of Hamiltonian mechanics. The Poincaré symmetry of the special theory of relativity can be understood as a deformation of the Galilei symmetry of non-relativistic physics; (conversely, the Galilei group can be obtained as a "contraction" of the Poincaré group). Essential elements of the mathematics needed to understand how to implement such deformations have been developed in [4] and, more recently, in [5]. In Section 3, we illustrate this point of view by showing that atomistic theories of matter can be obtained by deformation/quantization of theories treating matter as a continuous medium. This is a relatively recent observation made in [6] – perhaps more an amusing curiosity than a deep insight. It leads to the realization that the Hamiltonian mechanics of systems of identical point particles can be viewed as the quantization of a theory of dust described as a continuous medium (Vlasov theory).

In Section 4, we sketch a novel approach (called "*ETH approach*") to the foundations of quantum mechanics. We will only treat non-relativistic quantum mechanics, which is a theory with a globally defined time. (But a relativistic incarnation of our approach appears to be feasible, too.) Most people, including grown-up professors of theoretical physics, appear to have rather confused ideas about a theory of events and experiments in quantum mechanics. Given that quantum mechanics has been created more than ninety years ago and that it may be considered to be the most basic and successful theory of physics, the confusion surrounding its interpretation may be perceived as something of an intellectual scandal. In Section 4 we describe ideas that have a chance to lead to progress on the way towards a clear and logically consistent interpretation of quantum mechanics. For those readers who are able to follow our thought process, the presentation in Section 4 will show, I hope convincingly, how important the quest for (or search of) an appropriate *mathematical structure* is when one attempts to formulate and then understand and use new theories in physics. It will lead us into territory where the air is rather thin and considerable abstraction cannot be avoided. Clearly, neither the mathematical, nor the physical details of this story, which is subtle and lengthy, can be explained in this essay. But I believe it is sufficiently important to warrant the sketch contained in Section 4. Readers not familiar with the standard formulation of basic quantum mechanics and some functional analysis may want to stop reading this essay at the end of Section 3.

2 The Second Law of Thermodynamics

> *The thermal agency by which a mechanical effect may be obtained is the transference of heat from one body to another at a lower temperature.*
> – Nicolas Sadi Carnot

Nicolas Léonard Jonel Sadi Carnot was a French engineer who, after a faltering military career, developed an interest in Physics. He was born in 1796 and died young of cholera in 1832. In his only publication, "*Réflexions sur la puissance motrice du feu et sur les machines propres à développer cette puissance*", of 1824, Carnot presented a very general law governing *heat engines* (and steam locomotives): Let T_1 denote the absolute temperature of the boiler of a steam engine with a time-periodic work cycle, and let $T_2 < T_1$ be the absolute temperature of the environment which the engine is immersed in (coupled to). Carnot argued that the "degree of efficiency", η, of the engine, namely the amount of work, W, delivered by the engine during one work cycle divided by the amount of heat energy, Q, needed during one work cycle to heat the boiler and keep it at its (constant) temperature T_1 is always smaller than or equal to $1 - (T_2/T_1)$, i.e.,

$$\eta := \frac{W}{Q} \leq 1 - \frac{T_2}{T_1}, \tag{1}$$

a quantity always smaller than 1 – so, some of the energy used to heat the boiler is apparently always "wasted", in the sense that it cannot be converted into mechanical work but is released into the environment! Carnot's law can also be read in reverse: Unless the environment, which a heat engine is immersed in, has a temperature strictly smaller than the "internal temperature" of that heat engine (i.e., the temperature of its boiler), it is *impossible* to extract any mechanical work from the engine.

Carnot's law is unbelievably simple and unbelievably interesting, because it is universally applicable and because it has spectacular consequences. For example, it says that one cannot generate mechanical work simply by cooling a heat bath, such as the Atlantic Ocean, at roughly the same temperature as that of the atmosphere. In other words, it is impossible to extract heat energy from a heat bath in thermal equilibrium and convert it into mechanical work without transmitting some part of that heat energy into a heat bath at a *lower* temperature. One says that it is impossible to construct a *"perpetuum mobile"* of the second kind. This is very sad, because if "perpetua mobilia" of the second kind existed we would never face any energy crisis, and the climate catastrophe would not threaten us. Carnot's discovery gave birth to the theory of heat, *Thermodynamics*, and his law later led to the introduction of a quantity called *Entropy*, which I introduce and discuss below. This quantity is not only fundamental for thermodynamics and statistical mechanics, but, somewhat surprisingly, has come to play a crucial role in information theory and has applications in biology. Scientists have studied it until the present time and keep discovering new aspects and applications of entropy.

Actually, entropy was originally defined and introduced into thermodynamics by *Rudolf Julius Emanuel Clausius*, born Rudolf Gottlieb (1822–1888), who was one of the central figures in founding the theory of heat. He realized that the main consequences deduced from the so-called "Carnot cycle" (a mathematical description of the work cycle of a heat engine, alluded to above, leading to the law expressed in Eq. (1)) can also be derived from the following general principle: Consider two macroscopically large heat baths, one at temperature T_1 (the boiler) and the other one (the refrigerator) at temperature $T_2 < T_1$. Imagine that the two heat baths are connected by a thermal contact (e.g., a copper wire hooked up to the boiler at one end and to the refrigerator at the other end). Then – if there isn't any heat pump connected to the system that consumes mechanical work – *heat energy always flows from the boiler to the refrigerator*. This assertion has become known as the 2nd *Law of Thermodynamics* in the formulation of Clausius.

It led him to discover entropy. Once one understands what entropy is and what proper-ties it has, one can *derive* Clausius' law – at least for sufficiently simple heat baths – in the following precise form: Let $\mathcal{P}_i(t)$ be the amount of heat energy released per second by heat bath/reservoir i at time t, with $i = 1, 2$. Then, for sufficiently simple models of heat baths, one can show that

$$\mathcal{P}_1(t) + \mathcal{P}_2(t) \to 0, \quad \text{as } t \to \infty, \tag{2}$$

and that $\mathcal{P}_1(t)$ has a limit, denoted $\mathcal{P}_\infty$, as time $t \to \infty$. (This last claim is the really difficult one to understand; see [7] and references given there.) The 2nd Law of Thermodynamics in the formulation of Clausius then says that

$$\underbrace{\mathcal{P}_\infty \left(\frac{1}{T_2} - \frac{1}{T_1} \right)}_{>0} \geq 0, \tag{3}$$

i.e., after having waited for a sufficiently long time until the total system has reached a *stationary state*, heat bath 1 (the boiler) releases a *positive* amount of heat energy per second, $\mathcal{P}_\infty > 0$, while heat bath 2 (the refrigerator) absorbs/swallows an *equal* amount of heat energy.

Before I define entropy and present some remarks explaining what Carnot's Law (1) and Clausius' Law (3) have to do with entropy (-production), I would like to draw some *general lessons*. It has become somewhat fashionable among scientists not properly trained in mathematics and physics to try to export physical or chemical laws, such as Carnot's law (1), to other fields; e.g., to the social sciences. So, for example, inspired by Carnot's law, one might speculate that creative activities will be almost entirely absent in a completely just and harmonious society that is in perfect equilibrium, ($W = 0$ if all temperature differences vanish, i.e., $T_1 = T_2$). This might then be advanced as an argument *against* striving for social justice and harmony. One may go on and speculate that Carnot's law explains why the degree of efficiency of society's investment in various human endeavors, such as science, tends to be smaller than right-wing politicians would like it to be. Encouraged by such "insights", one starts to construct models describing the yield of society's investment in science that depend on hundreds of parameters and involve some "non-linear equations". Of course, these models turn out to be too complicated to be studied analytically, but are believed to describe "chaotic behavior". So they are put on a computer, which can produce misleading data if the models really describe chaotic behavior. But, after adjusting sufficiently many of those parameters, the models appear to describe reality, and they are then used to determine the allotment of funding to different groups of researchers. – And so on. Well, let me pause to warn against frivolous transplantations of concepts, such as entropy, non-linear dynamics, chaos, catastrophe theory, etc. from the context that has given rise to them, to entirely different contexts. Without the necessary caution this may lead to bad mistakes! For example, the degree of efficiency of society's investment in science and engineering has been much, much higher than one might reasonably and naïvely expect – Carnot's Law simply does *not* apply here!

I think that, in doing honest and serious science, one should be humble. Heat engines are highly complicated pieces of mechanical engineering. I admire the engineers who were able to see through the intricacies involved in designing such machines. That there is a *universal law* as simple as Carnot's Law (1) that applies to *all* of them should be viewed as a miracle. And, although this law is very, very simple, to *discover* it and *understand* why it applies to *all* heat engines, independently of their mechanical complexity, is a highly non-trivial accomplishment! Carnot's discovery was not published in 'Science' or 'Nature', and his *h-index*[5] equals 1. But the impact of his discovery has been truly enormous. The point I wish to make here is that the discovery of a reliable and universal Law of Nature, even of a very simple one, such as

---

5. Definition of the *h-index* – for "Hirsch index": Suppose a scientist has written $n + m$ papers of which n have been quoted (by other people) at least n times, while m have been quoted less than n times. Then the *h-index* of this scientist is $h = n$.

Carnot's, is a miracle that happens only relatively rarely. Physics is concerned with the study of phenomena that are so simple that one may hope to discover precise mathematical laws governing some of these phenomena – and, yet, the discovery of such laws is a rather rare event, and it is advisable not to expect that an interesting one is found every second year.

I now turn to some remarks about *entropy* and how one of its properties enables us to understand the origin of Carnot's and Clausius' laws; (see [8] for more details).

Let us consider a boiler at temperature T_1 and a refrigerator at temperature T_2 connected by a thermal contact. The quantity

$$\mathcal{P}_1(t)/T_1 + \mathcal{P}_2(t)/T_2$$

is an expression for the amount of "entropy production" per second at time t. If entropy production per second has a limit, as $t \to \infty$, then this limit is always *non-negative*! I will try to explain this a little later in this section. If the state of the system consisting of the boiler, the refrigerator and the thermal contact approaches a *stationary state*, as $t \to \infty^6$, then the "heat flows" $\mathcal{P}_i(t), i = 1, 2$, have limits, as t tends to ∞. Together with the simple fact (2), this implies the 2nd Law in the formulation of Eq. (3)!

Next I turn to Carnot's Law (1). Let $\Delta Q_1'(n)$ denote the amount of heat energy lost by the boiler of a heat engine, i.e., heat bath 1, in the nth work cycle, and let $\Delta Q_2'(n)$ be the amount of heat energy absorbed by the refrigerator, heat bath 2, during the nth cycle. By energy conservation, the amount of mechanical work, $W(n)$, produced by the heat engine in the nth cycle is then given by

$$W(n) = \Delta Q_1'(n) - \Delta Q_2'(n).$$

It turns out that the *"entropy production"* in the nth cycle is given by

$$\sigma(n) := -\frac{\Delta Q_1'(n)}{T_1} + \frac{\Delta Q_2'(n)}{T_2} \tag{4}$$

If the state of the total system, consisting of the boiler 1 and the refrigerator 2 connected to one another by the heat engine, approaches a *time-periodic state*, as $n \to \infty$, then the entropy production, $\sigma(n)$, per cycle approaches a *non-negative* limit, σ_∞. Then (4) implies that

$$\eta = \frac{W}{Q} \equiv \lim_{n\to\infty} \frac{W(n)}{\Delta Q_1'(n)} = \lim_{n\to\infty} \frac{\Delta Q_1'(n) - \Delta Q_2'(n)}{\Delta Q_1'(n)}$$

$$\leq 1 - \frac{T_2}{T_1} \equiv \eta_{\text{Carnot}}, \tag{5}$$

which is Carnot's law (1).

What is difficult to understand (and is only proven for simple, idealized model systems) is that, in the example considered by Clausius, the state of the total system approaches a stationary state, as time tends to ∞, while in the example of the heat engine considered by Carnot, the state of the total system approaches a time-periodic state, as the number of completed

---

6. A property that tends to be very difficult to prove and is understood only for rather simple examples; see [7].

work cycles approaches ∞. In fact, these properties can only be established rigorously for infinitely extended heat baths of a very simple kind [9]; although they are expected to hold in quite general models of heat baths in the thermodynamic limit. Real heat baths are finite, but macroscopically large. Then the laws of Clausius and Carnot are only valid *typically*, i.e., in most cases observed in the lab.

Next, we attempt to explain how entropy is defined in statistical mechanics. This may serve as a first illustration of the importance of the quest for (mathematical) structure in the natural sciences. For fun, and because, in Section 4, I will review a few facts about quantum mechanics, I choose to explain this within quantum statistical mechanics. However, the definitions and the reasoning are similar in classical statistical mechanics. Let S be a finitely extended physical system described quantum-mechanically. In standard quantum mechanics, states of S are described by so-called density matrices, ρ, acting on some Hilbert space $\mathcal{H}$. A density matrix ρ is a positive linear operator acting on $\mathcal{H}$ that has a finite trace, i.e.,

$$\text{tr}\,(\rho) := \sum_{n=1}^{\infty} \langle e_n, \rho e_n \rangle < \infty,$$

where $\{e_n\}_{n=1}^{\infty}$ is a complete system of mutually orthogonal unit vectors in $\mathcal{H}$, (i.e., an orthonormal basis in $\mathcal{H}$), $\langle \varphi, \psi \rangle$ is the scalar product of two vectors, φ and ψ, in the Hilbert space $\mathcal{H}$, and $\rho\,\psi$ is the mathematical expression for the vector in $\mathcal{H}$ obtained by applying the linear operator ρ to the vector $\psi \in \mathcal{H}$. In fact, for a density matrix, the trace is normalized to 1,

$$\text{tr}\,(\rho) = 1.$$

So-called *pure states* of S are described by orthogonal projections, P_ψ, onto vectors $\psi \in \mathcal{H}$. (Obviously, such projections are special cases of density matrices.)

The *von Neumann entropy*, $S(\rho)$, of a state ρ of S is defined by [7]

$$S(\rho) := -\,\text{tr}\,(\rho\,\ln\rho) \tag{6}$$

We note that $S(\rho)$ is non-negative, for all density matrices ρ, (because $0 < \rho < 1$), and vanishes if and only if ρ is a pure state. Moreover, it is a concave functional on the space of density matrices. (Finally, it is subadditive and "strongly subadditive" [10], a deep property with interesting applications in statistical mechanics and (quantum) information theory.) Von Neumann entropy plays an important role in statistical mechanics. However, in many applications, and, in particular, in thermodynamics, another notion of entropy is more important: *relative entropy*! This is a functional that depends on *two* states, ρ_1 and ρ_2, of S. The relative entropy of ρ_1, given ρ_2, is defined by

$$S(\rho_1\|\rho_2) := \text{tr}\,(\rho_1(\ln\rho_1 - \ln\rho_2)), \tag{7}$$

(assuming that ρ_1 vanishes on all vectors in $\mathcal{H}$ on which ρ_2 vanishes). Relative entropy has the following properties:

(i)   Positivity: $S(\rho_1\|\rho_2) \geq 0$.
(ii) Convexity: $S(\rho_1\|\rho_2)$ is jointly convex in ρ_1 and ρ_2.

---

7. For a strictly positive operator ρ, the operator $\ln\rho$ is well defined – one can use the so-called spectral theorem for self-adjoint operators to verify this claim.

(iii) Monotonicity: Let $\mathcal{T}$ be a trace-preserving, "completely positive" map on the convex set of density matrices on $\mathcal{H}$. Then

$$S(\rho_1 \| \rho_2) \geq S(\mathcal{T}(\rho_1) \| \mathcal{T}(\rho_2)).$$

See, e.g., [11] for precise definitions and a proof of property (iii). I don't think that it is important that all readers understand what is being written here. I hope those who don't may now feel motivated to learn a little more about entropy. To get them started, I include an appendix where property (i) – positivity of relative entropy – is derived. I think it is interesting to see how relative entropy and, in particular, the fact that it is *positive* can be applied to understand inequalities (1) (Carnot) and (3) (Clausius).

Let us start with (3). Let ρ_t be the true state at time t of the total system consisting of the two heat baths, 1 and 2, joined by a thermal contact; and let ρ_{eq} denote the state describing perfect thermal equilibrium of the heat baths 1 and 2 at temperatures T_1 and T_2, respectively, *before* they are coupled by a thermal contact; (the state of the thermal contact decoupled from the heat baths is unimportant in this argument). Then a rather straightforward calculation shows that

$$\frac{d}{dt} S(\rho_t \| \rho_{eq}) = \frac{\mathcal{P}_1(t)}{T_1} + \frac{\mathcal{P}_2(t)}{T_2}. \tag{8}$$

Now, if the state ρ_t of the total system approaches a stationary state, ρ_∞, as $t \to \infty$, then the right side of Eq. (8) has a limit, as t tends to ∞, and hence the time derivative, $dS(\rho_t \| \rho_{eq})/dt$, of the relative entropy $S(\rho_t \| \rho_{eq})$ has a limit, denoted σ_∞, as t tends to ∞. Since $S(\rho_t \| \rho_{eq})$ is non-negative, by property (1) above, σ_∞ must be *non-negative*; and this proves inequality (3)! Next we turn to the proof of (1). Let ρ_n denote the state at the beginning of the nth work cycle of a system consisting of the two heat baths 1 and 2 connected to one another by a heat engine that exhibits a time-periodic work cycle, and let ρ_{eq} be the state of the system with the heat engine removed (meaning that the heat engine is not connected to the heat baths and is in a state of very high temperature), which describes thermal equilibrium of the heat baths 1 and 2 at temperatures T_1 and T_2, respectively. It is quite simple to show that

$$\sigma(n) := S(\rho_{n+1} \| \rho_{eq}) - S(\rho_n \| \rho_{eq}) = -\frac{\Delta Q_1'(n)}{T_1} + \frac{\Delta Q_2'(n)}{T_2}.$$

If the total system approaches a time-periodic state, as $n \to \infty$, then the right side of this equation approaches a well-defined limit, as n tends to ∞, and hence

$$\lim_{n \to \infty} \sigma(n) =: \sigma_\infty$$

exists, too. Since $S(\rho_{n+1} \| \rho_{eq})$ is non-negative, for all $n = 1, 2, 3, \ldots$, by property (1) above, σ_∞ must be non-negative, too. This proves (5)!

Note that, apparently, the difference between the degree of efficiency, η, of a heat engine and the Carnot degree, $\eta_{\text{Carnot}} = 1 - T_2/T_1$, can be expressed in terms of the amount of entropy that is produced per work cycle.

A definition and a few important properties of (relative) entropy can be found in the appendix.

To summarize the message I have intended to convey in this section, let me first repeat my claim that the discovery of precise and universally applicable laws of Nature, such as Carnot's or Clausius' laws, is a miracle that only happens quite rarely. Second, we have just learned on these examples that a deeper understanding of the origin of laws of Nature emerges only once one has found the right mathematical structure within which to formulate and analyze them. In our examples, the key structure is the one of states of physical systems and their time evolution, and of a functional defined on the space of (pairs of) states, namely relative entropy.

3 Atomism and quantization

> *The crucial step was to write down elements in terms of their atoms ...*
> *I don't know how they could do chemistry beforehand, it didn't make any sense.*
> – Sir Harry Kroto

> *Hier (namely in Quantum theory) liegt der Schlüssel der Situation,*
> *der Schlüssel nicht nur zur Strahlungstheorie, sondern auch*
> *zur molekularen Konstitution der Materie.[8]*
> – Arnold Sommerfeld

Let me recall that, almost 500 years BCE, Leucippus and Democritus proposed the idea that matter is composed of "atoms". Although their idea played an essential role in the birth of modern chemistry – brought forward by John Dalton (1766-1844) and his followers – and in the work of James Clerk Maxwell (1831-1879) on the theory of gases, the existence of atoms was only unambiguously confirmed experimentally at the beginning of the 20th Century by Jean Perrin (1870-1942).[9] From the point of view of the mechanics known to scientists towards the end of the 19th Century, it must have looked appropriate to describe matter as a *continuous medium* – as originally envisaged for fluid dynamics by Daniel Bernoulli (1700-1782) and Leonhard Euler (1707-1783), the famous mathematicians and mathematical physicists from Basel. The atomistic structure of the Newtonian mechanics of point particles could have appeared as merely an artefact well adapted to Newton's $1/r^2$-law of gravitation, as already mentioned above. The most elegant and versatile formulation of classical mechanics known towards the end of the 19th Century was the one discovered by William Rowan Hamilton (1805-1865). In this formulation, physical quantities pertinent to a mechanical system are described as real-valued continuous functions on a space of pure states, Γ, the so-called "phase space" of the system, thought to be what mathematicians call a "symplectic manifold". The reader does not need to know what symplectic manifolds are. It is enough to believe me that if the space of pure states of a physical system has the structure of a symplectic manifold then the physical quantities of the system (i.e., the real-valued continuous functions on Γ) determine so-called Hamiltonian vector fields, which are generators of one-parameter groups of flows on Γ. As

---

8. Quantum Theory is the key not only for the theory of radiation but also for an understanding of the atomistic constitution of matter, in: "Das Plancksche Wirkungsquantum und seine allgemeine Bedeutung für die Molekularphysik".

9. As one finds in Wikipedia: In 1895, Perrin showed that cathode rays were negatively charged. He then determined Avogadro's number by several different methods. He also explained the source of solar energy as thermonuclear reactions of hydrogen.
   After Albert Einstein had published his explanation of Brownian motion of a "test particle" as due to collisions with atoms in a liquid, Perrin did experimental work to verify Einstein's predictions, thereby settling a century-long dispute about John Dalton's hypothesis concerning the existence of atoms.

such, they form a Lie algebra: To every pair, F and G, of real-valued, continuously differentiable functions representing two physical quantities of the system one can associate a real-valued continuous function, $\{F, G\}$, the so-called *Poisson bracket* of F and G. If $F = H$ is the Hamilton function of the system whose associated vector field generates the time evolution of the system, and if G is such that $\{H, G\} = 0$ then G is conserved under the time evolution determined by H – one says that G is a "conservation law". Furthermore G gives rise to a flow on Γ that commutes with the time evolution on Γ; i.e., the vector field associated with G generates a one-parameter group of *symmetries* of the system – *connection between symmetries and conservation laws.*

If one starts from a model of matter as a continuous medium and attempts to describe it as an instance of Hamiltonian mechanics one is necessarily led to consider infinite-dimensional Hamiltonian mechanics, or *Hamiltonian field theory.* Examples of Hamiltonian field theories are the *Vlasov theory* of material dust (such as large clusters of stars) often used in astrophysics and cosmology, Euler's description of *incompressible fluids* such as water, and *Maxwell's theory* of the electromagnetic field, including wave optics.

In 1925, *Heisenberg*[10] and, soon after, *Dirac*[11] discovered how one can pass from the classical Hamiltonian mechanics of a fairly general class of physical systems to the *quantum mechanics* of these systems. Their discoveries are paradigmatic examples of the importance of finding the *natural mathematical structure* that enables one to formulate a new law of Nature.

Heisenberg's 1925 paper on quantum-theoretical "Umdeutung" contains the revolutionary idea to associate with each physical quantity of a Hamiltonian mechanical system represented by a real-valued continuous function F on the phase space Γ of the system a "symmetric matrix" (more precisely, a self-adjoint linear operator), $\hat{F}$, representing the same physical quantity – *but in a quantum-mechanical description of the system!* Since matrix multiplication is non-commutative, two operators, $\hat{F}$ and $\hat{G}$, representing physical quantities of a quantum-mechanical system do generally *not* commute with one another. Dirac then recognized that one should replace the Poisson bracket, $\{F, G\}$, of two functions on phase space by $\frac{i}{\hbar} \times$ the *commutator*, $[\hat{F}, \hat{G}]$, of the corresponding matrices, where $\hbar$ is Planck's constant. Thus, the Heisenberg–Dirac recipe for the "quantization" of a Hamiltonian system reads as follows:

$$F \text{ (real function on } \Gamma) \mapsto \hat{F} \text{ (self-adjoint linear operator)}$$

$$\{F, G\} \text{ (Poisson bracket)} \mapsto \frac{i}{\hbar}[\hat{F}, \hat{G}] \text{ (commutator)} \qquad (9)$$

Remarks

- The commutator, $[A, B]$, between two matrices or linear operators A and B is defined by

$$[A, B] := A \cdot B - B \cdot A.$$

- Planck's constant $\hbar$ is sometimes replaced by another so-called "deformation parameter", such as Newton's constant G_N, or some other "coupling constant", etc.

---

10. Werner Heisenberg (1901–1976): Über quantentheoretische Umdeutung kinematischer und mechanischer Beziehungen, *Zeitschrift für Physik* **33** (1925), 879–893.
11. Paul Adrien Maurice Dirac (1902–1984): On the Theory of Quantum Mechanics, *Proc. Royal Soc.* (1926), 661–677.

■ The operators $\hat{F}$ are usually thought to act on a separable Hilbert space $\mathcal{H}$.

The map

$$\hat{} : F \mapsto \hat{F}$$

is *not* an algebra homomorphism, because the real-valued continuous functions on Γ form an abelian (commutative) algebra under point-wise multiplication, whereas matrix multiplication is *non-commutative*; moreover, the product, $\hat{F} \cdot \hat{G}$, of two self-adjoint operators, $\hat{F}$ and $\hat{G}$, is not a self-adjoint operator, unless the operators $\hat{F}$ and $\hat{G}$ commute.

Let me briefly digress into somewhat more technical territory. Readers not familiar with the notions discussed in the following paragraph are advised to pass to Eq. (13). For the purposes of a general discussion, one can always assume that the functions F on Γ are bounded and that the operators $\hat{F}$ are bounded operators on $\mathcal{H}$. In the analysis of systems with infinitely many degrees of freedom, such as the electromagnetic field, it is actually convenient to use a more abstract formulation, interpreting the operators $\hat{F}$ as elements of a C^*- algebra, C, that plays the role, in quantum mechanics, that the algebra, $C(\Gamma)$, of bounded continuous functions on phase space Γ plays in classical mechanics.

In classical mechanics, *states* are given by probability measures on phase space Γ. This is equivalent to saying that states are given by positive normalized linear functionals on the algebra $C(\Gamma)$ of continuous functions on Γ. This definition of states can immediately be carried over to quantum mechanics: A state of a quantum system whose physical quantities are represented by the self-adjoint operators in a C^*- algebra C is a positive normalized linear functional on C.

Definition: A positive normalized linear functional, ρ, on a C^*-algebra $\mathcal{A}$ containing an identity element $\mathbf{1}$,[12] (for example, $\mathcal{A} = C(\Gamma)$, where Γ is a compact topological space, or $\mathcal{A} = C$), is a $\mathbb{C}$- linear map,

$$\rho : \mathcal{A} \ni X \mapsto \rho(X) \in \mathbb{C}, \tag{10}$$

with the properties that

$$\rho(X) \geq 0, \text{ for every positive operator } X \in \mathcal{A}, \quad \rho(\mathbf{1}) = 1. \tag{11}$$

So-called *pure states* on $\mathcal{A}$ are states that cannot be written as convex combinations of at least two distinct states. In the example where $\mathcal{A} = C(\Gamma)$, a pure state is a Dirac delta function on a point of Γ. This means that pure states of a Hamiltonian mechanical system can be identified with points in phase space Γ (and, hence, the space of pure states of such a system does usually not have any relationship to a linear space, as would be the case in standard quantum mechanics).

From a pair, $(\mathcal{A}, \rho)$, of a C^*- algebra $\mathcal{A}$ and a state ρ on $\mathcal{A}$ one can always reconstruct a Hilbert space, $\mathcal{H}_\rho$, and a representation, π_ρ, of $\mathcal{A}$ on $\mathcal{H}_\rho$. This is the contents of the so-called Gel'fand-Naimark-Segal construction. (See, e.g., [12] for definitions, basic results and proofs.)

We recall that if the operators $\hat{F}$ and $\hat{G}$ representing two physical quantities of some system do not commute then they cannot be measured simultaneously: If the system is prepared in a state ρ the uncertainties, $\Delta_\rho \hat{F}$ and $\Delta_\rho \hat{G}$, in a simultaneous measurement of the quantities

---

12. This will always be assumed in what follows.

represented by the operators $\hat{F}$ and $\hat{G}$ satisfy the celebrated Heisenberg Uncertainty Relations

$$\Delta_\rho \hat{F} \cdot \Delta_\rho \hat{G} \geq \frac{1}{2} |\rho([\hat{F}, \hat{G}])|. \tag{12}$$

As a special case we mention that if x denotes the position of a particle moving on the real line $\mathbb{R}$ and p denotes its momentum then

$$\Delta x \cdot \Delta p \geq \frac{\hbar}{2}, \tag{13}$$

in an arbitrary state of the system.

The Heisenberg–Dirac recipe expressed in Eq. (9) can be applied to *Vlasov theory*[13] and *Maxwell's theory*[14], and to many other interesting examples of Hamiltonian systems or Hamiltonian field theories. In these examples, atomism always arises as a consequence of quantization.

In the following, I propose to sketch the example of Vlasov theory. This is a theory describing the mechanics of (star) dust viewed as a continuous material medium. A state of dust at time t is described by the density, $f_t(x, v)$, of dust with velocity $v \in \mathbb{R}^3$ observed at the point x in physical space $\mathbb{E}^3$. Clearly $f_t(x, v)$ is non-negative, and

$$\int d^3 x \int d^3 v \, f_t(x, v) = v,$$

where v is the number of moles of dust. This quantity is conserved, i.e., independent of time t. The density of dust at the point $x \in \mathbb{E}^3$ at time t is given by

$$n_t(x) := \int d^3 v \, f_t(x, v).$$

The equation of motion of the state f_t is given by the so-called Vlasov (collision-free Boltzmann) equation:

$$\frac{\partial}{\partial t} f_t(x, v) + v \cdot \nabla_x f_t(x, v) - \nabla V_{\text{eff}}[f_t](x) \cdot \nabla_v f_t(x, v) = 0, \tag{14}$$

where

$$V_{\text{eff}}[f_t](x) := V(x) + \int d^3 y \, \phi(x - y) \, n_t(y). \tag{15}$$

In this expression, $V(x)$ is the potential of an external force acting on the dust at the point $x \in \mathbb{E}^3$, $\phi(x - y)$ is a two-body potential describing the force between dust at point x and dust at point y in physical space.

In the following, I sketch how Vlasov theory can be *quantized* by applying the Heisenberg–Dirac recipe. Since my exposition is somewhat more technical than the rest of this essay, I want to disclose what the result of this exercise is: *The quantization of Vlasov theory is nothing but the Newtonian mechanics of an arbitrary number of identical point particles moving in physical space under the influence of an external force with potential given by the function V and interacting with each other through two-body forces whose potential is given by $N^{-1}\phi$, where N^{-1}*

---

13. First proposed by *Anatoly Alexandrovich Vlasov* (1908–1975) in 1938.
14. Named after the eminent Scottish mathematical physicist *James Clerk Maxwell* (1831–1879).

is a "deformation parameter" that plays the role of Planck's constant ℏ in the Heisenberg–Dirac recipe. – Readers not familiar with infinite-dimensional Hamiltonian systems or not very interested in mathematical considerations are encouraged to proceed to the material after Eq. (33).

Since $f_t(x, v) \geq 0$, it can be written as a product (factorized)

$$f_t(x, v) = \overline{\alpha_t(x, v)} \cdot \alpha_t(x, v), \tag{16}$$

where $\alpha_t(x, v)$ is a complex-valued function of (x, v), with

$$|\alpha_t(x, v)| = \sqrt{f_t(x, v)}.$$

Clearly, the phase, $\alpha_t / |\alpha_t|$, is *not* observable. Perhaps surprisingly, it appears to be a good idea to encode the time evolution of the density f_t into a dynamical law for α_t. Here is how this can be done: Let $\Gamma_1 := \mathbb{E}^3 \times \mathbb{R}^3$ denote the "one-particle phase space" of pairs, (x, v), of points in physical space and velocities. By $\Gamma_\infty := H^1(\Gamma_1)$ we denote the complex Sobolev space of index 1 over Γ_1. This space can be interpreted as an ∞- dimensional affine phase space. Functions $\alpha \in \Gamma_\infty$ and their complex conjugates, $\overline{\alpha}$, serve as complex coordinates for Γ_∞. The symplectic structure of Γ_∞ can be encoded into the Poisson brackets:

$$\{\alpha(x, v), \alpha(x', v')\} = \{\overline{\alpha}(x, v), \overline{\alpha}(x', v')\} = 0,$$
$$\{\alpha(x, v), \overline{\alpha}(x', v')\} = -i\delta(x - x')\delta(v - v'). \tag{17}$$

We introduce a Hamilton functional on Γ_∞:

$$\mathcal{H}(\overline{\alpha}, \alpha) := i \int \int d^3x d^3v \, \overline{\alpha}(x, v)[v \cdot \nabla_x - (\nabla V)(x) \cdot \nabla_v]\alpha(x, v)$$

$$- \frac{i}{2} \int \int d^3x d^3v \, \overline{\alpha}(x, v)\left(\nabla_x \int \int d^3x' d^3v' \, \phi(x - x')|\alpha(x', v')|^2\right) \cdot \nabla_v \alpha(x, v). \tag{18}$$

Hamilton's equations of motion are given by

$$\dot{\alpha}_t(x, v) = \{\mathcal{H}, \alpha_t(x, v)\}, \quad \dot{\overline{\alpha}}_t(x, v) = \{\mathcal{H}, \overline{\alpha}_t(x, v)\}. \tag{19}$$

It is a straightforward exercise [6] to show that these equations imply the Vlasov equation for the density $f_t(x, v) = |\alpha_t(x, v)|^2$.

Note that this theory has a huge group of local symmetry transformations: The "gauge transformations"

$$\alpha_t(x, v) \mapsto \alpha_t(x, v) \, e^{i\theta_t(x, v)} \tag{20}$$

where the phase $\theta_t(x, v)$ is an arbitrary real-valued, smooth function on Γ_1, are symmetries of the theory. The global gauge transformation obtained by setting $\theta_t(x, v) =: \theta \in \mathbb{R}$ form a continuous group of symmetries, $\simeq U(1)$, that gives rise to a conservation law,

$$\|\alpha_t\|_2^2 = \int d^3x \int d^3v \, f_t(x, v) = \text{ const. in time } t, \tag{21}$$

in accordance with *Noether's theorem.*

Next, we propose to quantize Vlasov theory by applying the Heisenberg–Dirac recipe (9) to the variables $\alpha, \overline{\alpha}$; i.e., we replace α and $\overline{\alpha}$ by operators,

$$\alpha \mapsto \hat{\alpha} =: a, \quad \overline{\alpha} \mapsto \hat{\overline{\alpha}} =: a^*, \tag{22}$$

and trade the poisson brackets in (17) for commutators:

$$[a(x,v),a(x',v')] = [a^*(x,v),a^*(x',v')] = 0,$$

and

$$[a(x,v),a^*(x',v')] = N^{-1} \cdot \delta(x-x')\,\delta(v-v'), \tag{23}$$

where the dimensionless number Nv is proportional to the number of "atoms" present in the system; i.e., the role of Planck's constant $\hbar$ is played by N^{-1}. The creation- and annihilation operators, a^* and a, act on Fock space, $\mathcal{F}$:

$$\mathcal{F} := \oplus_{n=0}^{\infty} \mathcal{F}^n, \tag{24}$$

where

$$\mathcal{F}^0 := \mathbb{C}|0\rangle, \quad \text{with } a(x,v)|0\rangle = 0, \forall x,v,$$

and

$$\mathcal{F}^n := \left\langle \int \cdots \int \varphi_n(x_1,v_1,...,x_n,v_n) \prod_{i=1}^{n} a^*(x_i,v_i) d^3x_i d^3v_i |0\rangle \right\rangle, \tag{25}$$

where $\langle \cdot \rangle$ indicates that the (linear) span is taken.

The physical interpretation of the "n-particle wave functions" φ_n is that

$$f_n(x_1,v_1,...,x_n,v_n) := |\varphi_n(x_1,v_1,...,x_n,v_n)|^2 \tag{26}$$

is the state density on n-particle phase space

$$\Gamma_n := \Gamma_1^{\times n}$$

for n identical classical particles moving in physical space. (The state of the system is obtained by multiplying the densities f_n by the Liouville measures $\prod_{i=1}^{n} d^3x_i d^3v_i$.)

The "Hamilton operator" generating the dynamical evolution of the states of the quantized theory is obtained by replacing the functions $\overline{\alpha}$ and α in the Hamilton functional $\mathcal{H}(\overline{\alpha},\alpha)$ introduced in (18) by the operators a^* and a, respectively, and writing all creation operators a^* to the left of all annihilation operators a; ("Wick ordering"). The time-dependent Schrödinger equation for the evolution of vectors in $\mathcal{F}$ then implies the Liouville equations for the densities defined in (26),

$$\dot{f}_t(x_1,v_1,...,x_n,v_n) = -\sum_{i=1}^{n}\left(v_i \cdot \nabla_{x_i} + F(X_i) \cdot \nabla_{v_i}\right) f_t(x_1,v_1,...x_n,v_n), \tag{27}$$

where

$$F(x_i) := -\nabla_{x_i}\left(V(x_i) + N^{-1}\sum_{j \neq i} \phi(x_i - x_j)\right)$$

is the total force acting on the ith particle, which is equal to the external force, $-(\nabla V)(x_i)$, plus the sum of the forces exerted on particle i by the other particles in the system; the strength of the interaction between two particles being proportional to N^{-1}. The equations (27) are equivalent to Newton's equations of motion for n identical particles with two-body interactions moving in physical space, (which are Hamiltonian equations of motion).

"Observables" of this theory are operators on Fock space $\mathcal{F}$ that are invariant under the symmetry transformations given by

$$a(x,v) \mapsto a(x,v)\, e^{i\theta_t(x,v)}, \quad a^*(x,v) \mapsto a^*(x,v)\, e^{-i\theta_t(x,v)}, \tag{28}$$

corresponding to the symmetries (20); (they are the elements of an infinite-dimensional group of local gauge transformations). These symmetries imply that the particle number operator

$$\widehat{\mathcal{N}} := \int d^3x \int d^3v\, a^*(x,v)\, a(x,v)$$

is conserved under the time evolution, and that (in the absence of an affine connection that gives rise to a non-trivial notion of parallel transport of "wave functions", φ_n) the observables of the theory are described by operators that are functionals of the densities $a^*(x,v)\, a(x,v)$. These operators generate an *abelian* (i.e., commutative) algebra. Together with the equations of motion (27), this means that the structure of observables and the predictions of this "quantum theory" are *classical*, in the sense that all observables can be diagonalized simultaneously and hence have objective values, and the time evolution of the system is *deterministic*. In fact, this theory is just a reformulation of the Newtonian mechanics of systems of arbitrarily many identical non-relativistic particles moving in physical space $\mathbb{E}^3$ under the influence of an external potential force and interacting with each other through two-body potential forces.

Thus, what we have sketched here is the perhaps somewhat remarkable observation (see [6], and references given there) that the classical Newtonian mechanics of the particle systems studied above, which treats matter as *atomistic*, can be viewed as the quantization of Vlasov theory, which treats matter as a *continuous medium* of dust. Conversely, Vlasov theory can be viewed as the "classical limit" of the Newtonian mechanics of systems of $\mathcal{O}(N)$ identical particles with two-body interactions of strength $\propto N^{-1}$, which is reached when $N \to \infty$. This has been shown (using different concepts) in [13]. Apparently, the parameter N^{-1} plays the role of Planck's constant $\hbar$.

To express these findings in words, it appears that a mechanics taking into account the atomistic structure of matter arises as the result of quantization of a mechanics that treats matter as a continuous medium.

Mathematical digression on "pre-quantization" of the one-particle phase space and on the passage to the quantum theory of systems of an arbitrary number of identical non-relativistic particles (bosons) with two-body interactions

Obviously the one-particle phase space Γ_1 carries a symplectic structure given by the symplectic 2-form

$$\omega := d\,x \wedge d\,v.$$

One-particle "wave functions", $\alpha(x,v)$, can be viewed as section of a complex line bundle over Γ_1 associated to a principal $U(1)$- bundle. We equip this bundle with a connection, $A = A_x\, dx + A_v\, dv$, (i.e., a gauge field, namely a mathematical object analogous to the well known vector potential in electrodynamics), whose curvature, i.e., the field tensor associated with A, is given by

$$d\,A = \omega.$$

In these formulae, dx and dv are differentials, and "d" denotes exterior differentiation. The connection A introduces a notion of parallel transport on the line bundle of "wave functions"

α. The symmetries (20) can then always be obeyed by replacing ordinary partial derivatives by *covariant* derivatives, i.e.,

$$(\nabla_x, \nabla_v) \mapsto (\nabla_x - iA_x, \nabla_v - iA_v),$$

and products

$$\overline{\alpha}(x, v)\, \alpha(x, v) \mapsto \overline{\alpha}(x, v) U_y(A) \alpha(x', v'), \tag{29}$$

where $U_y(A)$ is a complex phase factor describing parallel transport along a path y from the point $(x', v') \in \Gamma_1$ to the point $(x, v) \in \Gamma_1$. These replacements lead us to the theory of "pre-quantization" of one-particle mechanics formulated over the one-particle phase space Γ_1. By applying the Heisenberg–Dirac recipe (22), (23) and then using the connection A to define parallel transport of creation- and annihilation operators, a^*, a, and n-particle wave functions φ_n, we arrive at what is called "pre-quantization" of the mechanics of arbitrary n-particle systems.

In principle, the introduction of a connection A on the line bundle of one-particle "wave functions" α would allow one to consider vast generalizations of Vlasov dynamics, based on using (29), and, subsequently, of the quantized theory resulting from the replacements (22), (23). Some of these generalizations could be understood as Vlasov theories on a "non-commutative phase space", namely the non-commutative phase space obtained by applying the Heisenberg–Dirac recipe (9) to the Poisson brackets

$$\{x^i, x^j\} = 0 = \{p_i, p_j\},$$

and

$$\{x^i, p_j\} = -\delta^i_j, \qquad i, j = 1, 2, 3.$$

This leads us to the question whether standard quantum mechanics of systems of arbitrarily many identical non-relativistic particles could be rediscovered by appropriately extending the ideas discussed so far. One approach to answering this question is to pass from pre-quantization, as sketched above, to genuine quantization by following the recipes of geometric quantization, à la Kostant and Souriau; see, e.g., [14]. (An alternative is to consider "deformation quantization", see [15], which, however, is usually inadequate to deal with concrete problems of physics.) We cannot go into explaining how this is done, as this would take us too far away from our main theme. Instead, we return to Vlasov theory, whose states are represented by densities $f(x, v)$ on one-particle phase space Γ_1. We propose to replace the factorization (16) of $f(x, v)$ by the Wigner factorization

$$f_\hbar(x, v) = \frac{1}{(2\pi)^3} \int e^{-iv \cdot y} \overline{\psi}\left(x - \frac{\hbar y}{2}\right) \psi\left(x + \frac{\hbar y}{2}\right) d^3 y, \tag{30}$$

where the "Schrödinger wave function" ψ is an arbitrary function in $L^2(\mathbb{R}^3)$. Assuming that the time-dependent Schrödinger wave function ψ_t solves the so-called *Hartree equation*

$$i\hbar \partial_t \psi_t = \left(-\frac{\hbar^2}{2}\Delta + V + |\psi_t|^2 * \phi\right)\psi_t \tag{31}$$

one finds that $f_{\hbar,t}$ solves the Vlasov equation in the limit where $\hbar$ tends to 0.

To understand and prove this claim it is advisable to interpret $f_\hbar(x, v)$ as the Wigner transform of a general one-particle density matrix, ρ, i.e.,

$$f_\hbar(x, v) = \frac{1}{(2\pi)^3} \int e^{-ivy} \rho(x - \frac{\hbar y}{2}, x + \frac{\hbar y}{2}) \, d^3y.$$

Expression (30) is the special case where $\rho(x, y) = \overline{\psi}(x)\psi(y)$ is the pure state corresponding to the wave function ψ. The equation of motion for the density $f_\hbar$ is derived from the Liouville-von Neumann equation of motion for the density matrix ρ,

$$\hbar\dot{\rho} = -i[H_{\text{eff}}, \rho] \tag{32}$$

corresponding to the effective Hamiltonian

$$H_{\text{eff}} := -\frac{\hbar^2}{2}\Delta + V + n * \phi, \tag{33}$$

where $n(x) = \rho(x, x) = \int f_\hbar(x, v) \, d^3v$ is the particle density, and $(n * \phi)(x) := \int n(y)\phi(x - y) \, d^3y$. It is then not hard to see that, formally, the Liouville-von Neumann equation of motion (32), with H_{eff} as in (33), implies the Vlasov equation for $f_\hbar$, as $\hbar$ approaches 0.

The Hartree equation (31) for the Schrödinger wave function ψ turns out to be a Hamiltonian evolution equation on an infinite-dimensional phase space $\hat{\Gamma}_\infty$ with complex coordinates given by the Schrödinger wave functions ψ and their complex conjugates $\overline{\psi}$. Applying the Heisenberg–Dirac recipe to quantize Hartree theory (with the same deformation parameter, N^{-1}, as in Vlasov theory), one arrives at the theory of gases of non-relativistic Bose atoms moving in an external potential landscape described by the potential V and with two-body interactions given by the potential $N^{-1}\phi$. This is an example of a quantum-mechanical many-body theory. In the limiting regime where $N \to \infty$, i.e., in the so-called mean-field (or classical) limit, one recovers Hartree theory. Details of this story can be found in [6].

Vlasov theory has many interesting applications in cosmology and in plasma physics. As an example I mention the rather subtle analysis of Landau damping in plasmas presented in [16]. Hartree theory is often used to describe Bose gases in the limiting regime of high density and very weak two-body interactions, corresponding to $N \to \infty$. Another, somewhat more subtle limiting regime (low density, strong interactions of very short range) is the Gross-Pitaevskii limit considered in [17]. Hartree theory with smooth, attractive two-body interactions of short range features solitary-wave solutions. In a regime where the two-body interactions are strong, the dynamics of multi-soliton configurations is well approximated by the Newtonian mechanics of point particles of varying mass moving in an external potential V and with two-body interactions $\propto \phi$. However, whenever the motion of the solitons is not inertial they experience friction. This has been discussed in some detail in [18]. This observation may have interesting application in cosmology, as first suggested in [19].

To conclude this section, we mention that the atomistic nature of the electromagnetic field, which becomes manifest in the quanta of light or photons, can be understood by applying the Heisenberg–Dirac recipe to Maxwell's classical theory of the electromagnetic field (the deformation parameter being Planck's constant $\hbar$). Historically, this was the first example of a quantum theory. Its contours became visible in Planck's law of black-body radiation and Einstein's discovery of the quanta of radiation, the photons.

4 The structure of quantum theory

> *... Thus, the fixed pressure of natural causality disappears and there remains,*
> *irrespective of the validity of the natural laws, space for autonomous and causally*
> *absolutely independent decisions; I consider the elementary quanta of matter to be the*
> *place of these 'decisions'.*
> – Hermann Weyl, 1920

In Section 3, we have seen that the *atomistic constitution* of matter may be understood as resulting from Heisenberg–Dirac *quantization* of a "classical" Hamiltonian theory that treats matter as a *continuous medium*, such as Vlasov theory. In the following, we propose to sketch some fundamental features of quantum mechanics proper. It turns out that the deeply puzzling features of quantum mechanics arise from the *non-commutativity* of the algebra generated by the linear operators that represent physical quantities/properties characterizing a physical system. This non-commutativity turns out to be intimately related to the atomistic constitution of matter! In a sense, Hartree theory *is* a quantum theory – Planck's constant $\hbar$ appears explicitly in the Hartree equation that describes the time evolution of physical quantities of the theory. Hartree theory describes matter (more precisely interacting quantum gases) as a continuous medium. As a result, the algebra of physical quantities of this theory is *abelian* (commutative). When it is quantized according to the Heisenberg–Dirac recipe – as indicated in Section 3 – one arrives at a theory (namely non-relativistic quantum-mechanical many-body theory) providing an atomistic description of matter, and the algebra of operators representing physical quantities becomes *non-commutative*.

The purpose of this section is to sketch some general features of *non-relativistic quantum mechanics* related to its probabilistic nature and its fundamental irreversibility. Our analysis is intended to apply to a large class of physical systems; and it is based on the assumption that the linear operators providing a quantum-mechanical description of physical quantities and events of a typical physical system, S, generate a *non-abelian* (non-commutative) algebra. An example of an important consequence of this assumption is the phenomenon of *entanglement* (see below), which does *not* appear in classical physics.

In classical physics, the operators representing physical quantities always generate an *abelian* (commutative) algebra, $\mathcal{E}^c$, over the complex numbers invariant under taking the adjoint of operators and closed in the operator norm. By a theorem due to I. M. Gel'fand (see, e.g., [12]), such an algebra is isomorphic to the algebra of complex-valued continuous functions over a compact topological (Hausdorff) space, Γ,[15] i.e.,

$$\mathcal{E}^c \simeq C(\Gamma). \tag{34}$$

The operator norm, $\|F\|$, of an element $F \in \mathcal{E}^c$ is the sup norm of the function on Γ corresponding to F, which we also denote by F. The physical quantities of the system are described by the *real-valued* continuous functions on Γ, which are the self-adjoint elements of $\mathcal{E}^c$. *States* of the system are given by probability measures on Γ; *pure* states correspond to atomic measures, i.e., Dirac δ- functions, supported on points, ξ, of Γ. Thus, the pure states are "characters" of the algebra $\mathcal{E}^c$, i.e., positive, normalized linear functionals, δ_ξ, with the property that

$$\delta_\xi(F \cdot G) = \delta_\xi(F) \cdot \delta_\xi(G).$$

---

15. Of course, Γ is usually *not* a symplectic manifold – it is symplectic, i.e., a "phase space", only if the system is Hamiltonian.

Passing to a subsystem of the system described by the algebra $\mathcal{E}^c$ amounts to selecting some *subalgebra*, $\mathcal{E}^c_0$, of the algebra $\mathcal{E}^c$ invariant under taking adjoints and closed in the operator norm. Characters of $\mathcal{E}^c$ obviously determine characters of $\mathcal{E}^c_0$; i.e., *pure states of the system remain pure when one passes to a subsystem*. This implies that the phenomenon of entanglement is completely absent in classical physics.

Time evolution of physical quantities from time s to time t is described by *automorphisms, $\tau_{t,s}$, of $\mathcal{E}^c$, which form a one-parameter groupoid. This may sound curiously abstract. But it turns out that any such groupoid is described by flow maps,

$$y_{t,s} : \Gamma \to \Gamma, \qquad \Gamma \ni \xi(s) \mapsto \xi(t) = y_{t,s}(\xi(s)) \in \Gamma.$$

Under fairly general hypotheses on the properties of the maps $y_{t,s}$ they are generated by (generally time-dependent) vector fields, X_t, on Γ; i.e., the trajectory $\xi(t) := y_{t,s}(\xi)$ of a point $\xi \in \Gamma$ is the solution of a differential equation,

$$\dot{\xi}(t) = X_t(\xi(t)), \qquad \text{with } \xi(s) = \xi \in \Gamma.$$

These properties of time evolution are preserved when one passes from the description of a physical system to the one of a subsystem. Since all this may sound too abstract and quite incomprehensible, I summarize the main features of classical physics in words:

(A) The physical quantities of a classical system are represented by by self-adjoint operators that all commute with one another. They correspond to the bounded, real-valued, continuous functions on a "state space" Γ.

(B) Pure states of the system can be identified with points in its state space Γ.

(C) *All* physical quantities have objective and unique values in every pure state of the system. Conversely, the values of all physical quantities of a system usually determine its state uniquely. Thus, pure states have an "ontological meaning": They contain complete information on all properties of the system at a given instant of time.

(D) Mixed states are given by probability measures on Γ. Probabilities associated with such mixed states are expressions of ignorance, i.e., of a lack of complete knowledge of the true state of the system at a given instant of time.[16]

(E) Time evolution of physical quantities and states is completely determined by flow maps, $y_{t,s}$, from the state space Γ to itself specifying which pure states, $\xi(t)$, at time t correspond to initial states, $\xi(s)$, chosen at time s. Thus, the "Law of Causation" holds (as formulated originally by Leucippus and Democritus), and there is perfect determinism; (disregarding from the possibly huge problems of computation of dynamics for chaotic systems).

(F) All these properties of a classical description of physical systems are preserved upon passing to the description of a subsystem (that may interact strongly with its complement).

Well, for better of worse, *these wonderful features of classical physics all disappear when one passes to a quantum-mechanical description of reality!* One of the first problems one encounters when one analyzes general features of a quantum-mechanical description of reality is that

---

16. One should add that, pragmatically, mixed states play an enormously important role in that they often enable us to make concrete predictions on quantities that are defined as time-averages along trajectories of true states of which one expects that they are identical to ensemble averages. Often, only the ensemble averages are accessible to concrete calculations, using measures describing certain mixed states, such as thermal equilibrium states, while time-averages along trajectories of true states remain inaccessible to quantitative evaluation.

one does not know how to describe the time evolution of physical quantities of a system unless that system has interactions with the rest of the universe that are so tiny that they can be neglected over long stretches of time. Such a system is called "*isolated*". In this section, we limit our discussion to isolated systems; (but see, e.g., [20, 21].)

Here is a *pedestrian definition of an isolated physical system* – according to quantum mechanics:

Let S be an isolated physical system that we wish to describe quantum-mechanically.

1. The physical quantities/properties of S are represented by bounded self-adjoint operators. They generate a C^*- algebra $\mathcal{E}$, i.e., an algebra of operators invariant under taking adjoints and closed in an operator norm with certain properties; (see, e.g., [12]). For simplicity, we suppose that the spectra of all the operators corresponding to physical quantities of S are finite point spectra. Then every such operator $A = A^*$ has a spectral decomposition,

$$A = \sum_{\alpha \in \sigma(A)} \alpha \, \Pi_\alpha, \tag{35}$$

where $\sigma(A)$ is the spectrum of A, i.e., the set of all its eigenvalues, and Π_α is the spectral projection of A corresponding to the eigenvalue α, (i.e., the orthogonal projection onto the eigenspace of A associated with the eigenvalue α, in case A is made to act on a Hilbert space).

2. An *event* possibly detectable in S corresponds to an orthogonal projection $\Pi = \Pi^*$ in the algebra $\mathcal{E}$. But *not* all orthogonal projections in $\mathcal{E}$ represent events. Typically, a projection Π corresponding to an event possibly detectable in S is a spectral projection of an operator in $\mathcal{E}$ that represents a physical quantity of S.

3. So far, *time* has not appeared in our characterization of physical systems, yet. Time is considered to be a real parameter, $t \in \mathbb{R}$. All physical quantities of S possibly observable during the interval $[s, t] \subset \mathbb{R}$ of times generate an algebra denoted by $\mathcal{E}_{[s,t]}$.[17] It is natural to assume that if $[s', t'] \subset [s, t]$ $(s \le s', t \ge t')$

$$\mathcal{E}_{[s',t']} \subseteq \mathcal{E}_{[s,t]} \subseteq \mathcal{E}. \tag{36}$$

Events possibly detectable during the time interval $[s, t]$ are represented by certain self-adjoint (orthogonal) projections in the algebra $\mathcal{E}_{[s,t]}$.

4. *Instruments*: An "instrument", $\mathcal{I}_S[s, t]$, serving to detect certain events in S during the time interval $[s, t]$ is given by a family of mutually orthogonal (commuting) projections, $\{\Pi_\alpha\}_{\alpha \in I_S[s,t]} \subset \mathcal{E}_{[s,t]}$. Typically, these projections will be spectral projections of commuting self-adjoint operators representing certain physical quantities of S that may be observable/measurable in the time interval $[s, t]$. For the quantum mechanics describing a physical system S to make concrete predictions it is necessary to specify its list of instruments $\{\mathcal{I}_S^{(i)}[s_i, t_i]\}_{i \in \mathcal{L}_S}$, where $\mathcal{L}_S$ labels all instruments of S. It should be noted that instruments located in different intervals of time may be related to each other by the time evolution of S. (Thus, for autonomous systems, it suffices to specify all instruments $\mathcal{I}_S^{(i)}[0, \infty), i = 1, 2, 3, \dots$ All other instruments of S are conjugated to the ones in this list by

---

17. Technically speaking, this algebra is taken to be a von Neumann algebra, which has the advantage that, with an operator $A \in \mathcal{E}_{[s,t]}$, all its spectral projections also belong to $\mathcal{E}_{[s,t]}$.

time translation. Luckily, we do not need to go into all these details here.) We emphasize that the operators belonging to *different* instruments all of which are located in the same interval of times do, in general *not* commute with each other. For example, one instrument may measure the position of a particle at some time belonging to an interval $I \in \mathbb{R}$, while another instrument may measure its momentum at some time in I.

Remark: For most quantum systems, the set of instruments tends to be very sparse. There are many very interesting examples of idealized mesoscopic systems for which the set of instruments serving to detect events at time t consists of the spectral projections of a single self-adjoint operator $X(t)$, with

$$X(t) = U_S(s,t)X(s)U_S(t,s),$$

where $U_S(t,s)$ is the unitary propagator of the system S describing time translations of operators representing physical properties of S observable at time s to operators representing the same physical quantities at time t; (we use the Heisenberg picture – as one should always do).

The notion of an "instrument" is not intrinsic to the theory and may depend on the "observer", but *only* in the sense that the amount of information available on a given physical system depends on our abilities to *retrieve* information about it, (which may change with time). The situation is similar to the one encountered in a description of the time evolution of systems in terms of stochastic processes.

DEFINITION. *We define the algebras*

$$\mathcal{E}_{\geq t} := \overline{\bigvee_{t':t<t'<\infty} \mathcal{E}_{[t,t']}}, \qquad \text{for } t \in \mathbb{R}, \tag{37}$$

where $\overline{(\cdot)}$ represents completion in the operator norm of $\mathcal{E}$. The algebra $\mathcal{E}_{\geq t}$ is the algebra of all events possibly detectable at times $\geq t$, i.e., happening in the future of time t. [18] *By property (36) we have that*

$$\mathcal{E} \supseteq \mathcal{E}_{\geq t} \supseteq \mathcal{E}_{\geq t'} \supseteq \mathcal{E}_{[t',t'']}, \tag{38}$$

whenever $t < t' \leq t''$.

Next, we describe the *key idea* underlying our approach to quantum mechanics:

A necessary condition for a physical system S to feature events that may be detectable around or after some time t' (= the present), using suitable instruments $\mathcal{I}_S[t', \infty)$, is that

$$\mathcal{E}_{\geq t} \underset{\neq}{\supset} \mathcal{E}_{\geq t'}, \quad \text{for some past time } t < t'. \tag{39}$$

Property (39) expresses a fundamental *loss of access to information* concerning the past (in (39): before time t', but after time t) that occurs in systems featuring detectable events. A property similar to (39), but appropriate for local relativistic quantum theory, has been established for quantum electrodynamics (QED), formulated in the language of algebraic quantum field theory, by Detlev Buchholz and the late John Roberts in [22]. It is a consequence of Huygens' Principle[19] for theories with massless modes or particles, such as the photons of QED. It should be

---

18. Since we are interested in projections representing events possibly detectable at times $\geq t$, it may be advantageous to assume that the algebras $\mathcal{E}_{\geq t}$ are actually von Neumann algebras; see, e.g., [12].

emphasized that a property perfectly analogous to (39) can also be derived for *classical* relativistic field theories obeying Huygens' Principle. Simple models of non-autonomous systems for which property (39) can be proven for certain (discrete) times t' have been discussed in [23].

We must ask why property (39) may actually represent a fundamental property (an "axiom", if you will) of the quantum theory of events and experiments. Our explanation is based on exploiting the phenomenon of entanglement. Suppose that the system S has been prepared in a state ρ at some time t_0. (How a system can be prepared in a specific state at approximately a fixed time is a question that we cannot answer in this essay; but see [24], where it is discussed at length.) The state ρ may be a pure state on the algebra $\mathcal{E}$. We define a state ρ_t on the algebra $\mathcal{E}_{\geq t}$ by setting

$$\rho_t := \rho|_{\mathcal{E}_{\geq t}}, \qquad \rho_t(A) = \rho(A), \forall A \in \mathcal{E}_{\geq t}. \tag{40}$$

Because of Eq. (39), the state ρ_t may be a *mixed* state on the algebra $\mathcal{E}_{\geq t}$ *even* if it is a *pure* state on the algebra $\mathcal{E}$, assuming that these algebras are non-commutative. This is what *entanglement* is all about! Furthermore, because of loss of access to information as expressed in (39), the states ρ_t "evolve" in time. This means that, at certain times (which one can predict), one may be able to use an "instrument", in the sense of item 4 above, to detect an event, in the sense of item 2 above, of which there were no signs at earlier times. Indeed, it is precisely the fundamental property of "loss of access to information", as expressed in (39), that makes it possible to *gain* information about a system by detecting events happening in it! One may want to call this fact the "*Second Law of quantum measurement theory*". Here is a rough indication of how to understand these things somewhat more precisely:

Given that a system S has been prepared in a state ρ at some time t_0, it may happen that, around some later time t, the state ρ_t is an incoherent superposition of eigenstates of a family of commuting self-adjoint projections belonging to the algebra $\mathcal{E}_{\geq t}$ and representing *events detectable at time t* or later; see item (2), above. These projections may be those of an instrument $\mathcal{I}_S[t, \infty)$, in the sense of item (4) above. Mathematically, this means that

$$\rho_t(A) = \sum_{\alpha \in I_S[t,\infty)} \rho(\Pi_\alpha \, A \, \Pi_\alpha) + \rho(\Pi^\perp \, A \, \Pi^\perp), \qquad \sum_{\alpha \in I_S[t,\infty)} \Pi_\alpha = \mathbf{1} - \Pi^\perp, \tag{41}$$

where $\{\Pi_\alpha\}_{\alpha \in I_S[t,\infty)} = \mathcal{I}_S[t, \infty) \in \mathcal{E}_{\geq t}$ is an instrument, and $\Pi^\perp$ projects on whatever is *not* identifiable by this instrument.

Well, things are a little more subtle than that, as we will explain presently. Given a (C^*- or von Neumann) algebra $\mathcal{M}$ and a state ρ on $\mathcal{M}$, we define the adjoint action of an operator $A \in \mathcal{M}$ on the state ρ to be given by a bounded linear functional, $ad_A(\rho)$, defined as follows:

$$ad_A(\rho)(B) := \rho([A, B]), \quad \forall B \in \mathcal{M}. \tag{42}$$

---

19. After the celebrated scientist Christiaan Huygens (1629–1695), who explained many phenomena related to the wave properties of light with the help of the idea of light spheres emanating from all points in physical space already reached by light.

We define the "centralizer" of the state ρ to be the subalgebra

$$C_\rho := \{A \in \mathcal{M} : ad_A(\rho) = 0\} \tag{43}$$

of the algebra $\mathcal{M}$.[20] Furthermore, let $\mathcal{Z}_\rho$ denote the center of C_ρ.[21]

Given a state ρ on the algebra $\mathcal{E}$, we define C_{ρ_t} to be the centralizer of the state ρ_t on the algebra $\mathcal{E}_{\geq t}$, and we denote the center of C_{ρ_t} by $\mathcal{Z}_{\rho_t}$.

We are now prepared to say what it means, *quantum-mechanically*, that an event detectable by an instrument $\mathcal{I}_S[t, \infty)$ happens at a certain time, given that we know the state the system has been prepared in.

Axiom concerning events in quantum mechanics:

(I)   Given that the system has been prepared in state ρ, the first event after the preparation of the system, detectable by some instrument, $\mathcal{I}_S[t, \infty)$, of S, happens as soon as equation (41) holds true, provided all the projections $\Pi_\alpha \in \mathcal{I}_S[t, \infty)$ *and* the projection $\Pi^\perp$ belong to the center $\mathcal{Z}_{\rho_t}$ of the centralizer C_{ρ_t} of the state ρ_t.

(II)   The *probability* to detect the event $\Pi_\alpha \in \mathcal{I}_S[t, \infty)$, is given by *Born's Rule*:

$$\text{Prob}\{\Pi_\alpha \text{ happens}\} = \rho(\Pi_\alpha) \tag{44}$$

and $\rho(\Pi^\perp)$ is the probability that the instrument does not detect anything it can identify.

(III) If the event corresponding to the projection Π_α is detected then the state to be used for predictions after time t must be taken to be

$$\rho_{t,\alpha}(A) := \frac{\rho(\Pi_\alpha A \Pi_\alpha)}{\rho(\Pi_\alpha)}, \quad \forall A \in \mathcal{E}_{\geq t} \tag{45}$$

and if the instrument does not detect anything it can identify then the state

$$\rho_t^\perp(A) := \frac{\rho(\Pi^\perp A \Pi^\perp)}{\rho(\Pi^\perp)}, \quad \forall A \in \mathcal{E}_{\geq t} \tag{46}$$

must be used.

Item (III) of the axiom is sometimes called the *"collapse of the wave function"*, a terrible expression, because the "collapse" involved here is not a physical process, but the passage to a conditional expectation.

The formulation of the basic "Axiom concerning events" given above lacks certain elements of precision that cannot be provided here, because they involve concepts – such as conditional expectations defined on non-abelian algebras, etc. – and mathematical subtleties that one cannot explain on a page or two; (see, however, [25]). A precise formulation of this axiom shows that the approximate time ($\approx s_{i_0}$) at which the first event is detected after the preparation of the state of the system[22] and the instrument, $\mathcal{I}_S^{(i_0)}[s_{i_0}, t_{i_0}]$, for some $i_0 \in \mathcal{L}_S$, that detects this first event can be predicted if one knows the state the system has been prepared in; see [25].

---

20. It is an easy exercise that I recommend to the reader to show that C_ρ is an algebra contained in $\mathcal{M}$ and that ρ is a trace on C_ρ.

21. The center, $\mathcal{Z}$, of an algebra $\mathcal{N}$ consists of *all* operators in $\mathcal{N}$ that commute with *all* operators in $\mathcal{N}$. Note that $\mathcal{Z}$ is an abelian subalgebra of $\mathcal{N}$.

22. I.e., the approximate time at which "a detector clicks".

Loss of access to information, as formulated in property (39), together with items (II) and (III) of the basic Axiom are fundamental expressions of the *probabilistic nature of quantum mechanics* (i.e., of its *indeterminism*) and of its *fundamental irreversibility*.

Whenever an event happens, in the sense of item (I) of the basic Axiom, then we should pass to the corresponding conditional state given in Eq. (45) to make predictions of the future evolution of the system, whereas if the instrument does *not* detect any event it can identify then the state in Eq. (46) must be used to predict the future. The passage from the state ρ_t to one of the states in (45) and (46) is obviously *not* a linear process and *cannot* be derived from the solution of any Schrödinger equation. The statements that the time evolution of states in quantum mechanics is described by a Schrödinger equation and that the Heisenberg picture and the Schrödinger picture are equivalent are *not* tenable when one studies physical systems featuring events – and, ultimately, only such systems are interesting for physics.

"I leave to several futures (not to all) my garden of forking paths" – Jorge Luis Borges[23]

To summarize our findings, one may say that the time evolution of states of physical systems featuring events is described, in quantum mechanics, by a generalized *"branching process"*. At every fork of the process, an event detectable by some instrument of the system happens, or an event not identifiable by that instrument happens – as formulated in the basic Axiom. The probabilities of the different outcomes are given by *Born's Rule*. If one takes notice of the particular event happening at the fork one is advised to use the corresponding state, as given in (45) and (46), for improved predictions of the future. This is a new initial state, and one then studies whether the system will feature another event in the future, in the sense of the basic Axiom, when prepared in this new initial state, etc. The different possibilities form a tree-like structure (a little like the different descendants of a parent in population dynamics – but with the difference that, in quantum mechanics, only *one* "descendant", among all possible "descendants", is real), and the *actual trajectory* of the system corresponds to a path on this tree-like structure, called a *"history"*. This has motivated me to call our approach to quantum mechanics the *"ETH approach"* – for "Events", "Trees", and "Histories". In quantum mechanics, the "ontology" of a system S lies in its possible "histories" (the probabilities or "frequencies"[24] of which are predicted by the theory).

It should be emphasized that, in quantum mechanics, the notion of "conserved quantities", such as energy, momentum and angular momentum, becomes somewhat fuzzy in systems featuring events, because such quantities are actually not strictly conserved along "histories": If the instrument involved in the detection of an event does not commute with the operator corresponding to a conserved quantity this quantity is not conserved when the event is detected. This follows from the "collapse rules" (45) and (46).

I conclude this essay by drawing an analogy between quantum mechanics and the standard theory of stochastic (or branching) processes: The filtration of algebras $\{\mathcal{E}_{\geq t}\}_{t \in \mathbb{R}}$ in quantum mechanics is the analogue of a filtration of abelian algebras, $\{\mathcal{E}^c_{\geq t}\}_{t \in \mathbb{R}}$, of functions defined on the path space Ξ of a stochastic process with state space X, where the functions belonging to $\mathcal{E}^c_{\geq t}$ only depend on the part $\xi_{\geq t}(\cdot) := \{\xi(t') \in X : t' > t\}$ of the trajectory $\xi(\cdot) \in \Xi$ of

---

23. In: "El jardín de senderos que se bifurcan", Editorial Sur, 1941 – I thank P. F. Rodriguez for having drawn my attention to this story.

24. A notion due to Jacob Bernoulli (1655–1705), a member of the famous Bernoulli family of Basel.

the process at times t or later. Quantum-mechanical events are somewhat analogous to events featured by a stochastic process, (for example the event that a trajectory $\xi(\cdot)$ of a stochastic process visits a certain measurable subset Ω of Ξ whose definition only depends on the part $\xi_{\geq t}$ of the trajectory). In the case of standard stochastic processes, all possible events generate an *abelian* algebra, and one can therefore assume that the "true" state of the system at time t corresponds to a point $\xi(t) \in X$, for all times t. In quantum mechanics, this is *not* the case! It tends to be rare that an "event" detectable by some "instrument" happens. This is a consequence of the non-commutativity of the algebras $\mathcal{E}_{\geq t}, t \in \mathbb{R}$.

In contrast to the situation in classical theories, the state of a system does *not* have an ontological significance in quantum mechanics; (the word "state" may therefore be considered to be a misnomer). It merely enables us to make plausible bets on possible events that may (or may not) happen in the future. In quantum mechanics, the "ontology" lies in the *"histories of events"* of a system, (every event giving rise to a new initial state in the range of the projection that corresponds to the event, as expressed in item (III) – the "collapse postulate" – of the basic Axiom).

Appendix on Entropy

In this appendix I recall the definition of the von Neumann entropy of a density matrix and the definition of relative entropy for a pair of density matrices. I then state the most important properties of relative entropy and derive its positivity from an inequality due to O. Klein.

The von Neumann entropy of a density matrix ρ is defined by

$$S(\rho) := - \operatorname{tr}(\rho \ln\rho) \tag{47}$$

It is obviously non-negative and vanishes only if ρ is a pure state. It has various important properties among which one should mention that it is *concave, subadditive* and *strongly subadditive*; see [11].

More important for our considerations in Section 2 is another functional, called *"relative entropy"*, defined on pairs of density matrices: Let ρ and σ be density matrices on $\mathcal{H}$; the relative entropy of ρ given σ is introduced as follows:

$$S(\rho\|\sigma) := \operatorname{tr}\left(\rho(\ln \rho - \ln \sigma)\right), \tag{48}$$

and it is assumed that $\ker(\sigma) \subseteq \ker(\rho)$. Important properties of relative entropy are:

- Positivity:

$$S(\rho\|\sigma) \geq 0, \quad \text{with "=" iff } \rho = \sigma \text{ on } \ker(\rho)^{\perp}. \tag{49}$$

- Convexity: $S(\rho\|\sigma)$ is jointly convex in ρ and in σ.

For the material in Section 2, positivity and joint convexity of relative entropy are the crucial properties.

Next, we state and prove a general inequality, due to O. Klein,[25] which turns out to imply the positivity of relative entropy. Let f be a real-valued, strictly convex function on the real line, and let A and B be self-adjoint operators on $\mathcal{H}$. Then

$$\text{tr}\,(f(B)) \geq \text{tr}\,(f(A)) + \text{tr}\,(f'(A) \cdot (B - A)), \tag{50}$$

with "=" only if $A = B$.

Proof of inequality (50): Let $\{\psi_j\}_{j=0}^{\infty}$ be a complete orthonormal system (CONS) of eigenvectors of B corresponding to eigenvalues β_j, $j = 0, 1, 2, \ldots$ Let ψ be a unit vector in $\mathcal{H}$, and $c_j :=$ $\langle \psi_j, \psi \rangle$. Then

$$\langle \psi, f(B)\psi \rangle = \sum_j |c_j|^2 \, f(\beta_j) \geq f\Big(\sum_j |c_j|^2 \beta_j \Big) = f(\langle \psi, B\psi \rangle), \tag{51}$$

by convexity of f; which, moreover, also implies that

$$f(\langle \psi, B\psi \rangle) \geq f(\langle \psi, A\psi \rangle) + f'(\langle \psi, A\psi \rangle) \cdot \langle \psi, (B - A)\psi \rangle.$$

If ψ is an eigenvector of A then the R.S. is

$$= \langle \psi, [f(A) + f'(A) \cdot (B - A)]\psi \rangle. \tag{52}$$

Eq. (50) follows by summing Eqs. (51) and (52) over a CONS of eigenvectors of A. $\square$

As an application we set $f(x) = x \ln(x)$. Then

$$f'(x) = \ln(x) + 1, \quad \text{and} \quad f''(x) = \frac{1}{x} > 0, \text{ for } x > 0,$$

i.e., f is convex on $\mathbb{R}_+$. We set $A := \sigma$ and $B := \rho$. Then A and B are positive operators and hence, by the convexity of f on $\mathbb{R}_+$, Klein's inequality (50) implies that

$$
\begin{aligned}
\text{tr}\,(\rho \ln(\rho)) &= \text{tr}\,(f(B)) \\
&\geq \text{tr}\,(f(A)) + \text{tr}\,(f'(A) \cdot (B - A)) \\
&= \text{tr}\,(\sigma \ln(\sigma)) + \text{tr}\,([\ln(\sigma) + 1](\rho - \sigma)) \\
&= \text{tr}\,(\rho \ln(\sigma)),
\end{aligned} \tag{53}
$$

and we have used the fact that $\text{tr}\,(\rho) = \text{tr}\,(\sigma)\,(= 1)$, and the cyclicity of the trace. This proves the positivity of relative entropy. $\square$

The joint convexity of the relative entropy $S(\rho\|\sigma)$ in ρ and σ is a fairly deep property that we do not prove here. Instead, we show that the von Neumann entropy $S(\rho)$ is a *concave* functional of ρ. Let $\rho = p\rho_1 + (1 - p)\rho_2$. We apply Klein's inequality (50) twice, with the following choices:

---

25. Oskar Benjamin Klein (1894–1977) was an eminent Swedish theorist. For example, independently of Kaluza, he invented the Kaluza-Klein unification of gravitation and electromagnetism involving a compact fifth dimension of spacee-time, and, in 1938, he was first to propose a non-abelian gauge theory of weak interactions.

- $B_1 := \rho_1, A := \rho$
- $B_2 := \rho_2, A := \rho$

Taking a convex combination of the two resulting inequalities, we find that

$$
\begin{aligned}
p \operatorname{tr} \left(\rho_1 \ln(\rho_1)\right) + (1-p) \operatorname{tr} \left(\rho_2 \ln(\rho_2)\right) & \\
\geq \operatorname{tr} \left(\rho \ln(\rho)\right) &+ p(1-p) \operatorname{tr} \left((\rho_1 - \rho_2)[\ln(\rho) + 1]\right) \\
&+ (1-p)p \operatorname{tr} \left((\rho_2 - \rho_1)[\ln(\rho) + 1]\right) \\
&= \operatorname{tr} \left(\rho \ln(\rho)\right),
\end{aligned} \tag{54}
$$

which completes the proof of concavity of $S(\rho) = -\operatorname{tr} \left(\rho \ln(\rho)\right)$. $\qquad \square$

For deep and sophisticated entropy inequalities we refer the reader to [10, 11].

Acknowledgements. I am very grateful to numerous former PhD students of mine and colleagues for discussions and collaboration on various results presented in this essay. Their names can be inferred from the bibliography attached to this essay. Among them, I gratefully mention B. Schubnel, who was my companion on my journey through the landscape sketched in Section 4. Part of this paper was written while I was visiting the School of Mathematics of the Institute for Advanced Study at Princeton. I wish to acknowledge the financial support from the 'Giorgio and Elena Petronio Fellowship Fund', and I warmly thank my colleague and friend Thomas C. Spencer for generous hospitality at the Institute and many very enjoyable discussions.

References

[1] E. H. LIEB and J. L. LEBOWITZ, The constitution of matter: Existence of thermodynamics for systems composed of electrons and nuclei, *Adv. Math.* (1972), 316–398.

[2] R. JOST, Das Carnotsche Prinzip, die absolute Temperatur und die Entropie, in: "Das Märchen vom Elfenbeinernen Turm – Reden und Aufsätze", pp. 183–188, K. Hepp, W. Hunziker and W. Kohn (editors), Berlin, Heidelberg: Springer-Verlag 1995.

[3] M. FLATO, A. LICHNÉROWICZ and D. STERNHEIMER, Déformations 1-différentiables des algèbres de Lie attachées à une variété symplectique ou de contact, Compositio Math. **31** (1975), 47–82; F. BAYEN, M. FLATO, C. FRONSDAL, A. LICHNÉROWICZ and D. STERNHEIMER, Deformation theory and quantization I-II, *Annals of Physics* (NY) **111** (1978), 61–110; and (1978), 111–151.

[4] M. GERSTENHABER, The cohomology structure of an associative ring, *Ann. Math.* **78** (1963), 267–288.

[5] M. KONTSEVICH, Deformation quantization of Poisson manifolds, *Lett. Math. Physics* **66** (2003), 157–216; (see also [15]).

[6] J. FRÖHLICH, A. KNOWLES and A. PIZZO, Atomism and quantization, *J. Phys. A* **40** (2006), 3033–3045; J. FRÖHLICH, A. KNOWLES and S. SCHWARZ, On the mean-field limit of bosons with Coulomb two-body interaction, *Commun. Math. Phys.* **288** (2009), 1023–1059.

[7] J. FRÖHLICH, M. MERKLI and D. UELTSCHI, Dissipative Transport: Thermal contacts and tunneling junctions, *Ann. H. Poicaré* **4** (2003), 897–945; C.-A PILLET and V. JAKŠIC, Non-equilibrium steady states of finite quantum systems coupled to thermal reservoirs, *Commun. Math. Phys.* **226** (2002), 131–162.

[8] W. K. ABOU SALEM and J. FRÖHLICH, Status of the fundamental laws of thermodynamics, *J. Stat. Phys.* **126** (2007), 1045–1068.

[9] J. FRÖHLICH, M. MERKLI, S. SCHWARZ and D. UELTSCHI, Statistical mechanics of thermodynamic processes, in: *A Garden of Quanta (Essays in honor of Hiroshi Ezawa)*, J. Arafune et al. (editors), London, Singapore and Hong Kong: World Scientific Publ. 2003;
W. K. ABOU SALEM and J. FRÖHLICH, Cyclic thermodynamic processes and entropy production, *J. Stat. Phys.* **126** (2006), 431–466.

[10] E. H. LIEB and M. B. RUSKAI, A fundamental property of quantum-mechanical entropy, *Phys. Rev. Letters* **30** (1973), 434–436;
Proof of strong subadditivity of quantum-mechanical entropy, *J. Math. Phys.* **14** (1973), 1938–1941.

[11] D. SUTTER, M. BERTA and M. TOMAMICHEL, Multivariate trace inequalities, arXiv:1604.03023v2 [math-ph] 30 Apr 2016.

[12] O. BRATTELI and D. W. ROBINSON, *Operator Algebras and Quantum Statistical Mechanics I*, New York: Springer-Verlag 1979;
Operator Algebras and Quantum Statistical Mechanics II, New York: Springer-Verlag 1981.

[13] W. BRAUN and K. HEPP, The Vlasov dynamics and its fluctuations in the 1/N limit of interacting classical particles, *Commun. Math. Phys.* **56** (1977), 101–113.

[14] N. M. J. WOODHOUSE, *Geometric Quantization*, Oxford Mathematical Monographs, Oxford: Clarendon Press 1991;
A. A. KIRILLOV, Geometric Quantization, in: *Dynamical Systems IV: Symplectic Geometry and its Applications*, V. I. Arnol'd and S. P. Novikov (editors), Berlin, Heidelberg: Springer-Verlag 1990.

[15] A. S. CATTANEO and G. FELDER, A path integral approach to the Kontsevich quantization formula, *Commun. Math. Phys.* **212** (2000), 591–611, and references given there.

[16] C. MOUHOT and C. VILLANI, On Landau damping, *J. Math. Phys.* **51** (2010), 015204.

[17] L. ERDŐS, B. SCHLEIN and H.-T. YAU, Derivation of the cubic non-linear Schrödinger equation from quantum dynamics of many-body systems, *Inventiones Math.* **167** (2007), 515–614;
Derivation of the Gross–Pitaevskii equation for the dynamics of Bose-Einstein condensates, *Ann. Math.* 2(172) (2010), 291–370;
P. PICKL, Derivation of the time dependent Gross–Pitaevskii equation with external fields, *Rev. Math. Phys.* **27** (2015), 1550003;
A. KNOWLES AND P. PICKL, Mean-field dynamics: Singular potentials and rate of convergence, *Commun. Math. Phys.* **298** (2010), 101–139.

[18] J. FRÖHLICH, T. P. TSAI and H.-T. YAU, On a classical limit of quantum theory and the non-linear Hartree equation, GAFA – Special Volume GAFA 2000 (2000), 57–78;
On the point-particle (Newtonian) limit of the non-linear Hartree equation, *Commun. Math. Phys.* **225** (2002), 223–274;
J. FRÖHLICH, B. L. G. JONSSON and E. LENZMANN, Effective dynamics for Boson stars, *Nonlinearity* **32** (2007), 1031–1075;
W. K. ABOU SALEM, J. FRÖHLICH and I. M. SIGAL, Colliding solitons for the nonlinear Schrödinger equation, *Commun. Math. Phys.* **291** (2009), 151–176.

[19] W. H. ASCHBACHER, Large systems of non-relativistic Bosons and the Hartree equation, Diss. ETH Zürich No. 14135, Zürich April 2001;
W. H. ASCHBACHER, Fully discrete Galerkin schemes for the nonlinear and nonlocal Hartree equation, *Electron. J. Diff. Eqns.* **12** (2009), 1–22.

[20] E. B. DAVIES, *Quantum Theory of Open Systems*, New York: Academic Press 1976.

[21] W. DE ROECK and J. FRÖHLICH, Diffusion of a massive quantum particle coupled to a quasi-free thermal medium, *Commun. Math. Phys.* **303** (2011), 613–707.

[22] D. BUCHHOLZ and J. E. ROBERTS, New light on infrared problems: Sectors, statistics and spectrum, *Commun. Math. Phys.* **330** (2014), 935–972;
D. BUCHHOLZ, Collision theory for massless Bosons, *Commun. Math. Phys.* **52** (1977), 147–173.

[23] B. SCHUBNEL, Mathematical results on the foundations of quantum mechanics, Diss. ETH Zürich No. 22382, Zürich December 2014.

[24] J. FRÖHLICH and B. SCHUBNEL, The preparation of states in quantum mechanics, *J. Math. Phys.* **57** (2016), 042101.

[25] J. FRÖHLICH and B. SCHUBNEL, Quantum probability theory and the foundations of quantum mechanics, in: *"The Message of Quantum Science – Attempts Towards a Synthesis"*, Ph. Blanchard and J. Fröhlich (editors), Lecture Notes in Physics **899**, Berlin, Heidelberg, New York: Springer-Verlag 2015;
PH. BLANCHARD, J. FRÖHLICH and B. SCHUBNEL, A "Garden of forking paths" – the quantum mechanics of histories of events, *Nuclear Physics B* (in press), available online 12 April 2016;
J. FRÖHLICH, paper in preparation.

Geometry and freeform architecture

Helmut Pottmann and Johannes Wallner

During the last decade, the geometric aspects of freeform architecture have defined a field of applications which is systematically explored and which conversely serves as inspiration for new mathematical research. This paper discusses topics relevant to the realization of freeform skins by various means (flat and curved panels, straight and curved members, masonry, etc.) and illuminates the interrelations of those questions with theory, in particular discrete differential geometry and discrete conformal geometry.

1 Introduction

A substantial part of mathematics is inspired by problems which originate outside the field. In this paper we deal with outside inspiration from a rather unlikely source, namely *architecture*. We are not interested in the more obvious ways mathematics is employed in today's ambitious freeform architecture (see Figure 1) which include finite element analysis and tools for computer-aided design. Rather, our topic is the unexpected interplay of geometry with the spatial decomposition of freeform architecture into beams, panels, bricks and other physical and virtual building blocks. As it turns out, the mathematical questions which arise in this context proved very attractive, and the mundane objects of building construction apparently are connected to several well-developed mathematical theories, in particular discrete differential geometry.

The design dilemma. Architecture as a field of applications has some aspects different from most of applied mathematics. Usually having a unique solution to a problem is considered a satisfactory result. This is not the case here, because architectural design is *art*, and something as deterministic as a unique mathematical solution of a problem eliminates design freedom from the creative process. We are going to illustrate this dilemma by means of a recent project on the Eiffel tower.

The interplay of disciplines. We demonstrate the interaction between mathematics and applications at hand of questions which occur in practice and their answers. We demonstrate how a question Q, phrased in terms of engineering and architecture, is transformed into a specific

Figure 1. Freeform architecture. The Yas Marina Hotel in Abu Dhabi illustrates the decomposition of a smooth skin into straight elements which are arranged in the manner of a torsion-free support structure. The practical implication of this geometric term is easy manufacturing of nodes (image courtesy Waagner-Biro Stahlbau).

mathematical question Q^* which has an answer A^* in mathematical terms. This information is translated back into an answer A to the original question. Simplified examples of this procedure are the following:

Q_1: Can we realize a given freeform skin as a steel-glass construction with straight beams and flat quadrilateral panels?

Q_1^*: Can a given surface Φ be approximated by a discrete- conjugate surface?

A_1^*: Yes, but edges have to follow a conjugate curve network of Φ.

A_1: Yes, but the beams (up to their spacing) are determined by the given skin.

Q_2: For a steel-glass construction with triangular panels, can we move the nodes within the given reference surface, such that angles become $\approx 60°$?

Q_2^*: Is there a conformal triangulation of a surface Φ which is combinatorially equivalent to a given triangulation (V, E, F)?

A_2^*: Yes if the combinatorial conformal class of (V, E, F) matches the geometric conformal class of Φ.

A_2: Yes if the surface does not have topological features like holes or handles.

Overview of the paper. We start in Section 2 with freeform skins with straight members and flat panels, leading to the discrete differential geometry of polyhedral surfaces. Section 3 deals with curved elements, Section 4 with circle patterns and conformal mappings, Section 5 with the statics of masonry shells, and finally Section 6 discusses computational tools.

2 Freeform skins with flat panels and straight beams

Freeform skins realized as steel-glass constructions are usually made with straight members and flat panels because of the high cost of curved elements. Often, the flat panels form a watertight skin. Since three points in space always lie in a common plane, but four generic points do not, it is obviously much easier to use triangular panels instead of quadrilaterals. Despite this

Figure 2. Steel-glass constructions following a triangle mesh can easily model the desired shape of a freeform skin, at the cost of high complexity in the nodes. The *Złote Tarasy* roof in Warszaw (left) is welded from straight pieces and spider-like node connectors which have been plasma-cut from a thick plate (images courtesy Waagner-Biro Stahlbau).

difficulty, the past decade has seen much research in the geometry of freeform skins based on *quadrilateral* panels. This is because they have distinct advantages over triangular ones – fewer members per node, fewer members per unit of surface area, fewer parts and lighter construction (see Figure 2).

2.1 Meshes

We introduce a bit of terminology: A *triangle mesh* is a union of triangles which form a surface, and we imagine that the edges of triangles guide the members of a steel-glass construction. The triangular faces serve as glass panels. Similarly, *quad* meshes are defined, as well as general meshes without any restrictions on the valence of faces. We use the term *planar quad mesh* to emphasize that panels are flat. Dropping the requirement of planarity of faces leads to general meshes whose edges are still straight. We use V for the set of vertices, E for the edges, and F for the faces. The exact definition of "mesh" follows below.

Meshes from the mathematical viewpoint. While a triangle mesh is simply a 2D simplicial complex of manifold topology, a general mesh is defined as follows. This definition is engineered to allow certain degeneracies, e.g. coinciding vertices.

DEFINITION 1. *A mesh in $\mathbb{R}^d$ consists of a two-dimensional polygonal complex (V, E, F) with vertex set V, edge set E, and face set F homeomorphic to a surface with boundary. In addition, each vertex $i \in V$ is assigned a position $v_i \in \mathbb{R}^d$ and each edge $ij \in E$ is assigned a straight line e_{ij} such that $v_i, v_j \in e_{ij}$.*

We say the mesh is a polyhedral surface if it has planar faces, i.e., for each face there is a plane which contains all vertices v_i incident with that face.

2.2 Support structures

An important concept are the so-called torsion-free support structures associated with meshes [30]. Figure 3 shows an example, namely an arrangement of flat quadrilateral panels along the

Figure 3. Physical torsion-free support structures. The roof of the Robert and Arlene Kogod Courtyard in the Smithsonian American Art Museum exhibits a mesh with quadrilateral faces and an associated support structure. The faces of the mesh are not planar – only the view from outside reveals that the planar glass panels which function as a roof do not fit together.

edges of a quad mesh (V, E, F) (which does not have planar faces), such that whenever four edges meet in a vertex, the four auxiliary quads meet in a straight line. We define:

DEFINITION 2. *A torsion-free support structure associated with a mesh (V, E, F) consists of assignments of a straight line ℓ_i to each vertex and a plane π_{ij} to each edge, such that $\ell_i \ni v_i$ for all vertices $i \in V$, and $\pi_{ij} \supset \ell_i, \ell_j, e_{ij}$ for all edges $ij \in E$.*

A support structure provides actual support in terms of statics (whence the name), but also has other functions like *shading* [43]. In discrete differential geometry, support structures occur under the name "line congruences".

Benefits of virtual support structures. Figures 1 and 4 illustrate the Yas Marina Hotel in Abu Dhabi, which carries a support structure in a less physical manner: each steel beam has a plane of central symmetry, and for each node these planes intersect in a common node axis, guaranteeing a clean "torsion-free" intersection of beams. This is much better than the complex intersections illustrated by Figure 2.

Combining flat panels and support structures. It would be very desirable from the engineering viewpoint to work with meshes which have both flat faces and torsion-free support structures. They would be able to guide a watertight steel-glass skin and allow for a "torsion-free" intersection of members in nodes such as demonstrated by Figure 4. The following elementary result however says that in order to achieve this, we must essentially do without triangle meshes.

LEMMA 3. *Every mesh can be equipped with trivial support structures where all lines ℓ_i and planes π_{ij} pass through a fixed point (possibly at infinity).*

Triangle meshes admit only trivial support structures. More precisely this property is enjoyed by every cluster of generic triangular faces which is iteratively grown from a triangular face by adding neighbouring faces which share an edge.

Proof. For an edge ij, there exists the point $x_{ij} = \ell_i \cap \ell_j$ (possibly at infinity), because ℓ_i, ℓ_j lie in the common plane π_{ij}. If ijk is a face, then $x_{ij} = \ell_i \cap \ell_j = (\pi_{ik} \cap \pi_{jk}) \cap (\pi_{ij} \cap \pi_{jk}) = \pi_{ij} \cap \pi_{ik} \cap \pi_{jk}$ implying that $x_{ij} = x_{ik} = x_{jk} \implies$ all axes incident with the face ijk pass through

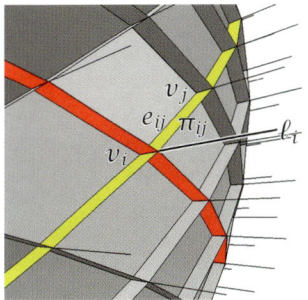 

Figure 4. Torsion-free support structure. For each edge ij and vertex i of a quadrilateral mesh, we have a plane π_{ij} and a line ℓ_i such that e_{ij}, ℓ_i, ℓ_j are contained in π_{ij} (at left, image courtesy Evolute). This support structure guides members and nodes in the outer hull of the Yas Marina hotel in Abu Dhabi, so that members have a nice intersection in each node (at right, image courtesy Waagner-Biro Stahlbau).

a common point. For faces sharing an edge that point obviously is the same, which proves the statement. □

Lemma 3 has far-reaching consequences since it expresses the incompatibility of two very desirable properties of freeform architectural designs. On the one hand frequently a freeform skin is to be watertight, acting as a roof, which for financial reasons imposes the constraint of planar faces. Unfortunately the planarity constraint is difficult to satisfy unless triangular faces are employed. On the other hand, triangle meshes have disadvantages: We already mentioned the large number of members. Lemma 3 states that only in special circumstances it is possible to reduce node complexity by aligning beams with the planes of a support structure.

2.3 Quadrilateral meshes with flat faces
Research related to meshes with planar faces is not new, as proved by the 1970 textbook [33] on *difference geometry* by R. Sauer which in particular summarizes earlier work starting in the 1930's. That work was pioneering for discrete differential geometry, which meanwhile is a highly developed area [11]. Relevant to the present survey, questions concerning quad meshes with planar faces ten years ago marked the starting point of a line of research motivated by problems in engineering and architecture [26], which again led to new developments in discrete differential geometry, see Section 2.4 below.

The meaning of "freeform". Research on quad meshes with flat faces has been rewarding mathematically, but unfortunately hardly any truly freeform meshes of that type have been realized as buildings. Their welcome qualities have nevertheless been used for impressive architecture, but meshes built so far enjoy special geometric properties (like rotational symmetry) which allows us to describe their shape using much less information than would be required in general, see Figure 5.

Smoothness limiting design freedom. A typical situation in the design process of freeform architecture is the following: A certain mesh has been created whose visual appearance fits the

Figure 5.   Not entirely freeform surfaces. Left: The hippo house in the Berlin zoo is based on a quadrilateral mesh with flat faces, but is not freeform in the strict sense. Its faces are parallelograms, so the mesh is generated by parallel translation of one polyline along another one. Mathematically, vertices $v_{i,j}$ have the form $v_{i,j} = a_i + b_j$.
Right: the Sage Gateshead building on the river Tyne, UK, is based on a sequence of polylines which are scaled images of each other, similar to a mesh with rotational symmetry.

 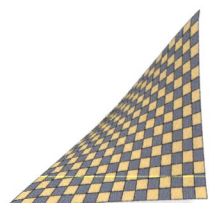 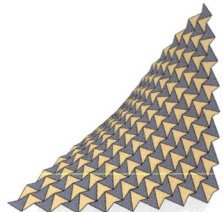

Figure 6.   Quad meshes. Left: The canopy at "Tokyo Midtown" is based on a quad mesh with planar faces. *Center:* This quad mesh has nonplanar faces, and all meshes nearby which have planar faces are far from smooth. *Right:* This mesh has planar faces and is no direct discrete analogue of a continuous smooth surface parametrization (for such regular patterns see [24]).

intentions of the designer and which eventually is to be realized as a steel-glass construction with flat panels. The designer therefore wants the vertex positions to be altered a little bit so that the faces of the mesh become planar, but its visual appearance does not change. Unfortunately this problem is typically not solvable. This is not because the nonlinear nature of this problem prevents a numerical solution – the reasons are deeper and of a more fundamental, geometric nature: For example, it is known that a "smooth" mesh of regular quad combinatorics which follows a smooth surface parametrization can have planar faces only if its edges are aligned with a so-called conjugate curve network of the reference surface. The network of principal curvature lines is the major example of that, cf. Figure 8. Since its singularities are shared (in a way) by all conjugate curve networks [45], the principal curvature lines already give a good impression of what a planar quad mesh approximating a given surface must look like. If the designer's mesh is not conjugate, there is no smooth mesh nearby which has planar faces – see Figure 6. There is no easy way out of this dilemma other than reverting to triangular faces, or redesigning the mesh entirely so that its edges follow a conjugate curve network, or forgoing smoothness.

Figure 7. A mesh with planar faces in the Louvre, Paris, by Mario Bellini Architects and Rudy Ricciotti. It has only as many triangular faces as are necessary to realize the architect's intentions (images courtesy Waagner-Biro Stahlbau and Evolute).

Example: The Cour Visconti in the Louvre. The "flying carpet" roof of the Islamic arts exhibition in the Louvre (Figure 7) provides a good illustration of the choices which have to be made when realizing freeform shapes with flat panels. In this particular case, the architects' design had triangular faces, but from the engineering viewpoint quadrilaterals would have been better (fewer parts, lighter construction, less complex nodes). Changing the visual appearance of the mesh was out of the question, for two reasons: Firstly conjugate curve networks of the flying carpet shape would not have led to meshes with sensible proportions of members, quite apart from their questionable aesthetics. Secondly there are both artistic and legal reasons not to change an architectural design after the decision to realize it has been made. In the end, many triangles (as many as it was possible to choose in a periodic way) were merged into flat quadrilaterals. This change is invisible since triangular shading elements were put on top of the glass panels.

Conjugate curve networks. As to the relation between quad meshes and curve networks, consider a mesh with vertices $v_{i,j}$, where i, j are integer indices. Using the forward difference operators $\Delta_1 v_{i,j} = v_{i+1,j} - v_{i,j}$ and $\Delta_2 v_{i,j} = v_{i,j+1} - v_{i,j}$, we express co-planarity of the four vertices $v_{i,j}, v_{i+1,j}, v_{i+1,j+1}, v_{i,j+1}$ by the condition

$$\Delta_1 \Delta_2 v \in \mathrm{span}(\Delta_1 v, \Delta_2 v).$$

One can clearly see the analogy to a parametric surface $x(t, s)$ with the property

$$\partial_1 \partial_2 x \in \mathrm{span}(\partial_1 x, \partial_2 x)$$

which is equivalent to $\mathrm{II}(\partial_1 x, \partial_2 x) = 0$, i.e., the parameter lines are conjugate w.r.t. the second fundamental form (this is the definition of a conjugate network). If in addition, $\mathrm{I}(\partial_1 x, \partial_2 x) = 0$ we get the network of principal curvature lines. In any case, Taylor expansion shows that for sequences $\{t_i\}, \{s_j\}$, the quad mesh with vertices $v_{i,j} := x(t_i, s_j)$ has faces which are planar up to a small error which vanishes as the stepsize diminishes. For the more difficult converse statement (convergence of meshes to smooth conjugate parametric surfaces) we refer to [11].

Multilayer constructions. In geometric research on freeform architecture, one focus was the problem of meshes at constant distance from each other [26, 30]. Like the support structures

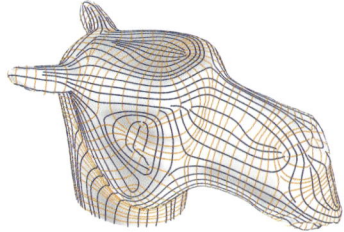 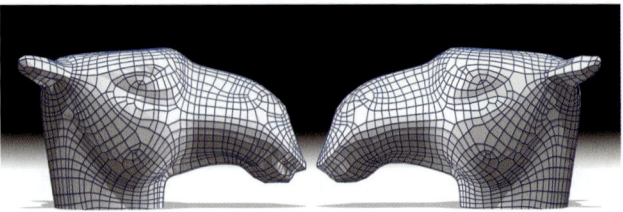

Figure 8.   A smooth surface can be approximated by a quad mesh with planar faces, if we first compute the principal curvature lines (at left), then align a mesh with those curves (center) and finally apply numerical optimization (right). One can hardly see the difference between the two meshes, which is why numerical optimization succeeds [26].

discussed above, such multilayer constructions can be physical or virtual – e.g., a steel-glass construction with members of constant height meeting cleanly in nodes is actually guided not by one mesh, but by two meshes whose edges are at constant distance (Figure 9). This topic has been very fruitful mathematically, and we discuss it in more detail in the next section.

2.4   Discrete differential geometry of polyhedral surfaces

The field of discrete differential geometry aims at the development of discrete equivalents of notions and methods of smooth surface theory. Discrete surfaces with quadrilateral faces can be treated as discrete analogues of parametrized surfaces. One focus of discretization is classical surface theory which is now associated with the theory of integrable systems. Remarkably this approach led to a better understanding of some fundamental structures [11] and has also impacted on the smooth theory.

Parallel meshes.   This concept is the basis of many constructions, from support structures to offsets to curvatures of polyhedral surfaces:

DEFINITION 4. *Two meshes M, M' with planar faces and the same combinatorics are parallel, if all corresponding edge lines e_{ij} and e'_{ij} are parallel.*

It is not difficult to see that parallel meshes and support structures are related: We can construct a support structure by connecting corresponding elements in a pair of parallel meshes, and for simply connected meshes also the converse is true.

Offset pairs.   If a pair of meshes has approximately constant distance from each other, they are interpreted as an offset pair. If one partner is approximating the unit sphere, they are seen as surface and Gauß image, in the spirit of relative differential geometry. There are several ways to exactify the notion of "approximating the unit sphere" (see also Figure 9):

DEFINITION 5. *A mesh M^* whose vertices are contained in the unit sphere is called inscribed. M^* is called circumscribed resp. midscribed if the planes resp. lines associated with faces resp. edges are tangent to the unit sphere. In any case, a parallel pair M, M^* with M^* approximating the unit sphere defines a parallel offset family of meshes $M^{(t)}$, by defining vertex positions $v_i^{(t)} = v_i + t v_i^*$.*

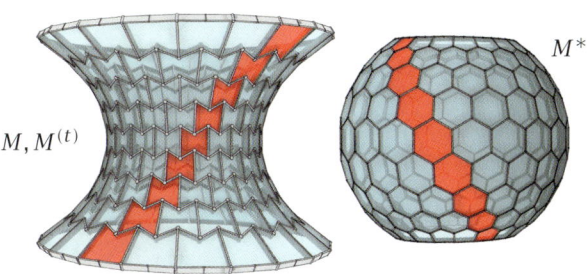
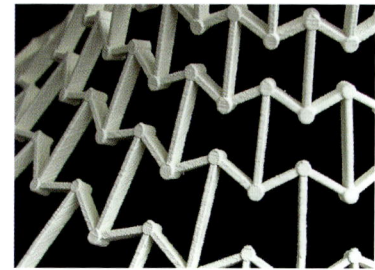

Figure 9. A discrete surface M, its offset $M^{(t)}$ and the Gauß image mesh M^*. Since M^* is midscribed to the unit sphere, corresponding edges of M and $M^{(t)}$ are at constant distance t from each other. The detail at right shows a physical realization of the mesh pair $M, M^{(t)}$ by members of constant height which fit together in the nodes.

The following is easy to see: In the inscribed (midscribed, circumscribed) case the points (lines, planes) assigned to the vertices (edges, faces) of both M and $M^{(t)}$ have distance t. Such meshes are therefore directly relevant to multilayer constructions in freeform architecture (see Figure 9 and [26, 30, 32]).

Discrete differential geometry of circular meshes, conical meshes, and Koebe polyhedra. There are some nice characterizations of meshes which possess a parallel mesh M^* with special properties. Quad meshes with inscribed M^* are circular, i.e., all faces have a circumcircle (the converse is true for simply connected quad meshes). Meshes with circumscribed M^* are conical, i.e., all vertices possess a cone of revolution tangent to the faces there (the converse is true for all meshes). A very interesting case is meshes where M^* is midscribed to the unit sphere. In that case M^* is uniquely determined, up to Möbius transformations, by its combinatorics alone [10]. It is known that the circular and conical quad meshes are discrete versions of principal parametric surfaces (for a convergence statement, see [6]). The Koebe polyhedron case corresponds the so-called Laguerre-isothermic parametrizations which do not exist on all surfaces.

Existence of meshes at constant distance. There are practical implications of this discussion, namely regarding the question if a given shape can be approximated by a quad mesh M with special properties. For circular and conical meshes this is possible in an almost unique way (see Figure 8 for a circular example), but only special surfaces can be approximated by a mesh where M^* is Koebe. This statement immediately translates to the existence of multilayer constructions: For a given shape, meshes at constant vertex-vertex distance or face-face distance exist, with the direction of their edges being essentially unique. Meshes with constant edge-edge distance exist only for special shapes.

A remark on discretization. Throughout its development, discrete differential geometry has studied discrete surfaces defined by properties analogous to properties which define classes of smooth surfaces. The latter may be equivalently defined by various properties, leading to several different discrete versions of it (so that still the resulting discrete surfaces converge to

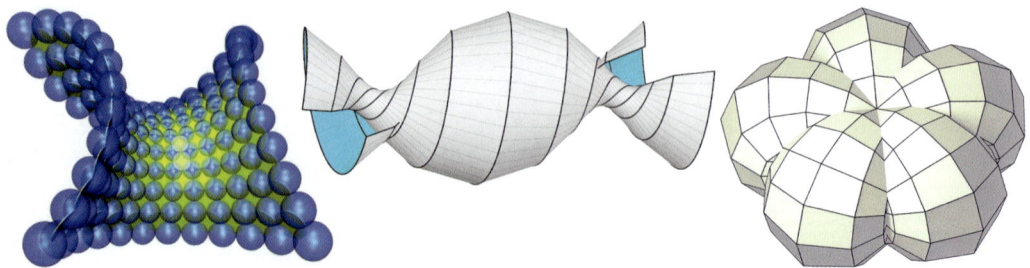

Figure 10. Discrete surfaces of constant mean curvature H. Left: An s-minimal surface by T. Hoffmann where Definition 6 yields $H = 0$. *Center:* Definition 6 does not apply to this cmc surface associated to an unduloid by W. Carl [12], but [22] yields $H = $ const. *Right:* This discrete "Wente torus" by C. Müller [27] has $H = $ const. w.r.t. both theories.

their smooth counterparts upon refinement). "Good" discretizations retain more than one of the original properties.

Integrable systems vs. curvatures. A major focus in discretization is the "integrable" PDEs that govern surface classes. For example, the angle ϕ between asymptotic lines in a surface of constant Gauß curvature obeys the sine-Gordon equation $\partial_t \partial_s \phi = \sin \phi$ after a suitable parameter transform. Consequently, Bobenko and Pinkall [8] based their study of discrete K-surfaces [44] on a discrete version of that equation. Other examples are furnished by *cmc surfaces* (defined by constant mean curvature) and other isothermic surfaces [19]. Apparently integrable systems provide the richest discrete surface theory. This approach, systematically presented in [11], creates discrete surface classes named after curvatures, whose actual definition however does not involve curvatures at all. It is therefore remarkable that notions of curvature have been discovered which assign the "right" values of curvature to such discrete surfaces after all.

Curvature and the surface area of offsets. In the smooth case, the surface area of an offset surface $M^{(t)}$ at distance t is expressed in terms of mean curvature H and Gauß curvature K, via the so-called Steiner formula. Since a similar relation holds in the discrete case, in [9,30] we gave a definition of curvatures of polyhedral surfaces which is directly inspired by the classical Steiner formula,

$$\text{area}(M^{(t)}) = \int_M (1 - 2tH + t^2 K)\, d\, \text{area}(M)$$

DEFINITION 6. *Assume that a mesh $M = (V, E, F)$ has planar faces, a Gauß image M^*, and an offset family $M^{(t)}$. For a face f with vertices $v_{i_1}, \ldots, v_{i_n}$, let*

$$\text{area}(f) = \frac{1}{2} \Big(\det(v_{i_1}, v_{i_2}, N_f) + \cdots + \det(v_{i_n}, v_{i_1}, N_f) \Big) = \langle f, f \rangle,$$

where N_f is a normal vector common to the face f, the corresponding face f^* in the Gauß image, and the face $f^{(t)} = f + t f^*$ in the offset mesh. We use the notation $\langle \, , \, \rangle$ to indicate the symmetric bilinear form induced by the quadratic form "area" in $\mathbb{R}^{3n}$. We define the mean *curvature H_f and the Gauß curvature K_f of the face f by comparing coefficients in the polynomials*

$$\text{area}(f^{(t)}) = \langle f, f \rangle + 2t \langle f, f^* \rangle + t^2 \langle f^*, f^* \rangle = (1 - 2tH_f + t^2 K_f)\, \text{area}(f)$$

$$\Rightarrow H_f = -\frac{\langle f, f^* \rangle}{\text{area}(f)}, \quad K_f = \frac{\text{area}(f^*)}{\text{area}(f)}.$$

Note that for convex faces, $\langle \, , \rangle$ is the well known *mixed area*. Interesting instances of Definition 6 are those where a canonical Gauß image M^* exists. Examples include the minimal surfaces of [7], the remarkable class of minimal surfaces dual to Koebe polyhedra of [5], and the cmc surfaces of [21].

Ongoing work on curvatures. Definition 6 is also the basis of further work, including discrete surfaces with hexagonal faces [27, 29]. Recently, Hoffmann et al. [22] developed a concept of curvature which applies to quad meshes (not necessarily planar) equipped with unit normal vectors in vertices. It coincides with Definition 6 in the circular mesh case, but also covers the K-surfaces of [8, 44] as well as other significant constructions like associated families of minimal and cmc surfaces [12, 22]. See Figure 10 for illustrations.

3 Freeform skins from curved panels

Curved beams and panels are employed in freeform architecture despite the fact that straight members and flat panels are cheaper and easier to handle. This has to do with artistic reasons and the required quality of the final surface: Discontinuities in the first and second derivative are clearly visible as discontinuities and kinks in reflection lines. Such requirements are known in the automotive industry, but in architecture there is no mass production and the cost of manufacturing curved elements is much higher. It is usually not easy to balance the required surface quality, the complexity of the shape, and the budget.

   Mathematical concepts which apply to curved elements are mostly of a differential- geometric nature and often are limit cases of notions known in discrete differential geometry. Optimization, both discrete and continuous, plays an important role. Let us start our discussion by listing a few manufacturing techniques:

- *Concrete* can be poured onto curved formworks. These can, e.g., be constructed by approximating the reference shape by a ruled surface. This allows us to use straight elements (Figure 11) or, on a smaller scale, hot wire cutting.
- *Sheet metal* can be bent into the shape of a developable surface, see Figure 12.
- *Double-curved glass panels* are expensive. They are made by hot bending using molds. On large structures, this technique has been employed only in situations were molds can be used to produce more than just one panel, see Figure 13.
- *Single-curved glass* panels can be made by bending in various ways (see Figure 12). In particular, cylindrical glass (see Figure 14) is comparatively cheap to make using machines which for the cooling process use the fact that cylinders permit a two-parameter group of rigid motions moving the surface into itself.

Opimization and manufacturing of panels. The shape of a freeform skin and its dissection into panels is usually very visible and thus is part of an architect's design. Manufacturing pan-

Figure 11. Formworks for concrete. Approximating surfaces of nonpositive Gauß curvature with ruled surfaces yields easily-built underconstructions. A union of ruled surface strips can be a C^1 smooth surface even if the rulings exhibit kinks (see marked area). This design by Zaha Hadid was intended for a museum in Cagliari (images courtesy Evolute).

Figure 12. Single-curved surfaces. Left: The outer skin of the Disney concert hall in Los Angeles by Frank Gehry is covered by metal sheets and consists of approximately developable surfaces (image courtesy *pdphoto.org*). *Right:* The glass canopy at the Strasbourg railway station exhibits "cold bent" glass panels (image courtesy RFR).

 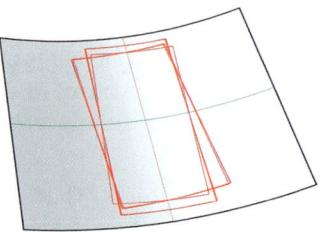 

Figure 13. Double-curved panels. Left: The entrance to the Paris metro at Saint Lazare station has rotational symmetry, which enables us to manufacture several glass panels with the same mold (image courtesy RFR). *Center:* symbolic image of a mold with panel outlines. *Right:* The 855 panels comprising the facade of the Arena Corinthians in Sao Paolo could be made with 61 molds [34].

els is an engineering responsibility. Typically, if the intended free form can be achieved by "simple" (e.g., cylindrical) panels, then the individual panels' geometric parameters are found via an optimization problem which usually is conceptually straightforward but might be cum-

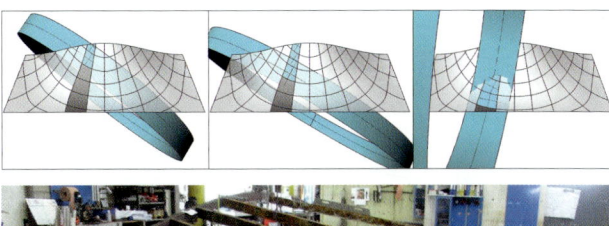

Figure 14. Panels and beams for the Eiffel tower pavilions. The facade of the pavilions on the first floor of the Eiffel tower feature cylindrical glass panels and curved beams with rectangular cross-section. *Top Right:* Best approximation of the original reference surface by cylinders ([2], image courtesy Evolute). *Bottom Right:* Since all four surfaces of each beam are manufactured by bending flat pieces of steel plate, they are developable surfaces. They also constitute a semidiscrete support structure (images courtesy RFR).

bersome to solve. Its target function involves gaps and kink angles between panels, as well as proximity to the original reference geometry, see Figure 14.

If custom panels are made in molds via hot bending, it is essential to re-use molds, by clustering similar panel shapes and determining a small number of molds capable of manufacturing all panels. The corresponding discrete-continuous optimization problems are hard. They have been studied by Eigensatz et al. [15] and are used for actual buildings, see Figure 13, right and [34].

The Eiffel tower refurbishment and a design dilemma. The first floor of the Eiffel tower exhibits three pavilions, completed in 2015, with a smooth glass facade consisting of cylindrical panels (see above) dissected by curved beams with rectangular cross-section. Figure 14 illustrates the fact that all four surfaces of such a beam are *developable*, since they are manufactured by bending flat sheets. Two of these surfaces are orthogonal to the facade, and it is known from elementary differential geometry that they must then follow principal curvature lines. This implies that the beams, up to spacing, are *uniquely determined* by the facade.

This relation presents a side-condition not easy to satisfy by a designer. In the case of the Eiffel tower pavilions, design freedom was restored by the fact that the principal curvature lines are very sensitive to small changes in the surface, and one could change the facade in imperceptible ways until the beam layout coincided with the original design intentions [36]. It should be mentioned that the Eiffel tower pavilions project benefitted from a cooperation between architects, engineers and mathematicians at an early stage, which possibly was responsible for its success.

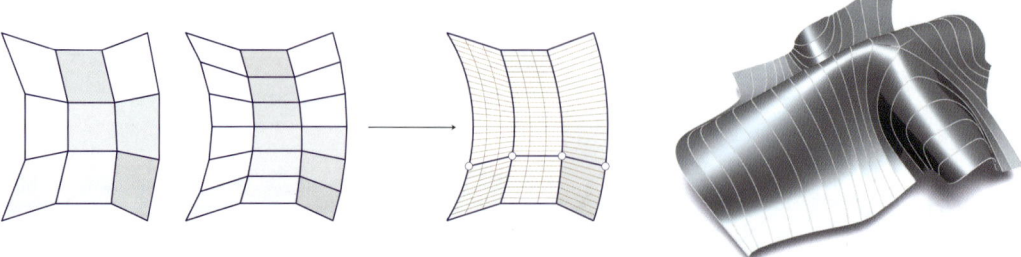

Figure 15. Left: Planar quad meshes upon refinement converge to a semidiscrete surface where each strip is developable. *Right:* Approximation by developable strips [31].

Differential geometry of semidiscrete surfaces.   Several geometric objects encountered in the discussion above can be seen as semidiscrete surfaces, which depend on one continuous and one discrete parameter. Figure 15 shows what is meant by that, how a semidiscrete surface arises as limit of discrete surfaces, and how it is visualized as a union of ruled strips. Various discrete surface classes have semidiscrete incarnations relevant to architecture:

- Developable ruled surfaces arise as limits of quad meshes with planar faces. They were investigated with regard to architecture (see Figure 15, right, and [26,31]). They also occur in semidiscrete versions of support structures (cf. our discussion of beams of the Eiffel tower pavilions).
- A union of ruled surfaces can be smooth, if it is a semidiscrete version of a mesh with planar vertex stars [16]. An application is shown by Figure 11.
- The geometry of semidiscrete surfaces is an active topic of research (cf. K-surfaces [42], curvatures [12,25,28], isothermic surfaces, etc.)

4 Regular Patterns

Regular patterns of geometric objects can mean different things, and there is hardly a limit to creativity: Figure 6 shows a mesh where the regularity lies in the repetitive features of edge polylines [24]. For the mesh consisting of equilateral triangles in Figure 20, regularity means

Figure 16.   Regular Patterns. Left: The facade of Selfridges, Birmingham (UK) is decorated with a circle pattern. *Right:* The "Kreod" pavilions, London, are derived from the support structure defined by a triangle mesh with incircle packing property.

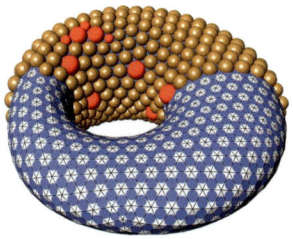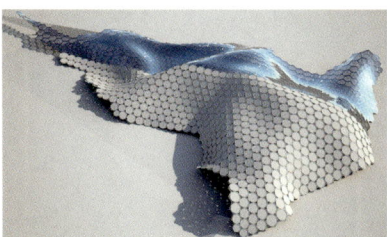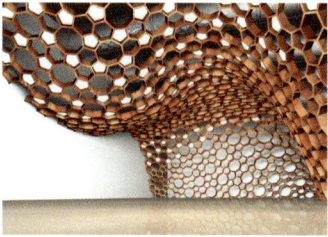

Figure 17.   Left: An incircle-packing (CP) mesh and its dual ball packing. *Center:* An existing freeform skin (Fiera di Milano) covered by a packing of circles dual to the incircle packing of Lemma 8, with circles having radius $\approx r_i$ [35]. *Right:* Torsion-free support structure made of the contact planes in the ball packing.

constant edgelength. For other examples see Figure 16. This section deals with a particular kind of patterns related to circle packings. We first discuss their connection to conformal mappings from the mathematical viewpoint, and then proceed with algorithms and applications.

Circle Packings and discrete conformal mappings.   The idea of *circle packing* is the following: Consider non-overlapping circles in a two-dimensional domain, take centers as vertices and put edges whenever two circles touch. If this graph is a triangulation, we have a circle packing. The natural correspondence between two combinatorially equivalent packings is a *discrete conformal mapping.* A Riemann mapping-type theorem states that in the simply connected case all packings can be conformally mapped to packings which fill either the disk or the unit sphere [38]. The proper statement is the following:

THEOREM 7 (Koebe-Andreev-Thurston). *Let K be a complex triangulating a compact surface S (possibly with boundary). Then there exists a unique Riemann surface S_K homeomorphic to S and a circle packing with contact graph K which is unique up to conformal automorphisms of S_K and which univalently fills S_K.*

This result (due to Koebe 1936, and more recently to Andreev and Thurston as well as Beardon and Stephenson [3]) justifies the definition of discrete conformal mappings together with the fact that they converge to smooth conformal mappings. We extend the concept of circle packing, starting with an elementary lemma [35]:

LEMMA 8. *For a triangle mesh (V, E, F) in $\mathbb{R}^n$, the following are equivalent:*

(1) *For faces $f = ijk$, $f' = ijl$, the incircles of triangles $v_i v_j v_k$, $v_i v_j v_l$ have the same contact point with $v_i v_j$ ("incircle packing property").*
(2) *Under the same assumptions, edgelengths obey*
$$|v_i - v_k| + |v_j - v_l| = |v_k - v_j| + |v_i - v_l|.$$
(3) *There are radii $r : V \to [0, \infty)$ with*
$$r_i + r_j = |v_i - v_j| \text{ for all } ij \in E.$$

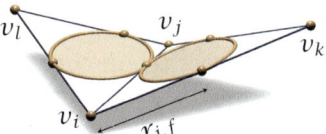

Proof. Using the notation of (1), define $r_{i,f}$ as the distance from v_i to the points where the incircle of f touches segments $v_i v_j$ and $v_i v_k$. Obviously, $|v_i - v_j| = r_{i,f} + r_{j,f}$. From $|v_i - v_j| = r_{i,f'} + r_{j,f'}$ we get $r_{i,f} - r_{i,f'} = r_{j,f'} - r_{j,f}$.

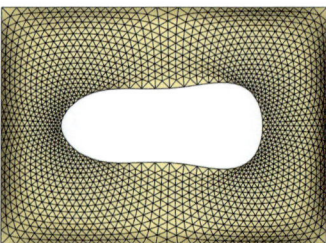

Figure 18. Optimization for the incircle-packing property of meshes which are not simply connected. The triangle mesh of the Great Court of the British Museum is only optimizable if the inner boundary is allowed to move. If one accepts Conjecture 9, this is caused by the geometric conformal class of the reference surface not coinciding with the combinatorial conformal class of the mesh.

Since (1) is expressed as $r_{i,f} = r_{i,f'}$, and (2) is expressed as $r_{i,f} - r_{i,f'} = r_{j,f} - r_{j,f'}$, the equivalence (1) $\iff$ (2) follows. Further, (1) $\implies$ (3) with $r_i = r_{i,f}$, for any face f containing the vertex i. The implication (3) $\implies$ (2) is obvious. $\qquad\qquad\square$

Circle patterns and discrete conformal mappings. The previous result implies that for a planar triangulation (V, E, F) with the incircle packing (CP) property, circles centered in vertices v_i, having radius r_i yield a packing whose contact graph is (V, E, F) (also the converse is true). These circles together with the original incircles form an *orthogonal circle pattern*. Such patterns (without the requirement of coming from a triangulation) define discrete conformal mappings in the same way packings do. This discrete approach to conformal mapping has been very fruitful: It has been used by He and Schramm for proving an extended Koebe's theorem [18], and by Bobenko et al. for determining the shape of a minimal surface from the combinatorics of its network of principal curves [5], based on a convergence result by O. Schramm [37].

Circle-Packing Meshes in freeform architecture. Meshes with the CP property of Lemma 8 are relevant to architecture in various ways. One reason is aesthetics, since such meshes tend to be regular in the sense of even distribution of angles and edgelengths [35].

Another application is geometric. By Lemma 8, the balls of radius r_i, centered in vertices v_i, intersect the incircles orthogonally, and the contact planes of balls constitute a *torsion-free support structure* associated with a mesh (V^*, E^*, F^*) whose vertices are incenters (Figure 17).

In [35], we optimized a triangle mesh for the CP property by moving its vertices tangential to it. This succeeds for meshes of disk and sphere topology, but not for others, see Figure 18. This behaviour can be explained by the following

CONJECTURE 9. *The canonical correspondence between combinatorially equivalent "incircle packing" meshes is a discrete version of conformal equivalence of general surfaces – in much the same way the correspondence between orthogonal circle patterns is a discrete version of conformal equivalence of Riemann surfaces.*

We do not attempt to give a precise statement here. If the conjecture is true, then the combinatorics of a triangle mesh $M = (V, E, F)$ define a "combinatorial" conformal class $[M]$, containing all CP meshes combinatorially equivalent to M. The shapes of those meshes are approximately related by continuous conformal equivalence. If the shape of M is not in $[M]$, there is no CP mesh combinatorially equivalent to M. Since for topological spheres and topological disks there is only one conformal class, optimization for the CP property works in these cases. So far, numerical evidence supports the conjecture, see also Figure 18.

5 Geometry and Statics

Statics is a very extensive and important field in building construction, and it is not surprising that it has connections to geometry. We discuss two specific topics here: One is the effort to include rudimentary statics in computational design (see Section 6). The other one is the geometry of freeform masonry, see Figure 19.

Self-supporting surfaces. We employ a simplified material model justified by the fact that failure of such structures is via geometry catastrophe, not by material failure. It is called Heyman's safe theorem [20] and postulates stability of a masonry shell S if there is a mesh whose vertices and edges are contained in S and whose edges carry compressive forces which are in equilibrium with the deadload. The geometry of such force systems has been investigated since J. C. Maxwell's time. P. Ash et al. in [1] summarize their geometry and connection to polyhedral bowls (see Figure 19). The use of such force systems ("thrust networks") in architectural design has been pioneered by P. Block, see, e.g., [4]. The method is further justified by the interpretation of force networks and polyhedral bowls as finite element discretizations of the shell equations and associated Airy potentials [17].

Discrete differential geometry of masonry shells. In [41] we established a differential-geometric context of freeform masonry together with algorithms for finding the nearest self-supporting shape to a given shape. The discrete shell equations have an interpretation in the framework of curvatures of polyhedral surfaces (Section 2.4) after a certain duality is applied, and there is also an interpretation in terms of geometric graph Laplacians. One result is the following: If a planar quad mesh is to have equilibrium forces in its edges, it must be aligned with the principal curvature lines in the sense of "relative" differential geometry, with

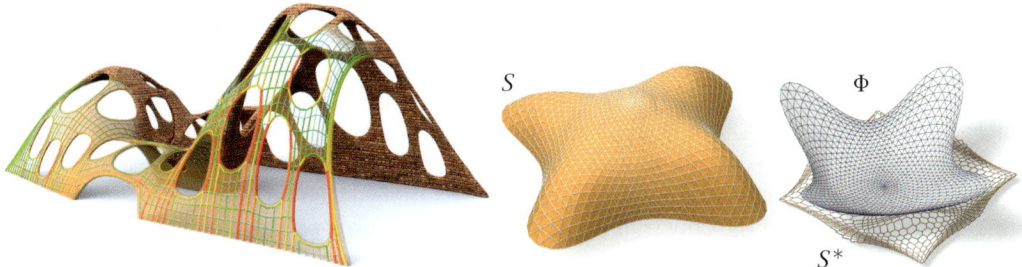

Figure 19. Self-supporting surfaces. A shell built from bricks without mortar in theory is stable, if it contains a fictitious network of compressive forces which is in equilibrium with the deadload. *Right:* Such a force system S has an associated polyhedral "Airy potential" bowl Φ (at least locally) whose face gradients are exactly Maxwell's force diagram S^* which is an orthogonal dual to the common projection of both S and Φ onto the horizontal plane.

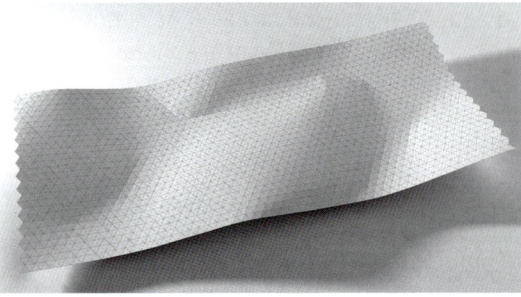  

Figure 20.   Design with nonlinear constraints. Left: A triangle mesh with equilateral faces approximates the "flying carpet" surface of Figure 7; it is a discrete model of crumpled paper [23]. *Center:* A union of developable strips whose developments fit together models a curved-crease sculpture capable of being folded from an annulus [39]. This model is inspired by sculptures by E. and M. Demaine (right). In both situations efficient design uses appropriately regularized Newton methods applied to quadratic equations.

the Airy potential as unit sphere. This is another instance of a problem which has a unique solution, effectively hindering the design process. The design and assembly of self-supporting masonry surfaces is still an active topic of research from both the theory and practice side. We exemplarily point to [13] and [14].

6 Computational Design

Interactive design often requires solving a large number of nonlinear constraints quickly. An example of that is design of meshes with planar faces. Another one is the inclusion of force systems during design, which can give a user important feedback (without replacing a full-blown statics analysis). A more complex example, not even discrete, is design of *developable* surfaces and curved-folding objects (Figure 20), where constraint equations are found by using spline surfaces and expressing geometric side conditions in terms of the spline control elements [39].

Setting up and solving constraints.   A system of n constraints on m variables can be written as $f(x) = 0$, with $x \in \mathbb{R}^m$ and $f : \mathbb{R}^m \to \mathbb{R}^n$. An iterative Newton method would have to solve $x_{j-1} + df_{x_{j-1}}(x_j - x_{j-1}) = 0$. Typically this linear system is under-determined (otherwise there would be no design freedom) and also contains redundant equations. These numerical difficulties can be overcome by appropriate regularization which geometrically amounts to "guided" projection onto the constraint manifold in $\mathbb{R}^m$. It turns out that *quadratic* equations give much better convergence than higher order ones, which is in part easy to achieve by introducing more variables. For some constraints like approximation of a reference surface there are special geometric ways to replace them with quadratic terms. For more details we refer to [40].

We see this line of research as a contribution to the longtime goal of interactive design tools capable of handling geometric constraints combined with statics.

Conclusion

We have reported on research in mathematics (discrete differential geometry) and computer science (geometry processing) which is either relevant to free forms in architecture, or is inspired by the geometric problems which occur there. This new area of applications is not exhausted yet; it continues to yield interesting problems and solutions. We emphasize that this field of applications has certain strange features: for artistic reasons it may happen that unique solutions to mathematical problems are are not acceptable in practice. Of course, this "design dilemma" is not restricted to architecture.

The research we reported on in this paper is focussed on architecture. Many ascpects of it are very relevant to other fields of application. This in particular applies to computational design, which lies at the *interface between technology and art*. The newly established doctoral college for computational design at Vienna University of Technology aims at addressing future challenges in this broad field.

Acknowledgments.   This research is supported by the Center for Geometry and Computational Design at TU Wien, and by the Austrian Science Fund via grants I-705 and I-706 (SFB-Transregio *Discretization in Geometry and Dynamics*).

References

[1] P. Ash, E. Bolker, H. Crapo and W. Whiteley, Convex polyhedra, Dirichlet tessellations, and spider webs. In *Shaping space (Northampton 1984)*, pages 231-250. Birkhäuser, 1988.

[2] N. Baldassini, N. Leduc and A. Schiftner, Construction aware design of curved glass facades: The Eiffel Tower Pavilions. In *Glass Performance Days Finland*, pages 406-410. 2013.

[3] A. F. Beardon and K. Stephenson, The uniformization theorem for circle packings. *Indiana Univ. Math. J.*, 39:1383-1425, 1990.

[4] P. Block and J. Ochsendorf, Thrust network analysis: A new methodology for three-dimensional equilibrium. *J. Int. Assoc. Shell and Spatial Structures*, 48(3):167–173, 2007.

[5] A. Bobenko, T. Hoffmann and B. Springborn, Minimal surfaces from circle patterns: Geometry from combinatorics. *Ann. Math.*, 164:231-264, 2006.

[6] A. Bobenko, D. Matthes and Yu. Suris, Discrete and smooth orthogonal systems: C^∞-approximation. *Int. Math. Res. Not.*, (45):2415-2459, 2003.

[7] A. Bobenko and U. Pinkall, Discrete isothermic surfaces. *J. Reine Angew. Math.*, 475:187-208, 1996.

[8] A. Bobenko and U. Pinkall, Discrete surfaces with constant negative Gaussian curvature and the Hirota equation. *J. Diff. Geom.*, 43:527-611, 1996.

[9] A. Bobenko, H. Pottmann and J. Wallner, A curvature theory for discrete surfaces based on mesh parallelity. *Math. Annalen*, 348:1-24, 2010.

[10] A. Bobenko and B. Springborn, Variational principles for circle patterns and Koebe's theorem. *Trans. Amer. Math. Soc.*, 356:659-689, 2004.

[11] A. Bobenko and Yu. Suris, *Discrete differential geometry: Integrable Structure.* American Math. Soc., 2009.

[12] W. Carl, On semidiscrete constant mean curvature surfaces and their associated families. *Monatsh. Math.*, 2016, to appear.

[13] F. de Goes, P. Alliez, H. Owhadi and M. Desbrun, On the equilibrium of simplicial masonry structures. *ACM Trans. Graph.*, 32(4):93, 1-10, 2013.

[14] M. DEUSS, D. PANOZZO, E. WHITING, Y. LIU, P. BLOCK, O. SORKINE-HORNUNG and M. PAULY, Assembling self-supporting structures. *ACM Trans. Graphics*, 33(6):214, 1-10, 2014.

[15] M. EIGENSATZ, M. KILIAN, A. SCHIFTNER, N. MITRA, H. POTTMANN and M. PAULY, Paneling architectural freeform surfaces. *ACM Trans. Graph.*, 29(4):45, 1-10, 2010.

[16] S. FLÖRY, Y. NAGAI, F. ISVORANU, H. POTTMANN and J. WALLNER, Ruled free forms. In L. Hesselgren et al., editors, *Advances in Architectural Geometry 2012*, pages 57-66. Springer, 2012.

[17] F. FRATERNALI, A thrust network approach to the equilibrium problem of unreinforced masonry vaults via polyhedral stress functions. *Mechanics Res. Comm.*, 37(2):198-204, 2010.

[18] Z. X. HE and O. SCHRAMM, Fixed points, Koebe uniformization and circle packings. *Annals Math.*, 137:369-406, 1993.

[19] U. HERTRICH-JEROMIN, T. HOFFMANN and U. PINKALL, A discrete version of the Darboux transform for isothermic surfaces. In A. Bobenko and R. Seiler, editors, *Discrete integrable geometry and physics*, pages 59-81. Clarendon Press, Oxford, 1999.

[20] J. HEYMAN, The stone skeleton. *Int. J. Solids Structures*, 2:249-279, 1966.

[21] T. HOFFMANN, Discrete rotational cmc surfaces and the elliptic billiard. In H.-C. Hege and K. Polthier, editors, *Mathematical Visualization*, pages 117-124. Springer, Berlin, 1998.

[22] T. HOFFMANN, A. O. SAGEMAN-FURNAS and M. WARDETZKY, A discrete parametrized surface theory in $\mathbb{R}^3$. *Int. Mat. Res. Not.*, 2016, to appear.

[23] C. JIANG, C. TANG, M. TOMIČIĆ, J. WALLNER and H. POTTMANN, Interactive modeling of architectural freeform structures – combining geometry with fabrication and statics. In P. Block et al., editors, *Advances in Architectural Geometry 2014*, pages 95-108. Springer, 2014.

[24] C. JIANG, C. TANG, A. VAXMAN, P. WONKA and H. POTTMANN, Polyhedral patterns. *ACM Trans. Graphics*, 34(6):172, 1-12, 2015.

[25] O. KARPENKOV and J. WALLNER, On offsets and curvatures for discrete and semidiscrete surfaces. *Beitr. Algebra Geom.*, 55:207-228, 2014.

[26] Y. LIU, H. POTTMANN, J. WALLNER, Y.-L. YANG and W. WANG, Geometric modeling with conical meshes and developable surfaces. *ACM Trans. Graph.*, 25(3):681-689, 2006.

[27] C. MÜLLER, On discrete constant mean curvature surfaces. *Discrete Comput. Geom.*, 51:516-538, 2014.

[28] C. MÜLLER, Semi-discrete constant mean curvature surfaces. *Math. Z.*, 279:459-478, 2015.

[29] C. MÜLLER and J. WALLNER, Oriented mixed area and discrete minimal surfaces. *Discrete Comput. Geom.*, 43:303-320, 2010.

[30] H. POTTMANN, Y. LIU, J. WALLNER, A. BOBENKO and W. WANG, Geometry of multi-layer freeform structures for architecture. *ACM Trans. Graph.*, 26:65, 1-11, 2007.

[31] H. POTTMANN, A. SCHIFTNER, P. BO, H. SCHMIEDHOFER, W. WANG, N. BALDASSINI and J. WALLNER, Freeform surfaces from single curved panels. *ACM Trans. Graph.*, 27(3):76, 1-10, 2008.

[32] H. POTTMANN and J. WALLNER, The focal geometry of circular and conical meshes. *Adv. Comp. Math*, 29:249-268, 2008.

[33] R. SAUER, *Differenzengeometrie*. Springer, 1970.

[34] A. SCHIFTNER, M. EIGENSATZ, M. KILIAN and G. CHINZI, Large scale double curved glass facades made feasible – the Arena Corinthians west facade. In *Glass Performance Days Finland*, pages 494-498. 2013.

[35] A. SCHIFTNER, M. HÖBINGER, J. WALLNER and H. POTTMANN, Packing circles and spheres on surfaces. *ACM Trans. Graph.*, 28(5):139, 1-8, 2009.

[36] A. SCHIFTNER, N. LEDUC, P. BOMPAS, N. BALDASSINI and M. EIGENSATZ, Architectural geometry from research to practice – the Eiffel Tower Pavilions. In L. Hesselgren et al., editors, *Advances in Architectural Geometry 2012*, pages 213-228. Springer, 2012.

[37] O. SCHRAMM, Circle patterns with the combinatorics of the square grid. *Duke Math. J.*, 86:347-389, 1997.

[38] K. STEPHENSON, *Introduction to Circle Packing*. Cambridge Univ. Press, 2005.

[39] C. Tang, P. Bo, J. Wallner and H. Pottmann, Interactive design of developable surfaces. *ACM Trans. Graphics*, 36(2):12, 1-12, 2016.

[40] C. Tang, X. Sun, A. Gomes, J. Wallner and H. Pottmann, Form-finding with polyhedral meshes made simple. *ACM Trans. Graphics*, 33(4):70, 1-9, 2014.

[41] E. Vouga, M. Höbinger, J. Wallner and H. Pottmann, Design of self-supporting surfaces. *ACM Trans. Graph.*, 31(4):87, 1-11, 2012.

[42] J. Wallner, On the semidiscrete differential geometry of A-surfaces and K-surfaces. *J. Geometry*, 103:161-176, 2012.

[43] J. Wang, C. Jiang, P. Bompas, J. Wallner and H. Pottmann, Discrete line congruences for shading and lighting. *Comput. Graph. Forum*, 32(5):53-62, 2013.

[44] W. Wunderlich, Zur Differenzengeometrie der Flächen konstanter negativer Krümmung. *Sitz. Öst.. Akad. Wiss. Math.-Nat. Kl.*, 160:39-77, 1951.

[45] M. Zadravec, A. Schiftner and J. Wallner, Designing quad-dominant meshes with planar faces. *Comput. Graph. Forum*, 29(5):1671-1679, 2010. Proc. SGP.

Some geometries to describe nature

Christiane Rousseau

Since ancient times, the development of mathematics has been inspired, at least in part, by the need to provide models in other sciences, and that of describing and understanding the world around us. In this note, we concentrate on the shapes of nature and introduce two related geometries that play an important role in contemporary science. Fractal geometry allows describing a wide range of shapes in nature. In 1973, Harry Blum introduced a new geometry well suited to describe animal morphology.

The range of shapes and patterns in nature is immense. At the larger scale, we can think of the spiral shapes of galaxies, which resemble those of some hurricanes. There are many other shapes on Earth, from mountains to dunes, to oceans, to clouds, and to networks of rivers. Coming to still smaller scales, we observe the many shapes of life, be it vegetation or animals. At still smaller scales, we explore the complexity of the capillary network and bronchial tubes in the lungs. We discover common features: for instance, the capillary network resembles a river network. Can we measure these resemblances? Can we explain how these complicated shapes are produced and why some similarities of shapes occur in many different situations?

Traditional geometry uses lines, circles, smooth curves and surfaces to describe nature. This is sufficient to describe remarkable shapes like the logarithmic spiral of a nautilus, or the spiral shapes of hurricanes and galaxies, and to explain why they occur. If we want to describe the DNA strands, then the help of knot theory is welcome since the DNA strands are so long that they require entanglement to be packed in the cell. The traditional geometry is not sufficient to describe the shapes of many plants like trees, ferns, the Romanesco cauliflower, or the branch of parsley. We need a geometry to describe the plants. This geometry is the *fractal geometry*, created by the visionary mathematician, Benoît Mandelbrot (see for instance [5]).

1 Fractal geometry

The fern, the Romanesco califlower, or the branch of parsley are very complicated objects. Well, they look very complicated if we observe them through our eyes ... Hence, let us put on mathematical glasses and discover their hidden simplicity: they are *self-similar* (Figure 1). This means that if we zoom to some part of an object, then we see the same pattern reappearing at different scales. In practice, this self-similarity means that the object looks complicated just

a The fern   b A branch of parsley

Figure 1. A fern and a branch of parsley

because we do not look at it from the right point of view. And indeed, self-similar objects can be reconstructed with extremely simple programs if we make use of their self-similar structure (see, for instance, [6] or [8]).

Mathematicians like to focus on simpler models. Hence, we will concentrate on the Sierpiński carpet (see Figure 2a) which is obtained from a plain triangle, deleting the middle triangle, and iterating. We already face a surprise. The Sierpiński carpet is really an object created from the imagination of mathematicians. Well, nature seemed to have borrowed this pattern on Cymbiola Innexa REEVE (see Figure 2b)! And if you look at the computer generated pattern in Figure 2c, it is very similar to the pattern on the shell! We will come back to this pattern later. When introducing fractal geometry to describe the shape of plants, we have done much better. This new geometry can also be used to describe many more shapes of nature including rocky

a Sierpiński carpet   b Cymbiola Innexa REEVE   c computer generated pattern

Figure 2. The Sierpiński carpet, the shell Cymbiola Innexa REEVE (Photo credit: Ian Holden, Schooner Specimen Shells), and a computer generated pattern

coasts, the river networks or the patterns drawn by water flowing on the sand, the deltas of
large rivers like Niger, clusters of galaxies or the Milky Way, clouds, frost, the bronchial tubes
or the network of capillaries, etc.

We want to understand the geometry of these objects and what characterizes them.

1.1 Measuring a fractal

How can we measure the length of a rocky coast? This seems a difficult operation. Indeed, each
time we zoom, we see new details appearing, thus adding to the length. Hence, when do we
stop zooming?

Again, as mathematicians, we will first study a simple model. This model is the von Koch
snowflake (see Figure 3a). Is is obtained by *iteration*, i.e., by repeating the same step(s) again
and again. At each step of the iteration, we replace each segment by a group of 4 segments
with length equal to $1/3$ of the length of the original segment. If L is the length of the original
triangle in Figure 3b, then $\frac{4}{3}L$ is the length of the star of Figure 3c, $\left(\frac{4}{3}\right)^2 L$ is the length of the
object in Figure 3d, etc. In particular, this means that at each step the length is multiplied by
$\frac{4}{3}$. Since there are an infinite number of steps in the construction, then the length of the von
Koch snowflake is infinite even though the curve fits into a finite region of the plane!

As an exercise, you could play the same kind of game and check that the area of the Sierpiński carpet is zero.

We have seen that a curve can have an infinite length even when it lies in a finite region of
the plane. In the same way, there exists fractal surfaces with infinitely many peaks that have
infinite area and fit in a finite volume. Such properties are useful for nature. For instance, while
the outer surface of the small intestine has an approximate area of $0.5\,\mathrm{m}^2$, the inner surface
has an approximate area of $300\,\mathrm{m}^2$. The fractal nature of the inner surface allows its large area
and favors the intestinal absorption.

The examples above show that length and area give little information on a fractal. The concept which will allow to give more relevant information is that of *dimension*.

1.2 Dimension of a fractal object

How do we mathematically define *dimension*? We have an intuitive idea of dimension. Indeed,
smooth curves are of dimension 1, smooth surfaces, of dimension 2, and filled volumes, of

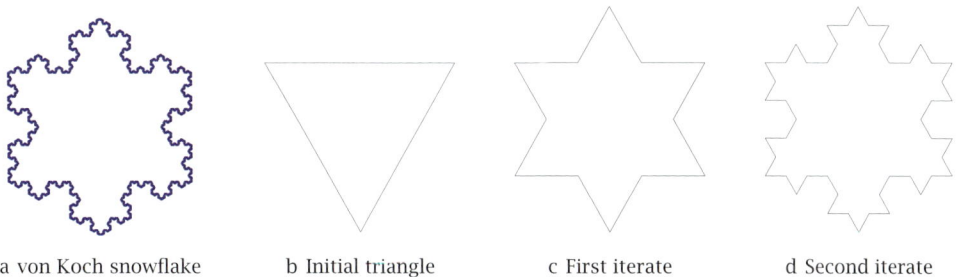

| a von Koch snowflake | b Initial triangle | c First iterate | d Second iterate |

Figure 3. Von Koch snowflake and the iteration process to construct it

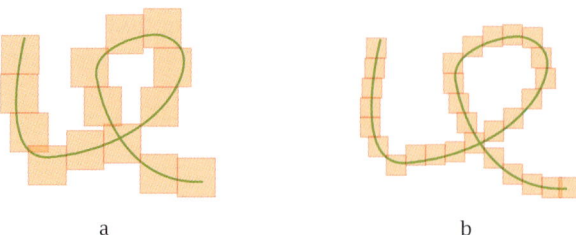

a   b

Figure 4. Calculating the dimension of a curve using squares of different sizes

dimension 3. Hence, we should give a mathematical definition of dimension, which yields 1 for smooth curves, 2, for smooth surfaces, and 3, for filled volumes. In the context of this paper, we will limit ourselves to dimensions 1 and 2. We want to cover an object in the plane with small squares. (If we would want to define dimension 3 we would use small cubes, but we could have used small cubes for curves and surfaces without changing the dimension!)

Case of a smooth curve. (see Figure 4)

- If we take squares with side of half size, then we approximately double the number of squares needed to cover the object.
- If we take squares with side one third of the size, then we approximately triple the number of squares needed to cover the object.
- …
- If we take squares with side n times smaller, then we approximately need n times more squares to cover the object.

Case of a surface. (see Figure 5)

- If we take squares with side of half size, then we approximately need four times more squares to cover the object.
- If we take squares with side one third of the size, then we approximately need nine times more squares to cover the object.

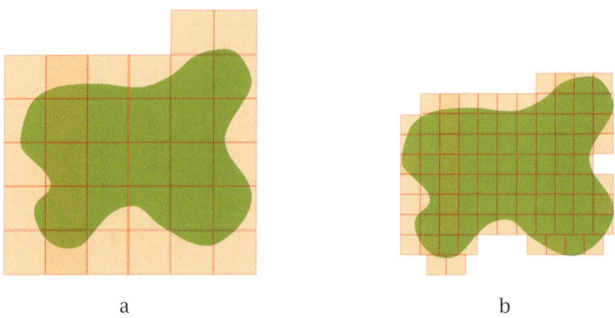

a   b

Figure 5. Calculating the dimension of a surface using squares of different sizes

- ...
- If we take squares with side n times smaller, then we approximately need n^2 times more squares to cover the object.

We can now give the (intuitive) definition of dimension.

DEFINITION. *An object has* dimension d *if, when we take squares (or cubes) with edge n times smaller to cover it, then we need approximately n^d more squares (or cubes) to cover it.*

Not all objects have a dimension. But, the self-similar objects have a dimension which, most often, is not an integer. We assert without proof that the fractal dimension of the von Koch snowflake is $\frac{\ln 4}{\ln 3} \sim 1.26$. Let us now calculate the dimension of the Sierpiński carpet (see Figure 6).

- Let us take a square with side equal to the length of the base (see Figure 6b). It covers the Sierpiński carpet of Figure 6a.
- If we take squares with side of half size, then we need three squares to cover the Sierpiński carpet. Note that $3 = 2^{\frac{\ln 3}{\ln 2}}$ (Figure 6c).
- if we take squares with side one fourth of the size, then we need nine squares to cover the Sierpiński carpet. Note that $9 = 4^{\frac{\ln 3}{\ln 2}}$ (Figure 6d).
- If we take squares with side one eight of the size, then we need 27 squares to cover the Sierpiński carpet. Note that $27 = 8^{\frac{\ln 3}{\ln 2}}$ (Figure 6e).

Hence, it is easy to conclude that the dimension of the Sierpiński carpet of Figure 6a is $d = \frac{\ln 3}{\ln 2} \sim 1.585$. We assert without proof that it is also the dimension of the Sierpiński carpet of Figure 2, even if that is not as obvious (see for instance [7]).

The dimension gives a "measure" of the complexity or density of a fractal. We feel that the Sierpiński carpet is denser than the von Koch snowflake, which looks more like a thickened

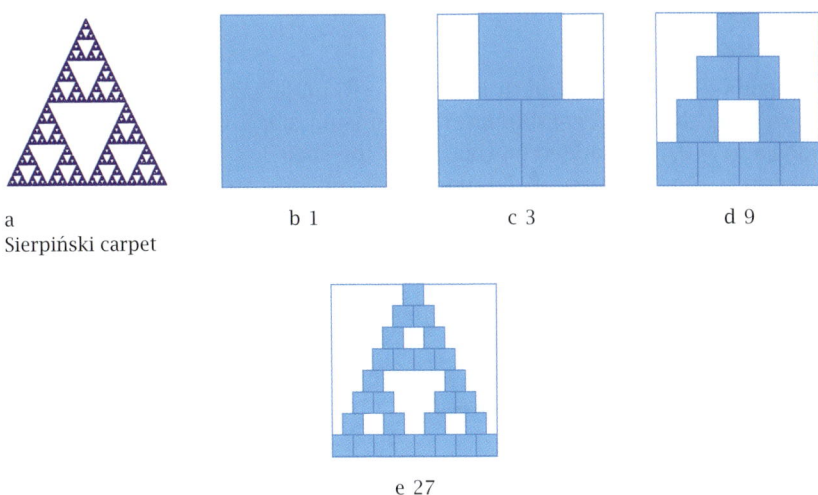

a
Sierpiński carpet

b 1

c 3

d 9

e 27

Figure 6. Number of squares to cover to cover the Sierpiński carpet appearing in a

curve. This is reflected by the fact that $\frac{\ln 3}{\ln 2} > \frac{\ln 4}{\ln 3}$, i.e., the dimension of the Sierpiński carpet is larger than the dimension of the von Koch snowflake.

1.3 Applications
There are many applications in a wide range of disciplines. Let us mention two in medicine.

The capillary network is not the same in the neighborhood of a tumor as elsewhere in the body. Research is carried on this, and in particular on its fractal dimension, in order to improve diagnosis from medical imaging.

High level athletes are more likely to suffer from asthma than the general population. The paper [3] studies the "optimal lung". There are 17 level of bronchial tubes before arriving to the terminal bronchioles followed by the acini involved in air exchange. If the bronchial tubes are too thin, then the pressure increases too much when the air penetrates in the next level of bronchial tubes. But if they are too wide, so that the volume never decreases from one level to the next, then the volume becomes too large. (It would become infinite if we had an infinite number of levels). Hence, the "optimal lung" would have the minimum volume without increasing the pressure. But the rate of variation of pressure close to the optimal lung is quite high, which means that a slight decrease in the diameter of the bronchial tubes induces a substantial increase of the pressure. To overcome this, the human lungs have a higher volume than the theoretical optimal lungs. This buffer provides a safety factor to protect in case of a pathology decreasing the diameter of the bronchial tubes, like asthma. Athletes have lungs generally closer to the theoretical optimal lungs, and hence, are more vulnerable.

1.4 Coming back to Cymbiola Innexa REEVE
How do we generate the pattern of Figure 2(c)? The scientists work on the hypothesis that a model of reaction-diffusion generates the pattern. The shell grows from its lower boundary, so the pattern should expand on the bottom side. We have two chemical reactants:

- one activator (colored),
- one inhibitor (uncolored).

The two reactants diffuse. The pattern is sensitive to initial conditions. We can obtain many patterns by varying the initial conditions. The diffusion starts from the top and goes to the bottom, when a new stripe is added on the side of the shell.

Let us present a simplification of the model. We approximate each layer by small squares (pixels). How is determined the color of a pixel? It depends on the color of the two pixels on the level just above, which are adjacent to it by the corner. If the adjacent two pixels are of the same type (both activators or inhibitors), then this generates a pixel with activator (colored). If the adjacent two pixels are of different types (one activator and one inhibitor), then this generates a pixel with inhibitor (uncolored).

This is how we write it mathematically. Activator (or color) is denoted by 1 and inhibitor by 0. The pattern is created following the three rules

$$\begin{cases} 0 + 0 = 0, \\ 1 + 0 = 0 + 1 = 1, \\ 1 + 1 = 0, \end{cases} \tag{1}$$

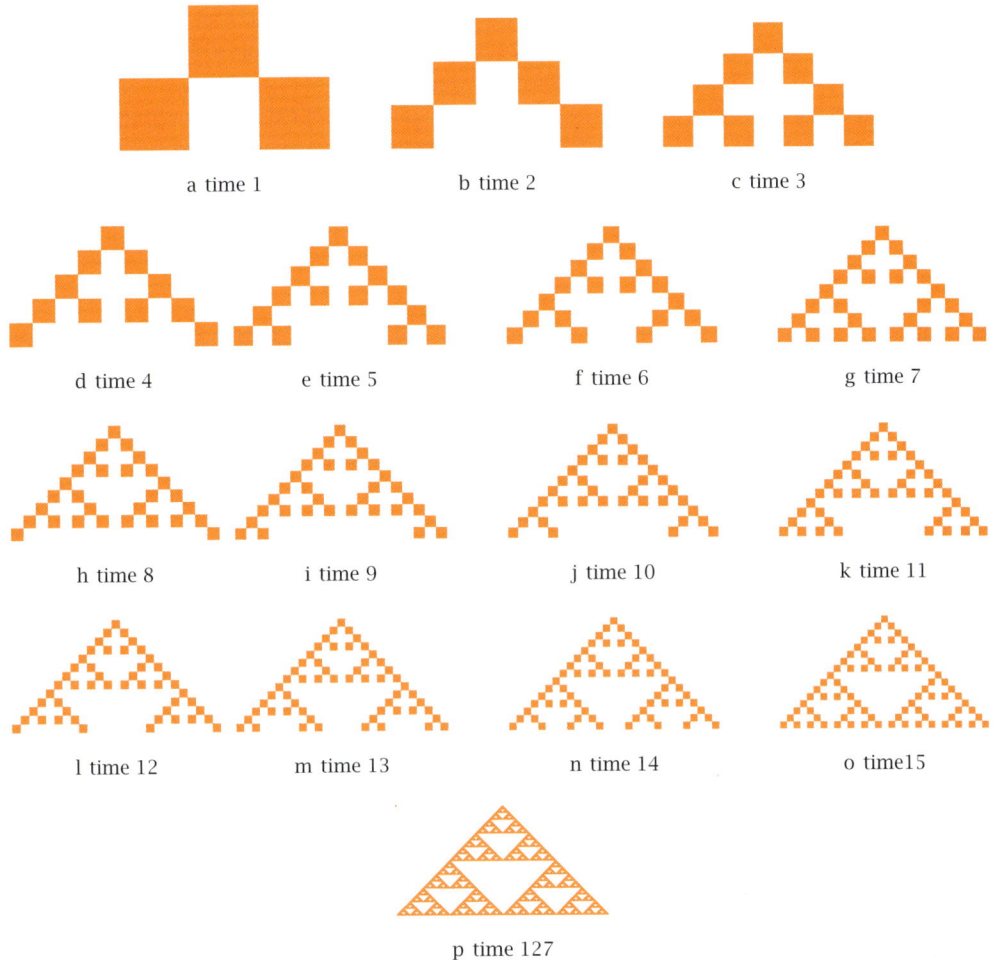

Figure 7. Pattern generated by the model (1) from a single nonzero entry 1 at time 0

which reflect the fact that activator needs inhibitor to produce color. Those who know arithmetic modulo 2 will recognize that this is addition modulo 2. Let us start with a single entry 1 on the first row at time 0 and observe in Figure 7 the pattern generated (we zoom out between each step, so that the figure does not grow too large). We see that we generate the Sierpiński carpet! We did the same to generate Figure 2c: on the top row we have generated random entries with activator, the others being filled with inhibitor. Then, we generated the lower rows, one by one, by applying the rules (1).

We do not claim to have given a complete explanation … Certainly, what strikes the mathematician is that the rules (1) generate the pattern of Cymbiola Innexa REEVE. This provides good motivation for interdisciplinary research to try to confirm biologically that this type of model is legitimate.

The pattern of the Cymbiola Innexa REEVE is very complex and yet, it is generated by an extremely simple rule, which could look like a computer instruction. And this rule is governed by chemical reaction. This is one example among many.

It is a very powerful idea that goes back to Turing: morphogenesis is generated by iterating simple rules, similar to computer instructions. Computers are very strong at iterating instructions. This is the principle of the *L-systems*.

2 The *L*-systems

The idea of the *L*-systems was introduced by the biologist Aristid Lindenmayer and he refined it together with the computer scientist Przemysław Prusinkiewicz. Have you already observed the growing of a fir tree? It grows once a year in spring. At that time, each bud produces one branch and several buds at the end of it. Hence, you could model it as a system performing exactly one operation, namely replacing a bud by a branch with several buds. If you want to be more realistic, then you could add a second operation, in which each previously formed branch would become wider by adding one growth ring.

Figure 9 presents an example of *L*-system with two symbols F and S and two rules presented in Figure 8. The symbol S corresponds to a stalk piece and the symbol F to a terminal branch. The first rule corresponds to replacing a terminal branch by a stalk of length 2 and three terminal branches. It is given by:

$$F \mapsto S + [F] - S + [S] - -[S].$$ (2)

The second rule doubles the length of the stalk:

$$S \mapsto SS.$$ (3)

The signs + and − in (2) corresponds to a rotation of a given angle to the left or to the right, and the notation [branch] means that we must come back to the beginning of the branch before starting the next instruction.

Rule 1 *Rule 2*

Figure 8. The two rules for the *L*-system of Figure 9

3 The geometry of Harry Blum

The fractal geometry described in Section 1 does not allow describing the shape of animals, for instance mammals or birds. Within a single species, there are huge variations between the

Figure 9. An *L*-system with two symbols, *F* and *S*, and the two rules in (2) and (3) (see Figure 8). We present the initial condition, a segment of type *F*, and the first six iterations.

shape of a tall skinny individual and that of another short fat individual. Describing some *invariants* of the shape for a given species is the purpose of morphology. In his paper [1], Harry Blum introduced a new geometry for this purpose. This geometry is based on the notion of *skeleton* (or *medial axis*) of a shape. The intuitive definition of skeleton is the following.

DEFINITION. *Let us imagine that the shape is made of flammable substance, and let us suppose that we ignite the fire simultaneously at all points of the boundary. Then the skeleton is the set of points where the fire dies because there is nothing more to burn.*

Figure 10 represents the skeleton of a square and that of an ellipse. The skeleton of a two-dimensional shape has the structure of a (non-oriented) *graph*. If the shape has no holes, then this graph is a *tree*, i.e., has no cycle starting from a vertex and coming back to it. The hypothesis of Harry Blum is that, for animals from a given species, the graphs of their skeleton are *equivalent*, i.e., have the same number and organization of vertices and edges, and that different numbers of branches, or different structures of the branch system could characterize different species. This requires some caution to be applied as such, but the idea is appealing. We show in Figure 11 several shapes that have the same skeleton.

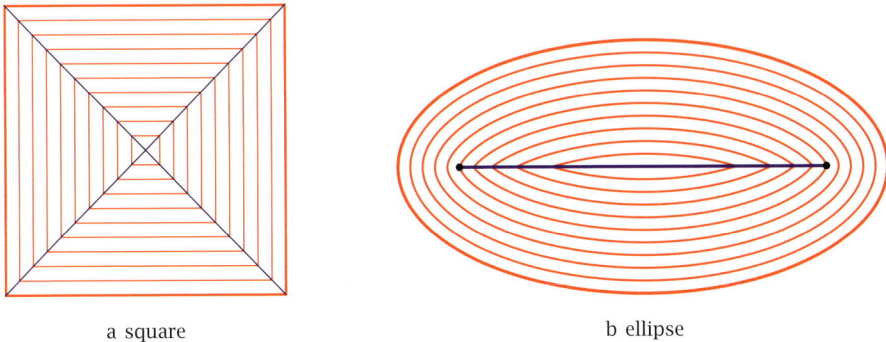

a square   b ellipse

Figure 10. Advance of the fire front and skeleton (in blue) for a square and an ellipse

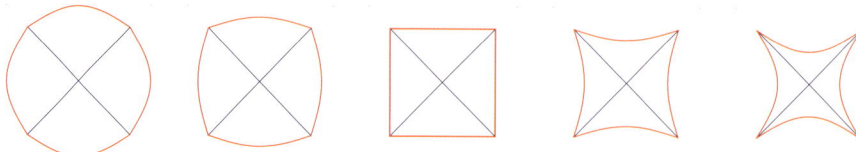

Figure 11. Some shapes with same skeleton

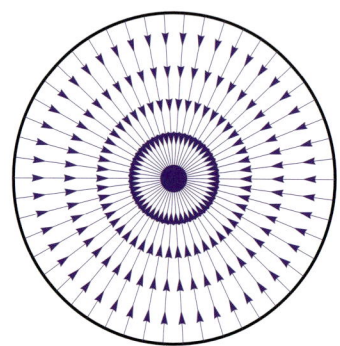

Figure 12. The skeleton of a circle is its center

Can we understand what is the mechanism behind the creation of a new branch of the skeleton? Let us start with a circle (see Figure 12). Then the fire propagates at constant speed towards the center of the circle, and the skeleton is only one point, namely the center of the circle. Of course, not all boundaries are circles. But we can try to approximate them by pieces of circles. Let us do that for the ellipse: see Figure 13. We take a circle tangent to the ellipse at one point and we let it grow until the limit position where it starts intersecting the ellipse near the tangency point (intersection points far from the tangency point are not relevant). This limit position is called the *osculating circle*. It is the circle that best approximates the ellipse close to the tangency point (see Figure 13). The skeleton of the ellipse starts precisely at the center of the osculating circle.

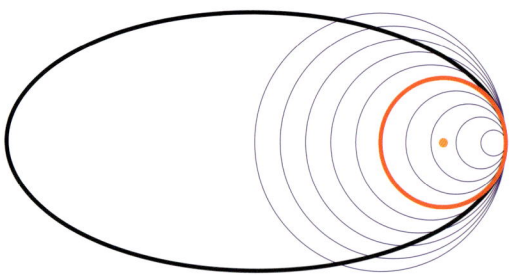

Figure 13. Among all circles tangent at a point, the osculating circle is the one which best approximates the boundary near the tangency point.

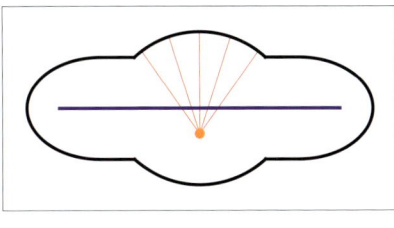 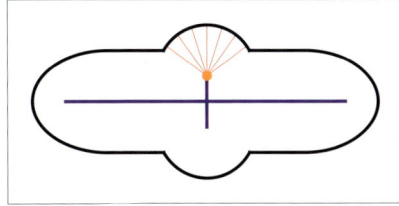

a no branch starts   b a branch starts

Figure 14. In a, no branch starts because the center of the osculating circle is passed the middle line where the fire has become extinct. In b a branch starts at the center of the osculating circle.

What we have discovered is a very general rule. The points where branches of the skeleton can start are most likely the centers of the osculating circles for different points of the boundary. A branch of the skeleton starts at the center of an osculating circle if the fire is not yet extinct. In Figure 14, we see two similar shapes, but with very different skeletons: in (a) the fire is extinct before reaching the center of the osculating circle, while in (b), the fire starts becoming extinct there.

Can this concept be useful in distinguishing species? Let us look try to distinguish the black bear from the brown bear, also called grizzly (see Figure 15), which are two different species of bears. The black bear can occur in all sorts of colors: black, brown, and even white for the Kermode bear … Hence, the color is not a distinctive feature. However, the brown bear always has a shoulder hump, while the black bear has none. The shoulder hump yields an additional

Figure 15. A black and a brown bear (Figure reproduced from http://commons.wikimedia.org/wiki/File:Black%26brownbears.JPG)

branch in the skeleton of the brown bear, compared to that of the black bear, thus allowing distinguishing the two species.

3.1 Application

An application is radio-active surgery with gamma-rays (see for instance [8]). In this technique, brain tumors are irradiated with doses of four different diameters, until all the volume of the tumor is irradiated up to some threshold ϵ. Since it is very lengthy to place the helmet for each dose, then the problem is to minimize the number of doses. For this, it is best to use doses centered either at the end of a branch of the skeleton, or at the intersection of several branches of the skeleton. The whole process is done via dynamic programming: once a few doses have been administered, one computes the skeleton of the remaining region to decide where to administer the next doses (see Figure 16).

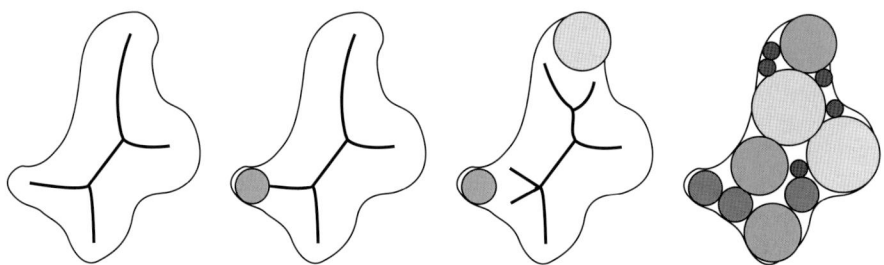

Figure 16.   A tumor, its skeleton, and different doses to irradiate it

4   A link between these two geometries

In the early stage, the embryo is spherical before morphogenesis begins: limbs, ears, nose, etc., start developing. What triggers morphogenesis? A potential explanation of the phenomenon goes back to the famous article "On the chemical nature of morphogenesis" (see [9]) of Alan Turing in 1952. In this article, Turing proposes a model of reaction-diffusion with chemical reactants to explain the phenomenon. His hypothesis is that the spherical form of the embryo is some kind of equilibrium, and that this equilibrium becomes unstable when the diffusion starts. This leads to a breaking of the spherical symmetry and limbs start growing. Turing gives the analogy with a flat surface of sand. When the wind starts blowing, the flat equilibrium becomes unstable and dunes began to grow. Since wind will always blow one day, this explains why we almost never observe flat stretches of sand on Earth. Reaction-diffusion models have also been proposed for explaining the variety of patterns of animal coatings.

We find here a great unifying principle: *The same type of reaction-diffusion models could explain the patterns of seashells, the animal coat markings, morphogenesis and phyllotaxy.*

References

[1] H. BLUM, Biological shape and visual science (Part I), *Journal of Theoretical Biology* **38** (1973), 205–287.

[2] P. PRUSINKIEWICZ and A. LINDENMAYER, *The Algorithmic Beauty of Plants*, The virtual Laboratory, Springer-Verlag, New York, 1990 (pdf version can be downloaded for free at http://algorithmicbotany.org/papers/#abop).

[3] B. MAUROY, M. FILOCHE, E.R. WEIBEL and B. SAPOVAL, An optimal bronchial tree may be dangerous, *Nature* **427** (2004), 633–636.

[4] H. MEINHARDT, *The Algorithmic Beauty of Sea Shells*, The virtual Laboratory, Springer-Verlag, Berlin, 1995.

[5] H.O. PEITGEN, H. JÜRGENS and D. SAUPE, *Chaos and Fractals: New frontiers of science*, Springer, New York, 2004.

[6] C. ROUSSEAU, Banach's microscope to find a fixed point, Klein vignette (http://blog.kleinproject.org/?p=162).

[7] C. ROUSSEAU, What means dimension?, Klein vignette (http://blog.kleinproject.org/?p=306).

[8] C. ROUSSEAU and Y. SAINT-AUBIN, *Mathematics and Technology*, Springer Undergraduate Texts in Mathematics and Technology, Springer, New York, 2008.

[9] A. TURING, The chemical basis of morphogenesis, *Philosophical Transactions Royal Society London, Series B*, **237** No. 641 (1952), 37–72.

Mathematics in industry

Helmut Neunzert

Industrial mathematics has become a fashionable subject in the last three decades. Today, there are many university groups and even research institutes dedicated to industrial and applied mathematics worldwide, proving that mathematics has indeed become a key technology. In this paper we first give a short account of the history of industrial mathematics. Using experiences from the Fraunhofer Institute for Industrial Mathematics (ITWM), we try to characterize the specific problem driven work of industrial mathematicians and take a look at future challenges.

There are some mathematicians who react very critically to a concept like "industrial mathematics". For them it doesn't exist. Mathematics may be pure, applied, or applicable (even these concepts are disputable), it may be good or bad (no doubt) – but industrial versus academic?

These people may be right, since all that depends strongly on the definition of mathematics itself. And such a definition may lead to endless discussions. I will try to explain my perception of mathematics shortly at the end of this article.

Although "industrial mathematics" may not exist, "industrial mathematicians" exist in large numbers. All lecturers and professors in mathematics educate many people, who become industrial mathematicians. I guess, the big majority of those students, who do not aim at becoming a high school teacher, will belong to this group after graduation. Few of our students will try to become professors themselves. But most of our master students will work for companies like Allianz (insurance) or Bayer (chemistry), Böhringer (Pharmaceutical) or Daimler (vehicles), Deutsche Bahn or Deutsche Bank, for IBM or Infineon, Linde or Lufthansa, RWE (energy) or SAP (software), Shell or Siemens (if you do not believe, please take a look into [9], it is worth reading anyway. Unfortunately, it has only been translated once, you have the choice between German and Japanese).

Even if they are educated mathematicians, their working contracts call them "software developers", "system analysts", "programmers" or (very rarely) "industrial mathematicians" (see, e.g., [6]. You will find that 28 % of all graduated mathematicians in Germany will become high school teachers, 17 % find jobs belonging to the so called STEM-professionals, but 41 % end up as IT-experts or in finances or in organization).

However, all these people, which I here call industrial mathematicians, use their mathematical knowledge, sometimes indirectly by applying their capacity in "logical thinking"; and sometimes directly when they are specialized in numerics, mathematical physics, statistics, optimization etc. What they most often do is mathematical modeling, scientific computing, optimization, control, data mining, visualization.

You find all these words as key words in the mathematics subject index; they belong to the codex of mathematics. And, sometimes these industrial mathematicians use and even develop new concepts, which belong to algebra, topology, probability theory, differential geometry etc.

But the big difference between an industrial and an applied mathematician is that the first one works mainly problem driven while the second one is in general method driven. Researchers at universities (and, maybe, institutes like Max Planck in Germany) are working in a special field of mathematics, they are algebraists or topologists or numerical analysts etc. Whether pure or applied, they develop and apply a certain methodology, they prove new theorems, they find new definitions and new algorithms.

In industrial mathematics, the situation is different: A practical problem is posed, e.g., how to cut a raw jewel to get, after polishing, gems of altogether highest value. How do we shape the geometry or how do we control a production unit in order to get very good products at low costs? What is the appropriate material for a filter and how do we design this filter in order to absorb as much of the damaging particles with as little energy costs as possible? These problems may not sound exciting, but they show that mathematics is literally everywhere. Not only in high tech areas, which you will find as examples in newspapers. Indeed, mathematics is in aerospace, semiconductors and even in cryptography, the standard example for the applicability of algebraic geometry, where it is often a real challenge for the mathematicians.

Today, mathematics is also needed in classical handicraft domains, e.g., in glass, steel, or even food production. "What is really difficult is to simulate the cooking of an omelette", said Jaques-Louis Lions already 30 years ago in a conference in Italy. Problems drive the work of an industrial mathematician: When he has understood the given problem (or when he believes he has understood it), he first makes a mathematical model of the system considered, then the model must be evaluated with the help of analytical or more often numerical methods. Already while modelling he or she has the algorithm to be used in mind. There is a strong interplay between modelling and computing, since the final goal is not a new model or a new algorithm, but the solution of "the" problem, in due time and with available hardware tools.

Of course, it is fun to construct a model, which includes all important details, to simplify it by tricky asymptotic methods (that's what my friends from the Oxford study groups loved to do), of course it is fun to develop an algorithm and to prove optimal order of convergence. But if all that cannot be used in the end, since the parameter, which should tend to zero is not really as small as necessary, if the order estimates are not telling how far the approximate solution you can achieve is away from the "true" solution, since your time and hardware resources are too small, then the industrial mathematician has to find another way. His goal to deliver a program, which simulates the behavior of the system, very often he tries to optimize this behavior. At the end, a visualization of the virtual system is most useful. And the customer, the problem poser, at the end wants to get the software to be able to do all this on his computer equipment. Often the customer is so happy and surprised about the benefits of the outcome that he wants more and more leading to further projects. Wonderful, at least if you need the money he hopefully pays for your work.

The whole process seems to be simple, but in general this is not the case. The steps are very much interwoven, going back and force many times. But, at the end, it is as satisfying as publishing a new proof and it adds a lot of surplus value to mathematics as a whole. Of course, it is fascinating for a non-mathematician to read about the solution of Fermat's conjecture –

especially many science journalists like to write about this kind of problems. But if a layman would know, how much mathematics is "contained" in almost every object of his daily use, he might be as fascinated. Some years ago, my, at that time rather small group, worked on the improvement of diapers, where nonlinear diffusion plays a crucial role. Not only my students but quite a lot of journalists were surprised and got interested: Mathematics and diapers! That doesn't seem to fit together well. They come from different worlds. Jewels and mathematics are obviously much closer, planes and mathematics belong together. But all that describes the past; today, I repeat, mathematics is everywhere. But is that true for all kinds of mathematics?

The methods which industrial mathematicians apply in modelling and computing reach deeply into many different mathematical fields – no, not in all of them, but in many. Industrial mathematicians should therefore be generalists, curious about the world around us, creative in inventing new concepts and new methods if necessary.

There is another characteristic feature of industrial mathematics versus academic mathematics: The topics of academic research are selected by a decision and/or the capability of one researcher or a small group of cooperating researchers. Industrial mathematics always starts with a dialogue between a person from industry who has to make or to support decisions – and one or better several mathematicians. The aim is most often to put the decision on a safer basis – not to make the decision. Mathematics should help, but it should not provide "the" solution. Mathematics with its "icy perfection", with its "absolute truth", which attracts so many people, often frightens or disappoints the decider. He still wants to use his experience, he wants to make the decision "by his gut feeling" – he hesitates to accept a proposal whose reasoning he has not fully understood. Problem finding here is a long process, needs many meetings between the industry representatives and the mathematicians. Even the perception of what is considered as "the" solution is the task of the problem poser, again in discussion with the "solver". I will give you a perfect example for that in the last of my 4 project descriptions, where we deal with multicriteria optimization.

We have to "humanize" mathematics if we want to become more successful in the real world. I will try to explain what I mean by humanization in the examples below. I will begin with some words about the history of "industrial mathematics". Of course, mathematics has always been applied in solving technical or organizational problems – even in the times of the pyramids or of the Indian Vedas. It started under the name of "Mathematica mixta" in the beginning of the 17th century and became "applied mathematics" a hundred years later. This name was used in France and Germany during the 19th century. Chairs for applied mathematics however were established in Germany not before the beginning of the 20th century, Carl Runge in Göttingen and Richard von Mises in Berlin were the first chair holders. At the same time the application of "modern mathematics" in industry started. Especially in electrotechnics, complex numbers and differential equations were used and graphical-numerical methods were developed. This new approach was sometimes called "to calculate instead of trying". A similar situation arose in optical and in aerospace industry. To name of few people: Balthasar van der Pol worked at Philips in the Netherlands, Charles Steinmetz at Bell Labs in the US and Iris Runge, the daughter of Carl Runge, worked at OSRAM and Telefunken in Germany (I recommend to read the book by Renate Tobies [18].)

A driving force behind closer contacts between industry and mathematics was Felix Klein who in 1898 founded the "Göttinger Vereinigung für angewandte Physik", which was extended to applied mathematics in 1901. This association had leading industry representatives as members, who sponsored the education of engineers, physicists, and mathematicians in order to

Figure 1. Iris Runge at the Osram Corporation (1929) (from [18])

create a new species of technoscientists (for example, see the concept of technoscience in the very interesting book of Steven Shapin [16], I believe that it also captures the profession of an industrial mathematician today quite well). These strong bridges between mathematics and industry contributed immensely not only to the quality of the staff, but also to the leading position of Göttingen before the nazis came to power. If you want to get the spirit of that time, take a look into volume 3 of Felix Klein's "Elementarmathematik vom höheren Standpunkt" [11]. Of course, the second world war triggered some mathematical research. But since it normally was classified as "secret", I have not enough information to give a proper description.

50 years later, during the 60s of the 20th century, academic groups dealing with industrial problems were created, e.g., in Oxford, where Alan Tayler and John Ockendon started the "Oxford Study Group with Industry". At the same time, in Claremont in California Jerome Spanier initiated the "Mathematical Clinic", treating industrial problems with students over longer periods. Ten years later, in the 70s, similar activities started in Austria (Linz), Germany (Kaiserslautern), the Netherlands (Eindhoven), in Italy (Bari and Florence), in Sweden (Stockholm) and, with a different organization, in France. This led to the foundation of the "European Consortium for Mathematics in Industry" (ECMI) in 1986, which still flourishes and has extended to almost all countries in Europe.

Today, many people in academia claim to do industrial mathematics, all over Europe, the US, Canada and Australia. I am not really sure that there is always a real industrial problem behind – the label "Industrial Mathematics" has become fashionable.

But there are institutions and institutes, which dedicate their work mainly to real world problems. To name a few: OCCAM in Oxford, ZIB and WIAS in Berlin, Radon in Linz, the three Fraunhofer-Institutes for Industrial Mathematics (ITWM, SCAI and MEVIS), the Fraunhofer-

Chalmers Center in Göteborg, the Institute for Applied Mathematics (IMA) in Minneapolis and many more.

To take "my own" institute ITWM as an example: Within 20 years, it has grown to about 250 full time staff positions, the majority being mathematicians. It has a budget of 22 million EURO of which more then 10 million come from industry. It permanently sponsors around 60 PhD students who do their PhD in mathematics at the TU Kaiserslautern, and it has 25 Felix Klein scholarships for excellent Bachelor and Master students. ITWM and TU have constructed a common roof for their activities, called Felix Klein Center (FKZ), which really takes Felix Klein as a role model, as someone who himself worked in pure and applied mathematics, interested in school and university education, building bridges between academia and industry. What we have learned from the activities at ITWM and at FKZ is that mathematics today is literally everywhere, almost each and every company needs mathematics, it is fair to say that it has become a key technology worldwide.

When we started in 1995, most people and especially the Fraunhofer Society were not convinced that mathematics is able to earn more than 40 % of the total budget as "hard money", i.e., directly from industry.

They gave us a test phase of five years, but with the support of our state minister an enthusiastic group of about ten young scientists solicited exciting projects, provided useful solutions, and at the end of the five years, Fraunhofer welcomed ITWM and with it mathematics as a proper subject for an institute. To date, it is one of the most successful IT-institutes in the Fraunhofer Society.

I shall now try to give four examples of genuine industrial mathematics. These examples are taken from projects at Fraunhofer-ITWM, completely different from all the examples in our recent book [14]. These examples show what we mean by problem driven in contrast to method driven research. I will describe how we found the problem, how we developed the method, how we dealt with the customer and what were the final results.

Example 1. Particle methods for flow problems. A big advantage of academic versus industrial research is the existence of worldwide networks, which are really cooperating. Active researchers often belong to several of these networks. My main network in the beginning of the 90th was "particle methods" – our group at the University of Kaiserslautern had the task to develop a particle code for the solution of the Boltzmann equation to simulate the flow of a dilute gas around a reentering space shuttle (called "Hermes" at that time). In high altitudes the gas is so thin that a normal aerodynamic description of this flow fails, the 6-dimensional nonlinear Boltzmann equation had to be applied (for the theoretical treatment of which Pierre-Louis Lions and Cedric Vilani were awarded with a Fields medal). For its approximate solution particle methods are most appropriate until today. At that time, we were one of the most active groups in that field and from the European Hermes project we got the task to do the calculations. When the shuttle comes down, the atmosphere is getting denser, the Boltzmann simulation becomes more and more time consuming and therefore, a code for solving the Euler equation had to be coupled in. The idea to use again a particle method, now for Euler and to couple it with the particle code for Boltzmann is very natural – particles here and there, just changing their behavior.

At that time, the best code for that purpose was Joe Monaghan's "smoothed particle hydrodynamics", SPH – in spite of the fact that there was no liquid but only air. Monaghan was in our network, he taught us about his code, and one of our PhD students studied and adopted it to

Figure 2. Re-entry of space shuttle Hermes (Simulation: University of Kaiserslautern, Working group "Technomathematics")

our situation. The change was substantial, so he called "his" method "Finite Pointset Method" (FPM) – the set of particles is considered as a very flexible grid, which moves in general with the flow. Flexible, movable grids are important when the domain of the flow changes rapidly. This is only a marginal aspect in the flight of the shuttle, but it is crucial, e.g., for the simulation of airbags when they are blown up. In milliseconds the bag is drastically enlarged and the airflow in it must be as well calculated as the forces on the bag. The coupling of both effects is necessary to simulate this blow-up. A French software company tried to make a simulation, failed and asked a very well known French mathematician, Pierre Arnaud Raviart for help. Raviart was a member of another of our networks and told the company to contact this German node. It is not usual to transmit such a project to another group, especially to a group in a foreign country. He did it, and we are still, 20 years later, grateful to him. Our FPM-method did the job very well. A very successful group at ITWM started, directed by the inventor of FPM, Jörg Kuhnert, who by now has run many projects for flows with free surfaces or with quickly changing containers.

In principle, FPM is rather simple and easy to understand: It is a finite difference method for conservation laws with moving gridpoints. In fluid problems, the grid moves with flow velocity – similar to particles immersed into the fluid. These particles carry the numerical values of mass, pressure, energy, the derivatives of the velocities, etc. Each particle has a domain of influence around itself. Far away from the boundary this is a ball whose radius defines an approximation length. The physical quantities named above are approximated by a weighted least–squares ansatz inside this ball. Here are the treatment of boundaries is crucial, free boundaries such as the surface of water waves, when curvatures or normals are needed, or given boundaries like (moving) walls. In the vicinity of the boundaries the shape of the influence zone changes, the particles even sometimes "look around the corner". The boundary conditions influence the motions of the gridpoints too: No slip conditions lead to "resting particles", slip conditions generate velocities of the particles in the neighborhood. There are no "weak formulations", everything is rather easy to understand, even time stepping. The time steps are explicit for compressible fluid flow, when the least square approximation is extended to an upwind formulation to provide stability. Implicit schemes are applicable for incompressible flows with a Chorin projection.

Why are these rather straightforward ideas so successful? Firstly, they work well, even in cases when the reality is complex. They work well in simulating a quick fuel filling of cars

Figure 3. Water crossing (Simulation: Fraunhofer ITWM)

(if you watch a Formula One race on TV, you know how important seconds in this process are) or of fuel sloshing in tanks, which influences the motion of the car and makes noise. It works for the simulation of a car passing a flooded road (the comparison with experiments shows excellent agreement), and it works for the simulation of mixing and stirring in food production.

Another reason for the success is that the method is directly "comprehensible" – the problem poser, often an engineer, can easily follow the steps, he understands how and why. It is, I repeat, not the icy perfection, not an indisputable solution, which has to be accepted. He may influence many parameters, many physical details such as the boundary conditions. There is room for discussion with the person who finally decides. Industrial problem posers do not like to deliver a problem and get back, after a while, a perfect solution, polished, correct, "true". Often they get no access to how it was achieved, no chance to understand why. This is one of my messages, which will be repeated several times: A true solution of a mathematical problem, with a nice elegant proof based on tricky ideas frightens the problem poser. What we mathematicians like, the indisputable truth, is not what the decider wants. And vice versa: Maybe that is why mathematicians rarely are deciders themselves?

Example 2. Simulation of lithium-ion batteries. Let us move to a second example, quite different from the first one: It is purely "problem driven" and originated in an urgent request from industry. It concerns the simulation of lithium ion batteries, especially their performance, and, later, their durability and their safety, too. The subject belongs to the field of electro chemistry and was surprisingly even new in physics when we started it ten years ago. The development here is so fast, that academia, in general, cannot keep up with this speed and is "behind". While many of our other problems belong to classical continuum mechanics, where physicists, engineers, mathematicians worked together for centuries, here is a problem, which comes from industry and is very new. Our Fraunhofer-Institute, in close contact with about 10 battery-producing companies, was one of the first facing these problems. A proof for this "priority statement" is that a habilitated physicist from our institute, after having worked for us for several years on modeling and simulating batteries, got a call to a chair for "electrochemical multiphysics modeling" at the university of Ulm, establishing a center for this field there. This

Figure 4. The Tower of Babel (Painting by Pieter Bruegel the Elder)

shows, that there is not only flow of science from academia to industry or research institutes, but sometimes also in the opposite direction.

What kind of research is needed for this storage of energy, a fashionable topic today? As often, it needs cooperation of physicists or chemists with mathematicians. But not physicists or chemists alone: Mathematics is really needed. Why is that? There are a lot of complicated multiphysics and multiscale tasks. In simulating a battery, very different scales must be considered: First, there is the microstructure of the porous electrodes and of the active particles. These are usually of micron or submicron size. But the battery cell itself is of millimeter or even centimeter size. The behavior on the microscale must be understood in order to get a reliable simulation of the whole cell. But it is hopeless to try to discretize the whole cell so finely that the microstructures are resolved. This would be far beyond the capacity even of supercomputers. Therefore, little pieces of the cell, which however are large enough to contain many pores and/or particles are considered and simulated. From this microsimulation, one gets input-output values for these pieces, considered as systems, which then are put together in order to simulate the whole cell. This process is called "homogenization". It has a strong mathematical basis in asymptotic analysis and is always a powerful tool, if – indeed if there is a clear separation of scales. By this we mean that we have, e.g., two quite different typical sizes, pores and the whole battery, so that this "micro-macro" game works (see, e.g., [7]). For those who are not sufficiently familiar with mathematics, they may think of paintings, which have a very fine microstructure, of the size of a "Pinselstrich" (a stroke of the brush), carefully executed by the artist. And if you now move away from the painting, you get an overall impression of this piece of art, not perceiving the details on the microscale, which however are very important. Just remember the work of impressionists or – with more then two scales – paintings of Pieter Bruegel the elder.

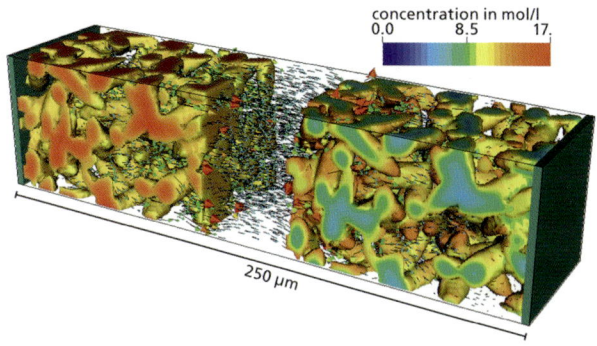

concentration in mol/l
0.0   8.5   17.

250 µm

Figure 5. Battery simulation (Simulation: Fraunhofer ITWM)

For batteries however, we have many different scales, especially with respect to time; and we have strong nonlinearities, which forbid simple superpositions. We do not have "scale separation". Other, new upscaling processes have to be invented and lead to coupled multiscale lithium-ion battery models. For that, one needs the exact asymptotic order of the densities of the interface currents. With this model, intense numerical simulations are undertaken, which prove that our upscaled model exactly captures the microscopic properties of the cell electrodes. The electrochemical process also finds a consistent and quite accurate macroscopic description (see, e.g., [10]).

As always in modeling, since decades there exist simpler models, which are regularly used, but now urgently need improvements. Industrial mathematics almost always means modeling – and it is computation as well. For the latter, Multiscale Finite Elements, MsFEM, are useful, allowing for solving microscale problems in subdomains in order to compute basis functions for the coarser grid. In comparison to a direct approach this reduces the dimension of the basis, but makes its computation rather elaborate. The subdomains are of the scale of the pores or the particles, and these objects are modeled as holes. MsFEM in perforated domains is an interesting research field, especially when the boundary conditions are not the most simple ones.

Convergence has still to be shown – even for linear systems. In our situation, with highly nonlinear battery models, there will be much to do in the future. And it is a field that forces closer cooperation between mathematicians and physicists. The interplay between different scales, appropriate model reductions and the search for optimal basis functions are exciting challenges for mathematical research, and, at the same time, they may have a strong influence on solving a crucial problem in energy conservation. Again, mathematics is fun for homo ludens – it is pleasure and satisfaction to play this fantastic and creative game of pure math. But mathematics is also fun for homo faber – it is an enormous satisfaction to realize that mathematics has solved, or at least essentially contributed to solving one of the most important problems mankind is facing now. The storage of energy seems to be one of these problems. We show a simulation of the motions of the ions between the electrodes, on a micro scale.

Example 3. Cosserat models for flexible cables in robotics. Let us pass over to our third example for industrial mathematics. With it we want to communicate a new message: Mathematics is a very old science. To solve very modern, industrial problems, the huge treasure of centuries of

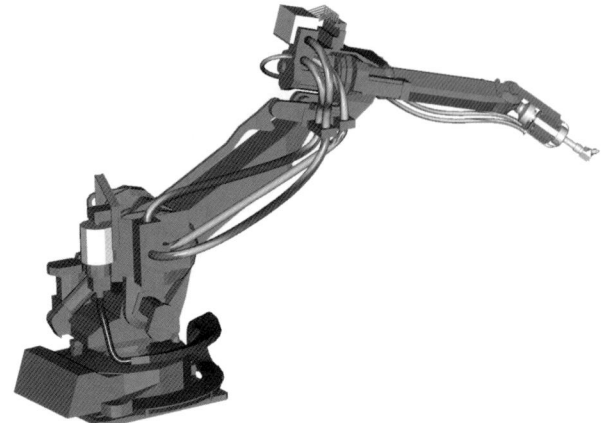

Figure 6. Industry robot with hydraulic hoses (© FCC Gothenburg (S))

mathematical research is beneficial, even unsubstitutable. The problem we treat is the motion of cables in a modern robot for industrial production.

What is our problem? You will easily understand it if you make a simple experiment: Take a 50 cm long piece of a garden hose in both hands and twist it; by gradually enforcing the twist you will see, after a while, a rather quick motion, in which the hose, or the cable, form a loop. Small changes in twisting create big differences in the shape of the cable as a whole – a typical nonlinear branching effect. To simulate this behavior, one needs nonlinear models of elasticity – and all this began with a book by Leonhard Euler from 1744 [8].

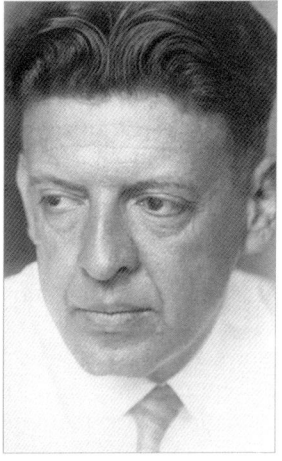

Figure 7.   Leonhard Euler, 1707–1783 (Painting by J. E. Handmann, 1753) and Wilhelm Blaschke, 1885–1962 (Source: Bildarchiv des Mathematischen Forschungsinstituts Oberwolfach, Photo: L. Reidemeister)

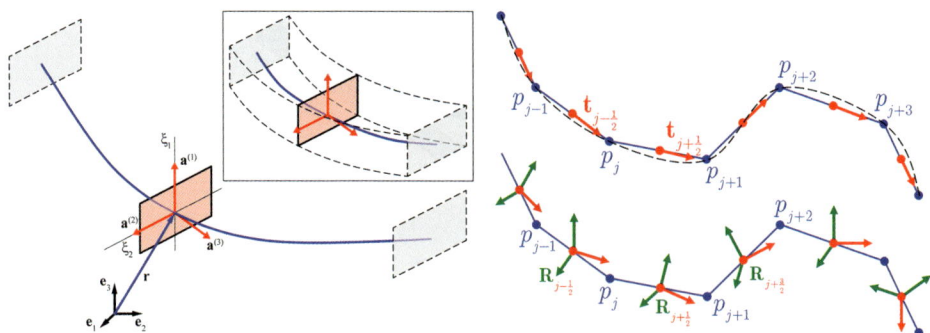

Figure 8. Cosserat rods (Graphic: J. Linn, S. Grützner, Fraunhofer ITWM)

Around 150 years later the French brothers François and Eugène Cosserat developed a theory of elastic rods [5], modeled by one-dimensional curves carrying an internal orientation given locally by a frame of three orthogonal unit vectors. Our cables are modeled as Cosserat rods [2]. The kinematics of these rods are closely related to the differential geometry of framed curves, with the differential invariants of rod configurations corresponding to the strain measures of the mechanical theory. These concepts were developed, again 50 years later and 200 years after Euler by a geometer from Hamburg, Wilhelm Blaschke [3]. Blaschke, a quite pure mathematician, has written a booklet with the title "Talks and Journeys of a Geometer" in 1956 [4] and I want to cite from this now 60 years old text:

> Science and technology need mathematics, without it they would be deprived of a powerful tool. But even vice versa: I am convinced, that a separation of mathematics from its application fields would be a calamity, it would create a Byzantine scholastic torpidity, a threat sometimes quite obvious today in spite of the fact that "old mathematics" still flourishes youthfully in a time period, when pictorial art seems to die away. Progress is often achieved at the border lines of science. Specialization is important but dangerous as well. Iron curtains are not favorable, neither in life nor in science. [I apologize for not being able to translate the poetic German of Blaschke into equivalent English.]

Our models are based on centuries of rather pure research – from Euler to Cosserat to Blaschke. And our numerics originated in research of R. Sauer [15] from before 1970. He was professor at TU München, when I studied at the LMU in the 50s. We use what is called geometric finite differences providing a discretization of the strain measures, that preserves their essential geometric properties. One may call that a "mimetic scheme", since it mimics the differential geometry of framed curves in Euclidean space in the discretized world. Combined with discrete Lagrangian mechanics, the discrete models constructed by this approach yield qualitatively correct and quantitatively good results even for very coarse discretization.

Figure 8 explains what we mean by framed curves and how we understand the "discretized world". The differential geometry of Cosserat curves provides the complete mathematical theory necessary for modeling the kinematics of geometrically exact rods.

The main point here is that curvatures and stresses are well defined for arbitrary large deformations (as we have in the motions of the cables), completely independent of the coarseness

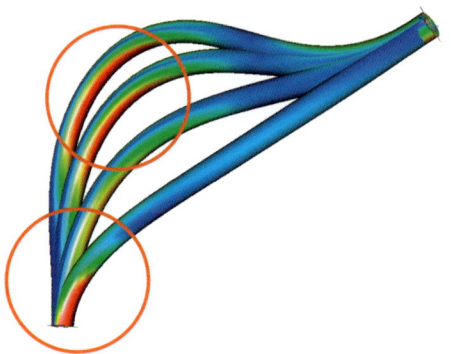

Figure 9. Deformations of hydraulic hoses (Simulation: Fraunhofer ITWM)

of the discretization. That means that the curvature increases when we bend the rod, up to the very limit of degenerate configurations.

The energy is only zero in the reference configuration – it does not possess any artificial minima otherwise. A detailed discussion of a proper discrete rod theory can be found in [13]. In Figure 9, the stresses of robot cables during the motion are shown, pointing to the maximal loads.

There are two messages in this third example: Even in industrial mathematics it is beneficial to study "old" papers of excellent mathematicians. And: It is not always advantageous to develop continuous models and discretize them afterwards, but it might be is better to do the modeling directly in the discrete world. This second message still leads to discussions even in our group at ITWM – is it better to model continuously and then discretize properly, or is it better to model directly in the discrete world? (I personally prefer the second way.)

Example 4. Optimal radiotherapy planning.   We are now coming to the fourth and last example, which shows in a perfect way how beneficial and successful industrial mathematics can be. However, sometimes one needs quite a long time and a lot of patience. 23 years ago, a PhD student from electrical engineering asked mathematicians for help with the optimization of radiotherapy: How do we control the radiation in such a way to get a maximal dose into the tumor, at the same time minimizing side effects on healthy parts of the body? It is obvious that this problem is of highest relevance. Every year almost 500000 people in Germany get a cancer diagnosis – and more than 50% of them are treated by radiotherapy.

The generation of a good therapy plan is a very complex task – there are many parameters to be determined, the simulation of the radiation effects in the tissue requests expensive computations. And one has to find a compromise between the different planning goals. Of course, this task is not only a problem for mathematicians – one also needs physicists, computer scientists, and of course medical doctors. The task is both multidisciplinary and multicriterial. What we finally want is to develop a decision support tool for physicians. Physicians often have a lot of experience and their decisions are based on this experience. They do not want to hand over the responsibility to a computer program. Finding and accepting the right compromise is case sensitive with regard to the patient and with regard to the experience of the doctor. They want to get help for their decision, but not to be pushed by a single solution of a computer system.

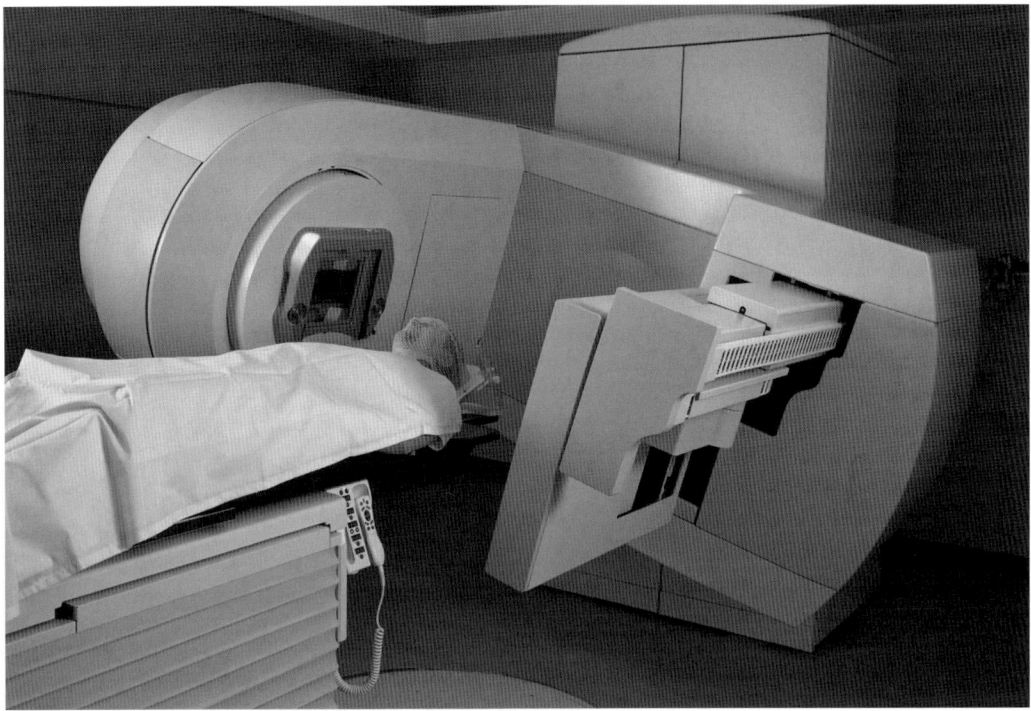

Figure 10. Radiation therapy (© M. Kostich, istockphoto.com)

Here, multicriteria optimization really helps. We want to find a solution that maximizes not only one objective function F, but several $F_1, F_2, \ldots, F_N$. But N-dimensional vectors cannot be strictly ordered, if $N > 1$. Here, the idea of the Italian-Swiss economist Vilfredo Pareto (1848–1923), who introduced the concept of so-called Pareto optima, comes into the game. A plan x is called a Pareto optimum, if you cannot improve $F_i(x)$ for one i, without worsening at least one of the other $F_j(x)$. That means, in a Pareto-optimal radiation plan, you cannot change the plan in such a way that the radiation of the tumor gets stronger without damaging at least one of the healthy organs of the body at the same time.

There are always many Pareto optima, together they form a Pareto set or, as it is sometimes called, a Pareto manifold. It is obviously lower dimensional, but it seems that there are not too many investigations about the geometrical structure of this manifold. But it is possible to construct approximations to this Pareto manifold and, therefore,we can be sure whether a point is close to a Pareto optimum or not. Navigation on this manifold (or at least in a close neighborhood) gives the physician the opportunity to use his experience: Maybe a bit less radiation into one organ – he reduces the dose for this organ by drawing the corresponding point closer to the origin. The navigation shows him immediately, how much this will reduce the dose into the tumor or how doses to other protected parts will be changed. This combination of mathematics with the experience and intuition of the user is a very good recipe to convince practitioners about the benefits of mathematics.

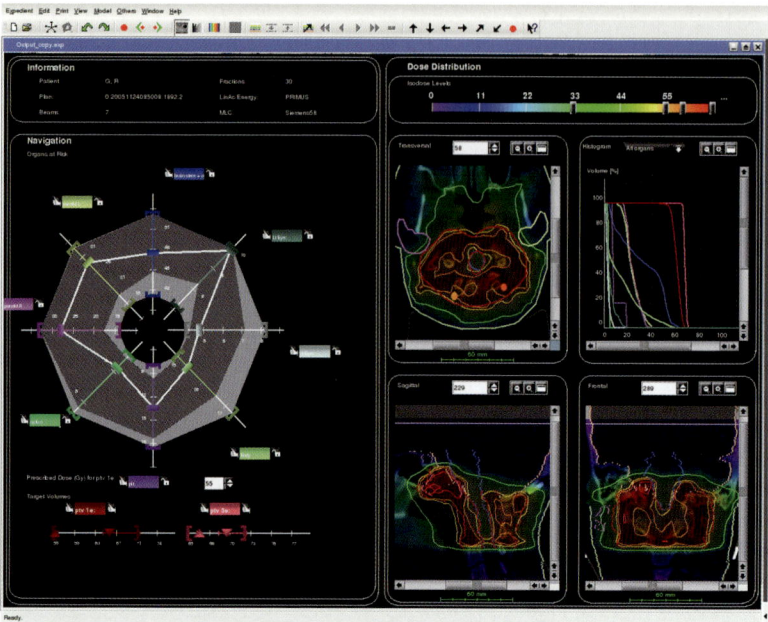

Figure 11. Navigation in radiotherapy (© Fraunhofer ITWM)

A close cooperation with the Heidelberg University Hospital and the German Cancer Research Institute (DKFZ) at Heidelberg and, later, with the Massachusetts General Hospital and the Harvard medical school at Boston led to very intense research during this decade, with 7 PhD theses, 30 peer reviewed publications, and many awards. In 2015, the world market leader for radiation software, Varian Medical Systems, bought the licenses for the software, which means that more than 60% of all radiation therapy patients worldwide will be supported with this product, "Mathematics saves lives" is not a euphemism [12]. In May 2016, the highest prize of the Fraunhofer Society, the so called "Preis des Stifterverbandes für die Deutsche Wirtschaft", which is awarded once every second year, was given to the radio therapy group at ITWM, another proof of the appreciation of mathematics.

The *future* of industrial mathematics will, of course, be dominated by industrial needs. Unfortunately, visions about the future of industrial mathematics are very unreliable. There are always fashions like, today, Industry 4.0 or Big Data. Sometimes, these buzzwords originate in mathematics, like fuzzy logic or evolutionary algorithms. When visiting companies 20 years ago, I remember almost everybody was asking about fuzzy logic, with a very vague understanding of the real concept. During that time many kitchen machines carried "fuzzy controlled" as a sign of quality.

Since it is really impossible to predict which mathematical tasks will dominate the future, which new mathematical methods will give new possibilities, which technical or organizatorial problems will dominate, I can only look for the present catchwords (politics is in general full of them, but this is not always very helpful, at least not for mathematics).

The only actual fashion that has definitely quite a lot of mathematics in it and may give some hints for the future is Big Data. I myself am not an expert in statistics or harmonic analysis, in statistical learning or data mining, the main areas related to this topic. But I tried during the last months to understand a bit.

I see huge clouds of points in high dimensional spaces and we have to search for patterns in these clouds. And patterns are, I believe, genuine subjects of mathematics. The points are most often perturbed by some kind of noise. The unperturbed points may belong to lower dimensional manifolds – the patterns mentioned above. If we have ideas about the mathematical structures describing the patterns, or, better: if we have mathematical models for it, we might try to identify the parameters of these models and "understand" the clouds.

But in the case of Big Data, the dimensions of the spaces are often very high and parameter identification is difficult or impossible. In other cases, the human behavior mainly influences the data (as in studying the purchase behavior of millions of customers), there are only statistical models, but none that explains the structure.

In a journal called "Wired Magazine" from 2008 I read

> Only models, from cosmological equations to theories of human behavior seemed to be able to consistently, if imperfectly, explain the world around us. Until now. Today companies like Google, which have grown up in an era of massively abundant data, don't have to settle for wrong models. Indeed, they don't have to settle for models at all.

And later:

> At the petabyte scale, information is not a matter of simple three- and four dimensional taxonomy and order but of dimensionally agnostic statistics ... It forces us to view data first mathematically and to establish a context for it later. [1]

Do we really understand, what the author means? And: What kind of mathematics is meant here? It seems that Chris Anderson talks about simple statistics, counting and comparing relative frequencies. He doesn't want to know why people do what they do. "The point is they do it, and we can track and measure it with unprecedented fidelity. With enough data, the numbers speak for themselves." I do not believe that it is as simple. There are at least advanced statistical models, e.g., random walks on graphs. Graphs are very natural data structures with nodes representing objects and edges representing the relation between them. One has to compare graphs in order to find relationships (Is a given protein an enzyme or not? Have different web pages a related content? ...). There are kernel methods, which start from linear algebra see [17], but are now applied for graph theoretical models with random walks and much more (see, e.g., [19]). In other cases, one needs feature selection and reduction, where concepts of sparsity, methods of compressed sensing etc may be used. There is a rather big mathematical arsenal available for Big Data, not at all simple, but not yet part of a traditional curriculum (and, to be honest, also far away from my competences).

The main application area for Big Data is bioinformatics. It seems to me that this is generally true for most future mathematics: Medicine is more and more becoming a science – it has been more like an art in the past with experience to be used for the case here and now. Therefore, we need more and more mathematical models for it. Indeed, there are models, for the functioning of the heart, the blood circulation, some crude approximations for some organs. But still, we are very far away from finding a good approximation for the system "human body", not to talk

Figure 12. Patterns (as a historical cake pan) (CC-BY-SA-3.0-AT Ailura)

about mental processes. I believe that we need completely new mathematics in order to deal with these problems – maybe we need somebody like Isaac Newton.

And other fields, for which new mathematics would be important? For example social sciences, economy or just industry (with or without 4.0)? Of course, we always look for patterns, in any kind of data, in order to "understand". We try to see patterns in images, or better, we try to separate different patterns, those, which are "background" and those, which carry information, patterns in texts, in economic data (that is what mathematical finance, still the most fashionable branch of mathematics, tries to do, with mixed success), patterns in medical data and many more. Mathematics is the science of order structures, into which we try to fit the real world – that way, they become patterns in real world data. I feel reminded of my childhood, when my mother used wooden cake forms – we called them "Springerle" – to fit in the dough of Christmas cakes. Sometimes it fitted, sometimes it did not. The cake, the real world, got a shape, a pattern.

Perhaps pure mathematics has not yet found the proper order structures where these problems fit in? We have not yet found the decent shape for the cake pan?

I do believe that pure and applied maths should not drift apart, they belong together. Mathematics is, has been, and will be the key science and the key technology. Mathematicians find order structures and try to use them for patterns, which order real world phenomena. And since the world gets more complicated, we will need more and more mathematicians, pure or applied or industrial ones (who cares), but creative and open minded.

References

[1] C. ANDERSON, The end of theory: The data deluge makes the scientific method obsolete, *Wired Magazine*, 2008 (http://www.wired.com/2008/06/pb-theory/)

[2] S. ANTMAN, *Nonlinear Problems of Elasticity*, Springer 2005.

[3] W. BLASCHKE, *Vorlesungen über Differentialgeometrie*, Springer 1930.

[4] W. BLASCHKE, Reden und Reisen eines Geometers. Deutscher Verlag der Wissenschaften, 1956.

[5] E. and F. COSSERAT, *Theorie des corps deformable*, Paris 1909.

[6] M. DIETER and G. TÖRNER, Der Arbeitsmarkt für Mathematiker, Teil I in *Mitteilungen der DMV*, Heft 22, 2014.

[7] Y. EFENDIEV and T. Y. HOU, *Multiscale Finite Elements Methods: Theory and Applications*, Springer 2009.

[8] L. EULER, *Methods inveniendi linear curvas maximi minimive proprietate gaudentes, sive solutio problematis isoperimetrici lattissimo sensu accepti*, 1744.

[9] G.-M. GREUEL, R. REMMERT and G. RUPPRECHT (eds.), *Mathematik – Motor der Wirtschaft*, Springer 2008.

[10] T. HOFMANN, R. MÜLLER, H. ANDRÄ and J. ZAUSCH, Numerical Simulation of Phase Separation in Cathode Materials of Lithium Ion Batteries, to appear in "Berichte des ITWM", 2016.

[11] F. KLEIN, C.H. MÜLLER and F. SEYFARTH, *Elementarmathematik vom höheren Standpunkte aus: III: Präzisions- und Approximationsmathematik*, Grundlehren der mathematischen Wissenschaften, Springer, 1968.

[12] K.-H. KÜFER, M. MONZ, A. SCHERRER, P. SÜSS, F. ALONSO, A. S. AZIZI SULTAN, T. BORTFELD and C. THIEKE, Multicriteria optimization in intensity modulated radiotherapy planning, in P.M. Pardalos, H.E. Romeijn (eds.): *Handbook of Optimization in Medicine* 5: 123–168, Kluwer 2009.

[13] J. LINN and K. DRESSLER, Discrete Cosserat Rod Models based on the Difference Geometry of Framed Curves for interactive Simulation of Flexible Cables, in L. Ghezzi, D. Hömberg, C. Landry, (eds.): *Math for the Digital Factory*, 29 pages, Springer 2016, submitted.

[14] H. NEUNZERT and D. PRÄTZEL-WOLTERS (eds.), *Currents in Industrial Mathematics*, Springer 2015.

[15] R. SAUER, *Differenzengeometrie*, Springer 1970.

[16] S. SHAPIN, *The Scientific Life: A Moral History of a Late Modern Vocation*, The University of Chicago Press 2008.

[17] J. SHAWE-TAYLOR and N. CRISTIANI, *Kernel Methods for Pattern Analysis*, Cambridge University Press 2003.

[18] R. TOBIES, *Iris Runge – A Life at the Crossroads of Mathematics*, Science and Industry, Birkhäuser 2012.

[19] S. V. N. VISHWANATHAN, N. N. SCHRAUDOLPH, R. KONDOR and K. M. BORGWARD, Graph Kernels, Journal of Machine Learning Research 11 (2010), 1201–1242.

Mathematics of signal design for communication systems

Holger Boche and Ezra Tampubolon

Orthogonal transmission schemes constitute the foundations of both our present and future communication standards. One of the major drawback of orthogonal transmission schemes is their high dynamical behaviour, which can be measured by the so called Peak-to-Average power value – the ratio between the peak value (i.e. L^∞-norm) and the average power (i.e., L^2-norm) of a signal. This undesired behaviour of orthogonal schemes has remarkable negative impacts on the performance, the energy-efficiency, and the maintain cost of the transmission systems. In this work, we give some discussions concerning to the problem of reduction of the high dynamics of an orthogonal transmission scheme. We show that this problem is connected with some mathematical fields, such as functional analysis (Hahn-Banach Theorem and Baire Category), additive combinatorics (Szeméredi Theorem, Green-Tao Theorem on arithmetic progressions in the primes, sparse Szeméredi type Theorems by Conlon and Gowers, and the famous Erdős problem on arithmetic progressions), and both trigonometric and non-trigonometric harmonic analysis.

1 Introduction

The rapid development of technologies and the astronomic growth in data usage over the past two decades are inter alia the driving force for the development of flexible and efficient transmission technologies. The latter can certainly not be imagined without the development of the *orthogonal transmission scheme*, which can roughly be described as the technique with which several data can be transmitted instantaneously orthogonally (orthonormally) in one single shot within a given time frame. Specifically, given a duration $T_s > 0$ of a transmit signal, and given finite transmit data $\{a_k\}_{k=1}^N$, which constitute simply a sequence in $\mathbb{C}$. The transmit signal of an orthogonal transmission scheme has the form:

$$s(t) = \sum_{k=1}^N a_k \phi_k(t), \quad t \in [0, T_s], \tag{1}$$

where $\{\phi_n\}_{n=1}^N$ constitutes an *orthonormal system* (ONS), in the space of square integrable functions on $[0, T_s]$. Each function in the collection $\{\phi_n\}_{n=1}^N$ is also referred to as *wave function*, and the expression (1) as *waveform*. In the literature on wireless communications, each

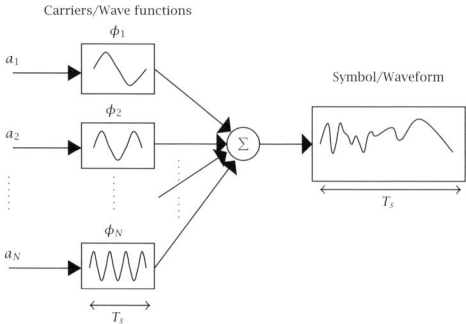

Figure 1. The block diagram of the information modulation unit of the orthogonal transmission scheme for the OFDM case

$\{\phi_n\}_{n=1}^{N}$ is also, according to its purpose, viz. to carry the information-bearing coefficients $\{a_n\}_{n=1}^{N}$, referred to as *carrier*. In the context of communications engineering, the way to process the information-bearing data by means of functions (in the space of square integrable functions on $[0, T_s)$) in the manner (1) for transmission purposes is also called *information modulation*. Quite popular choices of the wave functions in communications engineering are $\phi_n(\cdot) = e^{i2\pi(n-1)(\cdot)}$, and the Walsh functions (see Definition 30). The former case is referred to as *Fourier/OFDM case*, and the latter as *Walsh/CDMA case*. The origin of the terms OFDM and CDMA shall be introduced in the next section. To give a better understanding, the block diagram of the information modulation unit of the orthogonal transmission scheme for the OFDM case is provided in Figure 1. The coefficients $\{a_n\}_{n=1}^{N}$ in (1) might for instance be information sequences, which are available serially in time, or for each $k \in \{1, \ldots, N\}$, a_k corresponds to one of the information symbols of a user. For convenience, we mostly consider the time duration $T_s = 1$, since any other cases can be derived by simple rescaling.

In this work, we are mainly interested in the dynamical behaviour of the waveforms formed by orthonormal systems, which is measured by the ratio between the peak-value and the average power, called *peak-to-average-power ratio (PAPR)*. It is well-known, both theoretically and practically, that the waveforms of orthogonal transmission schemes possess high dynamical behaviour. A more detailed discussion on this aspect shall be given informally in the next section and formally in Section 4.1. The so-called *tone reservation method* [31–33] is without doubt one of the canonical ways to reduce the PAPR value of a waveform. There, the (indexes of) available wave functions ($\{1, \ldots, N\}$, $N \in \mathbb{N}$) are separated into two (fixed) subsets, one which is reserved for those, which carry the information data (call $\mathcal{I}$), called *information set*, and another which is reserved for those, which should, by means of the choice of the coefficients, reduce/compensate the peak value of the resulted waveforms, s.t. it is uniformly below a certain threshold constant $C_{Ex} > 0$, called *extension constant*. A more formal introduction of this method shall be given in Section 4.3. We refer to the applicability of tone reservation method with extension constant C_{Ex} simply as *the solvability of the PAPR problem with extension constant C_{Ex}* (see Definition 3).

Further, in this work, we mostly keep the option open, that the information data and the wave functions, used for the compensation of the peak value, are of infinite number. We shall give in Section 4.4 a discussion on the optimal extension constant, with which the PAPR reduction problem is solvable. Furthermore, we shall see in Section 4.5 that the solvability of the PAPR problem is connected to an embedding problem of a certain closed subspace of $L^1([0,1])$ (Definition 8) into $L^2([0,1])$ (Theorem 10 and Proposition 13). In turn, that fact shall give a relation between the non-solvability of the PAPR problem for a given extension constant, and the existence of certain combinatorial objects in the information set. We will observe, that in the OFDM case, the corresponding object is the so-called arithmetic progression (Definition 5.1), which leads us to involve the famous Szeméredi Theorem on arithmetic progressions (see Theorem 25 and Theorem 28), and some asymptotic and infinite tightening due to Green and Tao, and Conlon and Gowers (see Theorem 26 and Theorem 29). Further, it shall be obvious that the corresponding combinatorial object in the CDMA case is the subset, which indexed a so-called perfect Walsh sum (Definition 36). We will give a condition on the size of the information set, s.t. such a combinatorial object exists therein (Theorem 39). By means of the former, we are able to derive an asymptotic statement concerning to the existence of that combinatorial object (Theorem 41). Those results give in turn some statements concerning to the non-solvability of the PAPR problem, both in the finite – (Theorem 31), asymptotic – (Theorem 43, Theorem 44)), and infinite case (Theorem 46)

2 Motivation of the PAPR reduction problems

In this work, we consider two transmission schemes, which use basically the previous mentioned idea, namely the *orthogonal frequency division multiplexing (OFDM)* and the *direct sequence code division multiple access (DS-CDMA)*.

The OFDM constitutes a transmission scheme dominating, both the present, and future communication systems. It has become an important part of various, both wireless and wire line, current, and future standards, such as DSL, IEEE 802.11, DVB-T, LTE, and LTE-advanced/4G, and 5G. The wave functions of OFDM are basically complex sines of the form $\phi_n(\cdot) = \exp(2\pi i(n-1)(\cdot))$, $n \in \mathbb{N}$. One of the reasons for the attractiveness of OFDM for wireless applications is its robustness against multipath fading, which constitute major characteristic of wireless transmission channel. Roughly, the multipath fading of the signal transmitted through a wireless channel is caused by the fact that the signal will reach the receiver not only via the direct path, but also as a result of reflections from non-moving objects such as buildings, hills, ground, water, and moving objects such as automobiles and people, that are adjacent to the main direct path. Furthermore, the implementation of OFDM waveforms, and both the equalization and the decoding of the information contained in a received OFDM waveform, are fairly easy, since they all require only fast Fourier transform (FFT), inverse fast Fourier transform (IFFT), and simple multiplications. The latter is justified by the fact that the wave functions of OFDM are the eigenfunctions of a linear-time-invariant channel, which give a mathematical model of a (wireless) transmission channel. Notice that there are some minor technicalities, such as cyclic prefix insertion, and synchronization, to be considered. Also, making use of those techniques, OFDM communication scheme is proven to be robust against intersymbol interference (ISI), i.e. interference between consecutive waveforms/symbols. The occurrence of ISI in a communication system might impact the reliability of that system negatively. For details on previous mentioned aspects, we refer to standard textbooks on wireless communications. A rough sketch of OFDM transmission scheme, and the corresponding signal processing decoding at the receiver is given by the block diagram depicted in Figure 2.

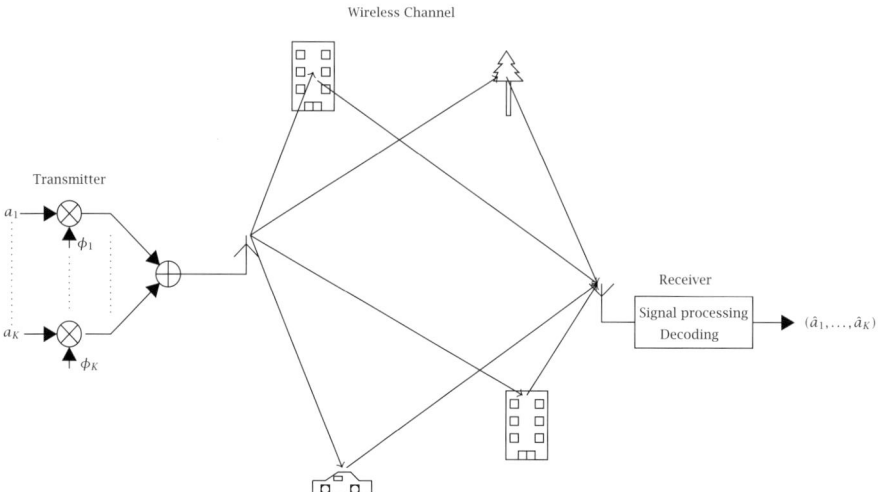

Figure 2. The block diagram of orthogonal transmission scheme and the corresponding signal processing at the receiver for multipath propagation channel. The sequence $(\hat{a}_1, \ldots, \hat{a}_K)$ denotes the estimates of the transmitted symbols $(a_1, \ldots, a_K)$.

DS-CDMA is certainly also a transmission scheme, which plays an important role in numerous present communication systems. This scheme has become an indispensable part of several communication standards, such as 3G, UMTS, GPS, and Galileo. DS-CDMA is used mainly for uplink communications, i.e. from users to a base station. There, a user n obtains an individual signature pulse ϕ_n, which basically a train of rectangular pulses. All of those pulses have to satisfy the orthonormality condition, which enables the base station receiver to separate each of the users separately. The principle mentioned previously works also for the downlink communication between base station and several users. The combined signal (1) is noise alike, so that a potential jamming is aggravated to intercept a certain user. To overcome the effect of multipath propagation, one may use the so-called RAKE-receiver, with which the multipath components can even be detected. A detailed treatment about those aspects is without scope of this work. Thus, we refer to standard textbooks on wireless communications.

In this work we are rather interested in the so-called *Peak-to-Average-Power-Ratio (PAPR)* behavior of orthogonal transmission schemes, than in the technical implementations of that scheme. The PAPR of a waveform (or more generally: a signal) gives a proportion between the peak value of a waveform, which is measured by the essential supremum, and its energy, which is measured by the L^2-norm of the waveform. In this case, the energy can also be interpreted as the average (quadratic) value of the waveform. Thus, the PAPR of a waveform gives an insight into the behaviour of its peaks. In particular, a high PAPR value indicates the existence of extreme peaks in the considered waveform. Thus in some literature, PAPR is denoted more vividly by *crest factor*. High PAPR value might have negative impacts to the reliability of a transmission scheme: Commonly, a transmission signal is amplified, before sent through a channel (see Figure 3). However, every (non-ideal) amplifier in practice has a certain magnitude threshold $M \in \mathbb{R}^+$, beyond which the input signal is not linearly amplified, but distorted or

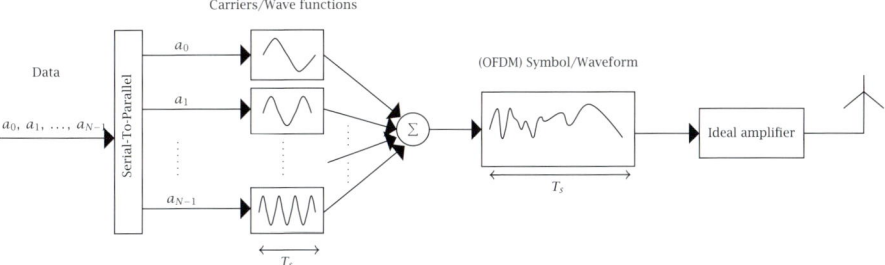

Figure 3. Block diagram for the generation of OFDM transmission signals

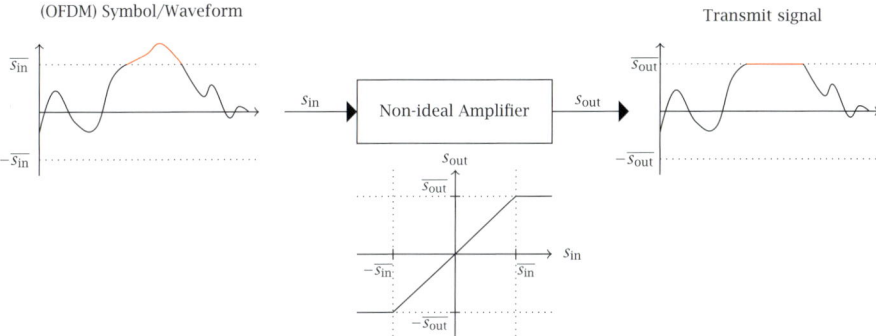

Figure 4. Impact of a realistic non-ideal amplifier to a waveform to be transmitted (clipping)

clipped. By *clipping*, we mean the operation, which leaves the part of the signal undisturbed, if its magnitude is below M, and which sets the magnitude of another part of the signal to the value M, if its magnitude is over M, while the phase is remained unchanged. For a depiction of the occurrence of clipping caused by a non-ideal amplifier, see Figure 4. Thus in case that an orthogonal transmission scheme having waveforms with high PAPR value, and an amplifier with low threshold value is used, then clipping might occur, which results in the alteration of the waveforms, and in particular in the destruction of the desired structure of the waveforms – the orthonormality of the wave functions. Another negative effects of clipping is for instance the occurrence of out-of-band radiation. For more detailed discussions on those aspects, we refer to standard textbooks on wireless communications.

A naive solution to that problem is to use another amplifier with a higher threshold value. However, this is practically not the best solution, since an amplifier with a high linear range is expensive, not only to purchase, but also to maintain. Such amplifier would in general have inefficiently high power consumption, and accordingly, require batteries with high capacity and long lifetime. Energy efficiency of a communication system, which is partially reflected in the low power consumption, is without doubt an important issue both for the present and future. The importance of this issue can be seen in the reports [3, 22] of consulting firms, which estimates, that 2 % of global CO_2 emissions are attributable to the use of information and communication technology, which is comparable to the CO_2 emissions due to avionic activities. Energy efficiency of communications systems is not only of environmental interests, but also of financial interests: Nowadays, energy cost of network operation can even make up to 50 % of the total operational cost.

An orthogonal transmission scheme tends to possess waveforms with high PAPR value. This disadvantage might be caused by the fact that such waveforms are generated by a superposition of large numbers of wave functions. For instance, there are up to 2048 wave functions for the downlink communication in the LTE standard [1], which uses OFDM as a basic transmission scheme. There are several methods proposed for reducing the PAPR [14,17]. However, the so-called *tone reservation method* [31–33] is without doubt a PAPR reduction method, which might have the potential to be popular, since it is canonical and robust, in the way that the only information required on the receiver's side is the indexes of the information-bearing coefficients. To understand this, let us first describe the tone reservation method. There, the (indexes of) available carriers ($\{1, \ldots, N\}$, $N \in \mathbb{N}$) are separated into (fixed) two subsets, one which is reserved for those, which carry the information data, and one which is reserved for those, which should, by means of the choice of the coefficients, reduce/compensate the peak value of the resulted signal. A more formal description of the task shall be given in Definition 3. The auxiliary coefficients may simply be ignored by the receiver. Thus there is no need for additional overhead in the transmission symbols. In the practical scenario, some further requirements for the compensation set have to be considered. There, the compensation set can also be used for channel estimation purposes [34].

At last, for some further discussions on PAPR, we recommend the recent comprehensive overview article [36], which gives for instance a discussion on alternative metrics beyond the PAPR, which are relevant to the behaviour of the energy consumption of a transmission system, and new mathematical concepts aiming to overcome the high PAPR behavior of an orthogonal transmission scheme.

3 Basic notions

For $N \in \mathbb{N}$, we denote the set $\{1, 2, \ldots, N\}$ simply by $[N]$. Let $\mathcal{I} \subset \mathbb{N}$ be an index set, we denote the space of square-summable sequences in $\mathbb{C}$ indexed by $\mathcal{I}$ by $l^2(\mathcal{I})$. For ease of notations, we denote sequences $\{c_n\}_{n \in \mathcal{I}}$ in $\mathbb{C}$, simply by bold letters $\mathbf{c}$, and vice versa.

Operators between Banach spaces
We call a mapping between vector spaces an *Operator*. For an operator $\Phi : \mathcal{X} \to \mathcal{Y}$ between Banach spaces, we define the norm by:

$$\|\Phi\|_{\mathcal{X} \to \mathcal{Y}} := \sup_{\|x\|_{\mathcal{X}} \leq 1} \|\Phi x\|_{\mathcal{X}}.$$

Clearly, in case Φ is linear, we can write:

$$\|\Phi\|_{\mathcal{X} \to \mathcal{Y}} := \sup_{\substack{x \in \mathcal{X} \\ x \neq 0}} \frac{\|\Phi x\|_{\mathcal{X}}}{\|x\|_{\mathcal{X}}} = \sup_{\|x\|_{\mathcal{X}} = 1} \|\Phi x\|_{\mathcal{X}}.$$

Lebesgue spaces
For $T > 0$, we denote the *Lebesgue space (Banach space) of p-integrable functions on $[0, T]$ by* $L^p([0, T])$. As usual, the Lebesgue space has to be understand as equivalence classes of func-

tions, which differ in a set of (Lebesgue-)measure zero, i.e. almost everywhere (a.e.). $L^p([0,T])$ is as usual equipped with the norm $\|f\|_{L^p([0,T])} := [(1/T)\int_0^T |f(t)|^p]^{1/p}$, in case $p \in [1,\infty)$, and in case $p = \infty$, the norm is defined by $\|f\|_{L^\infty([0,1])} = \operatorname{ess\,sup}_{t\in[0,T]} |f(t)|$. As already mentioned in the introduction, it is sufficient only to consider $T = 1$, since any other cases can be treated by means of simple rescaling.

Given a sequence $\{\phi_n\}_{n\in\mathcal{I}}$, where $\mathcal{I} \subset \mathbb{N}$ is an index set, of functions in $L^2([0,1])$. $\{\phi_n\}_{n\in\mathcal{I}}$ is said to be an *orthonormal system (ONS)* in $L^2([0,1])$, if $\int_0^1 \phi_k(t)\phi_l(t)$ is 0, if $k \neq l$, and 1, if $k = l$, where $k, l \in \mathcal{I}$. A collection of functions $\{\phi_n\}_{n\in\mathcal{I}}$ in $L^2([0,1])$ is said to be a *complete orthonormal system (CONS)* for $L^2([0,1])$, if $\{\phi_n\}_{n\in\mathcal{I}}$ is an ONS in $L^2([0,1])$, and if every function $f \in L^2([0,1])$ can be represented as the sum $f = \sum_{k\in\mathcal{I}} c_k\phi_k$, where $\mathbf{c} \in l^2(\mathcal{I})$ is given by $c_n = \int_0^1 f(t)\overline{\phi_n(t)}dt$, $\forall n \in \mathcal{I}$. The convergence of the sum has to be interpreted in the sense of $L^2([0,1])$, i.e., w.r.t. $\|\cdot\|_{L^2([0,1])}$.

4 PAPR problem for orthogonal transmission schemes

4.1 Basic bounds for PAPR of orthogonal transmission schemes

Given a signal $f \in L^2([0,1])$. The *Peak-to-Power-Average-Ratio* of f is defined generally as the ratio between peak value and the energy of the signal f:

$$\mathrm{PAPR}(f) := \frac{\|f\|_{L^\infty([0,1])}}{\|f\|_{L^2([0,1])}}.$$

Since we only consider signals that are basically the linear combination of orthonormal functions, we emphasize the following definition of PAPR for the corresponding subclass of $L^2([0,1])$. Further, we allow in the definition that the considered signal is generated by an infinite number of orthonormal functions.

DEFINITION 1. *Let $\{\phi_k\}_{k\in\mathcal{K}}$ be a set of orthonormal functions, where $\mathcal{K} \subset \mathbb{N}$. We define the Peak-to-Average Power Ratio (PAPR) of a set of coefficients $\mathbf{a} \in \mathbb{C}^{\mathcal{K}}$, $\mathbf{a} \neq 0$ (w.r.t. $\{\phi_k\}_{k\in\mathcal{K}}$) as the quantity:*

$$PAPR(\{\phi_k\}_{k\in\mathcal{K}},\mathbf{a}) = \operatorname*{ess\,sup}_{t\in[0,1]} \frac{\left| \sum_{k\in\mathcal{K}} a_k\phi_k(t) \right|}{\|\mathbf{a}\|_{l^2(\mathbb{N})}}$$

By the orthonormality of $\{\phi_k\}$, it is obvious that the PAPR of a sequence $\mathbf{a} \in l^2(\mathcal{K})$ is equal to the PAPR of the signal $s \in L^2([0,1])$, given by $s = \sum_{k\in\mathcal{K}} a_k\phi_k$. The behaviour of the peak value of orthonormal systems might, as already discussed informally in the introduction, be worse. Indeed, one can show [4, 7], that for finite number of orthonormal functions, the following worst-case behaviour holds:

$$\sqrt{N} \leq \sup_{\|\mathbf{a}\|_{l^2([N])}\leq 1} \mathrm{PAPR}(\{\phi_k\}_{k\in[N]},\mathbf{a}). \tag{2}$$

Furthermore, a corresponding sequence $\mathbf{a}$, with $\|\mathbf{a}\|_{l^2([N])} = 1$, for which the above inequality holds, can easily be constructed.

Such a behavior is of course not tolerable, since the waveforms of an orthogonal transmission scheme typically consists a large number of wave functions. One of the canonical and robust way to reduce the PAPR of a signal is the so-called tone reservation method. We will formalize the method and discuss it soon. But, let us first give some remarks concerning to the PAPR behaviour of an ONS in the following subsection.

Furthermore, it is also interesting to ask whether such a bad behaviour can occur for orthonormal single-carrier systems, for instance the systems with which the information-bearing signals are carried by shifted kernels. For single-carrier systems generated by N number of mutually distinct sinc-kernels, it was recently shown in [5] (Theorem 2.1), that in case that the information coefficients are chosen i.i.d. by Gaussian normal distribution with zero expectation, and a given variance, then the expected value of the PAPR of resulted single-carrier signal is comparable with $\sqrt{\log N}$. By some additional requirements on the information coefficients, one can even have the result, that the expectation the PAPR of the resulted signal is comparably with $\log \log N$. Those statements might asserts, that the PAPR behaviour of single-carrier systems (compared to the multi-carrier orthogonal transmission systems) is fairly good.

4.2   General remarks on coefficients of ONS

For ONS in $L^2([0,1])$, we have already mentioned in the previous subsection, that PAPR of some non-zero coefficients in $l^2([N])$ is bigger than $\sqrt{N}$, which is surely not tolerable. This asserts that some efforts have to be done to prevent such an undesired behaviour. From the mathematical point of view, the problem does not look so helpless, since by the Nazarov's solution of the coefficients problem [19], we have the following statement:

THEOREM 2 (Nazarov [19]). *Let $\{\phi_n\}_{n \in \mathbb{N}}$ be an ONB, for which $\|\phi_n\|_{L^1([0,1])} \geq C_1$, $\forall n \in \mathbb{N}$, for a constant $C_1 > 0$. Then there exists a constant $C > 0$, such that for every coefficients $\mathbf{a} \in l^2(\mathbb{N})$, there exists a function $f_* \in L^\infty([0,1])$, with:*

1. $\|f_*\|_{L^\infty([0,1])} \leq C \|\mathbf{a}\|_{l^2(\mathbb{N})}$.
2. $\left| \int_0^1 f_*(t)\overline{\phi_n(t)}dt \right| \geq |a_n|$, $n \in \mathbb{N}$.

Thus, in case that the considered ONS is in addition complete, one can construct for arbitrary $\mathbf{a} \in l^2(\mathbb{N})$, another sequence $\mathbf{b} \in l^2(\mathbb{N})$, with $|b_n| \geq |a_n|$, $\forall n \in \mathbb{N}$, by means of suitable enlargement of the absolute value – and phase change of each of its members, such that another waveform[1] f_* is obtained, whose peak value can be controlled by a certain factor $C > 0$, depending only on the choice of ONS $\{\phi_n\}_{n \in \mathbb{N}}$:

$$f_* := \sum_{k=1}^\infty b_n \phi_n \quad \text{and} \quad \|f_*\|_{L^\infty([0,1])} \leq C \|\mathbf{a}\|_{l^2([0,1])}.$$

However, this result is not applicable for communication systems, since the inverse transformation of the information-bearing coefficients $\mathbf{a}$ from the signal formed by $\mathbf{b}$ after a transmission for instance through a noisy channel is generally not possible.

For other applications in electrical engineering, Nazarov's solution might be very interesting: In some cases, it is crucial to construct waveforms (1), formed by means of a given ONB

---

1. Actually, Nazarov's theorem ensures the existence of a function $f_* \in L^\infty([0,1])$. But since $\{\phi_n\}_{n \in \mathbb{N}}$ is an ONB for $L^2([0,1])$, and $L^\infty([0,1]) \subset L^2([0,1])$, f_* can be represented as the series $f = \sum_{n=1}^\infty b_n \phi_n$ converges w.r.t. $\|\cdot\|_{L^2([0,1])}$, with $b_n := \int_0^1 f_*(t)\overline{\phi_n(t)}dt$.

$\{\phi_n\}_{n\in\mathbb{N}}$, s.t. the modulus of its coefficients is lower bounded, and its L^∞-norm can be controlled. For instance, to identify parameters of an unknown (wireless) channel, which can be seen as a linear-time invariant system, one may send suitable known waveforms, called *pilot signals*. In particular, the identification of channel parameters constitute an important step for a reliable communications. Based on the knowledge of the those – and the resulting received signals, one hopes to estimate those unknowns (see, e.g., [34], which also gives some novel ideas). To fulfill that task, the pilot signals have of course to possess some desired properties, i.a.: their peak values have to lie within an admissible range, which is in particular determined by the hardware of the transmitter, i.e. amplifier, filter, and antennas. In Section 2, we have already discussed about the effect of the distortion of the amplified signal, in case that the peak value of it does not lie within the linear range of the amplifier. The occurrence of such effect in the process of channel measurement is clearly undesired. Furthermore, it is in some cases desired, that the energy, i.e., the L^2-norm, is spread over all the coefficients of the pilot signal, see, e.g., the notion of *block-type pilot model* in the textbooks on wireless engineering. A necessary condition for the fulfillment of the latter condition is certainly, that the modulus of the coefficients of the pilot signal lies uniformly under a certain threshold. For instance, if the pilot signals are desired to have the energy of 1, one may require, that the modulus of their coefficients are lower bounded by $1/\sqrt{N}$ (in case that N wave functions are available). Nazarov's result shows, that this is basically possible.

The Nazarov's solution might also be interesting for designing test signals: After preparation of a system (e.g., circuits, chips, amplifier, etc.), one aims to examine, whether the implemented system possesses the desired property. This task is done by injecting input signals, which are admissible for later applications of the system. For instance, to check whether an implemented broadband amplifier fulfills the given requirements, one may inject a broadband signal, which is in our context a signal of the form (1), whose coefficients are non-zero for a large index set. Furthermore, it is desired that the peak value of the test broadband signal lies within the linear range of the designed broadband amplifier. A similar approach is also suitable for the testing of the small-signal behaviour of certain circuits.

Furthermore, in measurement-technological applications, it is often desired to construct waveforms f of the form (1), for a specific ONS $\{\phi_n\}_{n=1}^N$, fullfilling certain conditions on the coefficients (for instance they should be of modulus 1), such as flatness property, in the sense that the modulus of f is nearly constant. For the ONS $\left\{e^{i2\pi(n-1)(\cdot)}\right\}_{n=1}^N$, this problem is connected to the Littlewood's flatness problem [18], which can be stated as follows:

Given a polynomial s_N of the form (1), with $\phi_n(\cdot) = e^{i2\pi(n-1)(\cdot)}$, $n \in [N]$, whose coefficients are either complex and fulfill $|a_n| = 1$, $\forall n \in [N]$ (such a polynomial is called *complex unimodular polynomial*), or $a_n \in \{-1,1\}$, $\forall n \in [N]$ (such a polynomial is also called *real unimodular polynomial*). How close can s_N come to satisfying $|s_N(t)| = \sqrt{N}$, $\forall t \in [0,1]$? Specifically, for a given possibly small $\epsilon > 0$, we seek for an unimodular polynomial s_N, which is ϵ-flat, i.e., s_N fulfills the condition:

$$(1 - \epsilon)\sqrt{N} \le |s_N(t)| \le (1 + \epsilon)\sqrt{N}.$$

Equivalently, one may also seek for a sequence of polynomials $\{s_{N_k}\}_{k\in\mathbb{N}}$, whose members are each unimodular, and of degree $N_k \in \mathbb{N}$, $\forall k \in \mathbb{N}$, and which fulfills:

$$(1 - \epsilon_k)\sqrt{N_k} \le |s_{N_k}(t)| \le (1 + \epsilon_k)\sqrt{N_k}, \quad \forall k \in \mathbb{N},$$

for a sequence $\{\epsilon_{N_k}\}_{k\in\mathbb{N}}$ tending to 0. Such a sequence of unimodular polynomials is also called $\{\epsilon_{N_k}\}_{k\in\mathbb{N}}$ – *ultraflat.* Erdős conjectured in 1957 [9], that such a task is impossible, in the sense that every unimodular polynomials s_N of degree N have the property:

$$\max_{t\in[0,1]} |s_N(t)| \geq (1+\epsilon)\sqrt{N},$$

where $\epsilon > 0$ is a constant, which is independent on N. In the case that the considered uni-modular polynomials are complex, Kahane [15] showed the existence of a sequence of com-plex unimodular polynomials $\{s_N\}_{N\in\mathbb{N}}$, which is $\{\epsilon_N\}_{N\in\mathbb{N}}$-ultraflat, where ϵ_N tends basically not faster to zero than $N^{-1/17}\sqrt{\log(N)}$, as N goes to infinity. This statement disproves the Erdős conjecture, and sheds light to the solvability of the Littlewood's flatness problem, in the case, where the considered unimodular polynomials are complex. Further discussions on Kahane's result is given in [2]. For real unimodular polynomials, Erdős conjecture is still un-solved.

Kahane's result asserts both the possibility to reduce the peak value of an OFDM wave-form, and to obtain another waveform having nearly constant envelope, only by changing the phase of its coefficients. To see this, consider for instance the normalized OFDM wave-forms f, viz. $\|f\|_{L^2([0,1])} = 1$. Furthermore, we consider such signals, whose energy are spread equally among those coefficients, i.e. the coefficients $\{a_n\}$ in the representation (1) yields, $|a_n| = 1/\sqrt{N}$, $\forall n \in [N]$. By the triangle inequality, and by the fact that complex exponentials are of modulus 1, one can show that the peak value of such signals is upper bounded by $\sqrt{N}$. If we set $a_n = 1/\sqrt{N}$, for each $n \in [N]$, we obtain a waveform, which achieves such a worst case behaviour. As asserted in the previous paragraph, for each $k \in \mathbb{N}$, we can find $\{\beta_n^{(k)}\}_{n\in[N_k]}$, such that the L^∞-norm of the following sequence of waveforms:

$$f_{N_k}(t) = \frac{1}{\sqrt{N_k}} \sum_{n=1}^{N_k} e^{i\beta_n^{(k)}} e^{i2\pi(n-1)t}, \quad k \in \mathbb{N},$$

tends to 1, i.e., $\lim_{k\to\infty} \|f_{N_k}\|_{L^\infty[0,1]} = 1$. Furthermore, as it has already been discussed in pre-vious paragraph, the existence of such a sequence, whose minimum value tends to 1, as the index increases, i.e. $\lim_{k\to\infty} \min_{t\in[0,1]} |f_{N_k}(t)| = 1$. Summarily, we obtain OFDM waveforms $\{f_{N_k}\}_{k\in\mathbb{N}}$, each possessing envelopes approximately constant, by only changing the phase of their coefficients. An interesting application of this result is certainly the design of pilot sig-nals in OFDM system, which we have already discussed in the beginning of this section. The obtained OFDM waveforms might serve as a suitable pilot signals. Furthermore, a broadband amplifier for sending the pilot signal can be designed relatively easy, since the pilot signal has nearly constant envelope.

4.3 PAPR reduction problem and tone reservation method

As already mentioned in the introduction, the strategy considered in this paper is to reserve one subset of orthonormal functions for carrying the information-bearing coefficients, and to determine the coefficients for the remaining orthonormal functions, s.t. the resulted sum of functions has a peak-value smaller than a given threshold:

DEFINITION 3 (PAPR reduction problem). *Given $\mathcal{K} \subset \mathbb{N}$. Let $\{\phi_n\}_{n\in\mathcal{K}}$ be an orthonormal sys-tem, and $\mathcal{I} \subset \mathcal{K}$. We say the PAPR reduction problem is solvable for the pair $(\{\phi_n\}_{n\in\mathcal{K}}, \mathcal{I})$ with* *constant $C_{Ex} > 0$, if for every $\mathbf{a} \in l^2(\mathcal{I})$, there exists $\mathbf{b} \in l^2(\mathcal{I}^c)$ (the complementation is of course w.r.t. $\mathcal{K}$), satisfying $\|\mathbf{b}\|_{l^2(\mathcal{I}^c)} \leq C_{Ex}\|\mathbf{a}\|_{l^2(\mathcal{I})}$, for which the following holds:*

$$\operatorname*{ess\,sup}_{t \in [0,1]} \left| \sum_{k \in \mathcal{I}} a_k \phi_k(t) + \sum_{k \in \mathcal{I}^c} b_k \phi_k(t) \right| \leq C_{Ex}\|\mathbf{a}\|_{l^2(\mathcal{I})} \tag{3}$$

We further refer to $\mathcal{I}$ as *information set*, $\mathcal{I}^c$ as *compensation set*, $\{\phi_n\}_{n \in \mathcal{I}}$ as *information tones*, and respectively $\{\phi_n\}_{n \in \mathcal{I}^c}$ as *compensation tones*. We refer to the quantity $|\mathcal{I}| / |\mathcal{K}|$ as the density of information set (in $\mathcal{K}$). Unless otherwise stated, $\{\phi_n\}_{n \in \mathbb{N}}$ is a ONS for $L^2([0,1])$, $\mathcal{K}$ is a subset of $\mathbb{N}$, $\mathcal{I} \subset \mathcal{K}$, and $\mathcal{I}^c$ is complemented w.r.t. $\mathcal{K}$. We shall always ignore the uninteresting case, where $\mathbf{a} = 0$. A necessary condition for the solvability of the PAPR reduction problem is surely, that $\{\phi_n\}_{n \in \mathcal{I}}$ is uniformly bounded, in the sense that $\phi_n \in L^\infty([-\pi, \pi])$, for all $n \in \mathcal{I}$. Otherwise, one can easily construct a sequence $\mathbf{a} \in l^2(\mathcal{I})$, for which the peak-value of the signal $\sum_{k \in \mathcal{I}} a_k \phi_k$ is unbounded. To avoid any further undesirable behaviour, we assume not only that the restricted orthonormal system $\{\phi_n\}_{n \in \mathcal{I}}$ is bounded, but that all of the considered orthonormal systems are bounded. In practice, the compensation tones might also be used for another purposes, such as the estimation of the transmission channel (e.g., [34]).

Sometimes, we refer to the PAPR reduction problem simply as *PAPR problem*. Notice that if (3) is fulfilled, then the PAPR of the combined signal is below the given threshold value C_{ex}. To see this, notice that the L^2-norm of the combined signal is simply $\sqrt{\|\mathbf{a}\|^2_{l^2(\mathcal{I})} + \|\mathbf{b}\|^2_{l^2(\mathcal{I}^c)}}$. Correspondingly, we have:

$$\frac{\operatorname*{ess\,sup}_{t \in [0,1]} \left| \sum_{k \in \mathcal{I}} a_k \phi_k(t) + \sum_{k \in \mathcal{I}^c} b_k \phi_k(t) \right|}{\sqrt{\|\mathbf{a}\|^2_{l^2(\mathcal{I})} + \|\mathbf{b}\|^2_{l^2(\mathcal{I}^c)}}} \leq \frac{\operatorname*{ess\,sup}_{t \in [0,1]} \left| \sum_{k \in \mathcal{I}} a_k \phi_k(t) + \sum_{k \in \mathcal{I}^c} b_k \phi_k(t) \right|}{\|\mathbf{a}\|_{l^2(\mathcal{I})}} \leq C_{Ex},$$

by assumption, that the PAPR reduction problem is solvable with constant C_{Ex}. The choice of the extension constant is dependent on the design of the communication scheme, which is in particular constrained by some factors, such as the admissible maximum energy consumption, and the linear range of the amplifier. In case that the PAPR reduction problem is solvable with a certain extension constant, the corresponding compensation coefficients carried by the compensation tones might be computed by means of linear program [31–33].

The condition $\|\mathbf{b}\|_{l^2(\mathcal{I}^c)} \leq C_{Ex}\|\mathbf{a}\|_{l^2(\mathcal{I})}$ might seem at first sight to come out of nothing. However, it can be shown [6], that if (3) holds for some coefficients $\mathbf{b}$, then $\|\mathbf{b}\|_{l^2(\mathcal{I}^c)}$ has to be less or equal than $C_{Ex}\|\mathbf{a}\|_{l^2(\mathcal{I})}$. Thus, the requirement $\|\mathbf{b}\|_{l^2(\mathcal{I}^c)} \leq C_{Ex}\|\mathbf{a}\|_{l^2(\mathcal{I})}$ serves in some sense as a restriction of the possible solutions of the PAPR reduction problem. Notice also, that we allow infinitely many carriers for the compensation of the PAPR value. This is not only of mathematical -, but also of practical interests, since the solvability of the PAPR reduction problem in this setting is a necessary condition for the solvability of the PAPR reduction problem in the setting, where the available compensation tones are of finite number.

In some cases, it is advantageous to consider the following restricted form of the PAPR reduction problem:

DEFINITION 4 (Restricted PAPR (RPAPR) reduction problem). *Given $\mathcal{K} \subset \mathbb{N}$. Let $\{\phi_n\}_{n \in \mathcal{K}}$ be an orthonormal system, and $\mathcal{I} \subset \mathcal{K}$. We say the restricted PAPR reduction problem is solvable* *for the pair* $(\{\phi_n\}_{n\in\mathcal{K}}, \mathcal{I})$ *with constant* $C_{Ex} > 0$, *if for every* $\mathbf{a} \in l^2(\mathcal{I})$, *with* $\|\mathbf{a}\|_{l^2(\mathcal{I})} \leq 1$, *there exists* $\mathbf{b} \in l^2(\mathcal{I}^c)$, *satisfying* $\|\mathbf{b}\|_{l^2(\mathcal{I}^c)} \leq C_{Ex}$, *for which it holds:*

$$\operatorname*{ess\,sup}_{t\in[0,1]} \left| \sum_{k\in\mathcal{I}} a_k \phi_k(t) + \sum_{k\in\mathcal{I}^c} b_k \phi_k(t) \right| \leq C_{Ex} \|\mathbf{a}\|_{l^2(\mathcal{I})}. \tag{4}$$

Clearly, if the PAPR reduction problem is solvable, then the restricted PAPR reduction problem is also solvable. To give a clear distinction between the restricted - and the PAPR reduction problem, we sometimes refer to the latter as the *general PAPR reduction problem*. By reason of energy efficiency, practitioners might be interested rather in the restricted PAPR reduction problem, than in the general PAPR reduction problem, since they would ensure that the energy of the transmission waveforms, resulted from the designed scheme, is below a certain threshold. Thus, it makes sense, only to consider information-bearing waveform under a certain energy threshold w.l.o.g. 1, i.e. $\mathbf{a} \in l^2(\mathcal{I})$, with $\|\mathbf{a}\|_{l^2(\mathcal{I})} \leq 1$. The restricted PAPR reduction problem is indeed related with the general PAPR reduction problem in the following manner:

REMARK 5. *The solvability of the restricted PAPR problem implies also the solvability of the general PAPR problem. To see this, suppose that the restricted problem is solvable for* $(\{\phi_n\}_{n\in\mathcal{K}}, \mathcal{I})$, *with an extension constant* $C_{Ex} > 0$. *Let* $\mathbf{a} \in l^2(\mathcal{I})$, $\|\mathbf{a}\|_{l^2(\mathcal{I})} \neq 0$, *be arbitrary. Now consider the sequence* $\mathbf{a}_1 := \mathbf{a}/\|\mathbf{a}\| \in l^2(\mathcal{I})$. *By the solvability of the restricted PAPR reduction problem, we can find a sequence* $\mathbf{b} \in l^2(\mathcal{I}^c)$, *for which (4) holds, with* $\mathbf{a}_1$ *instead of* $\mathbf{a}$. *Thus we have:*

$$\operatorname*{ess\,sup}_{t\in[0,1]} \left| \sum_{k\in\mathcal{I}} a_k \phi_k(t) + \sum_{k\in\mathcal{I}^c} b_k \|\mathbf{a}\|_{l^2(\mathcal{I})} \phi_k(t) \right| \leq C_{Ex} \|\mathbf{a}\|_{l^2(\mathcal{I})}.$$

Since $\mathbf{a}$ *is arbitrary, previous observation implies that the general PAPR problem is solvable for* $(\{\phi_n\}_{n\in\mathcal{K}}, \mathcal{I})$, *with extension constant* C_{Ex}. *The fact that the solvability of the general PAPR reduction problem implies the solvability of the restricted PAPR reduction problem with the same extension constant is trivial.*

4.4   On the optimal extension constant

In this subsection we aim to give some discussion about the extension constant. An elementary observation is that the extension constant with which the PAPR reduction problem is solvable is always bigger than 1. To see this, suppose that the PAPR reduction problem is solvable for $(\{\phi_n\}_{n\in\mathcal{K}}, \mathcal{I})$ with $C_{Ex} > 0$. Then for an information-bearing coefficients $\mathbf{a} \in l^2(\mathcal{I})$, $\mathbf{a} \neq 0$, we can obtain $\mathbf{b} \in l^2(\mathcal{I}^c)$, s.t. the combined signal has the peak value less than $C_{Ex} \|\mathbf{a}\|_{l^2(\mathcal{I})}$. Thus, we have:

$$\frac{\sum_{k\in\mathcal{I}} |a_k|^2 + \sum_{k\in\mathcal{I}^c} |b_k|^2}{\|\mathbf{a}\|_{l^2(\mathcal{I})}^2} \leq \frac{\operatorname*{ess\,sup}_{t\in[0,1]} \left| \sum_{k\in\mathcal{I}} a_k \phi_k(t) + \sum_{k\in\mathcal{I}^c} b_k \phi_k(t) \right|^2}{\|\mathbf{a}\|_{l^2(\mathcal{I})}^2} \leq C_{Ex}^2,$$

where the first inequality follows from the usual estimation of the integral by means of essential supremum, and the orthonormality of $\{\phi_n\}_{n\in\mathbb{N}}$. The above inequality shows that $C_{Ex} \geq 1$, as desired.

Suppose that the restricted PAPR reduction problem is solvable for the pair $(\{\phi_n\}_{n\in\mathcal{K}}, \mathcal{I})$ with a given constant $C_{Ex} \geq 1$. It is natural to ask, whether there is a lowest possible extension constant $\underline{C}_{Ex}$, less than C_{Ex}, with which the PAPR reduction problem for $(\{\phi_n\}_{n\in\mathcal{K}}, \mathcal{I})$ is still solvable. It is also natural to ask the same question for the general PAPR reduction problem, where the optimal constant is denoted by C_{Ex}^*. Furthermore, it is interesting to see whether those quantities differ, or are exactly the same. Before we answer those questions, let us first give the following preliminaries.

We define the operator, $E_{\mathcal{I}} : l^2(\mathcal{I}) \to L^\infty([0,1])$, which assign each $\mathbf{a} \in l^2(\mathcal{I})$ a waveform:

$$E_{\mathcal{I}}\mathbf{a} := \sum_{k\in\mathcal{I}} a_k \phi_k + \sum_{k\in\mathcal{I}^c} b_k \phi_k,$$

where $\mathbf{b}$ is a suitable sequence in $\mathbb{C}$ indexed by $\mathcal{I}^c$ (where the complement is w.r.t $\mathcal{K}$), as an *extension operator (EP)*. By means of that operator, we may say that the restricted PAPR reduction problem is solvable for $(\{\phi_n\}_{n\in\mathcal{K}}, \mathcal{I})$, with extension norm $C_{Ex} \geq 1$, if there exists an extension operator $E_{\mathcal{I}}$, fulfilling:

$$\|E_{\mathcal{I}}\mathbf{a}\|_{L^\infty([0,1])} \leq C_{Ex} \|\mathbf{a}\|_{l^2(\mathcal{I})}, \quad \forall \|\mathbf{a}\|_{l^2(\mathcal{I})} \leq 1. \tag{5}$$

Notice that such an operator is not necessarily unique, and in general not linear. By means of Remark 5, we can give the following observation:

REMARK 6. *Furthermore, by Remark 5, an operator $E_{\mathcal{I}}$, for which (5) holds, induces another operator $E_{\mathcal{I}}' : l^2(\mathcal{I}) \to L^\infty([0,1])$ assigning each $\mathbf{a} \in l^2(\mathcal{I})$ to the wave form:*

$$E_{\mathcal{I}}'\mathbf{a} := \sum_{k\in\mathcal{I}} a_k \phi_k + \sum_{k\in\mathcal{I}^c} b_k \|\mathbf{a}\|_{l^2(\mathcal{I})} \phi_k,$$

which gives the solution of the general PAPR problem for $(\{\phi_n\}_{n\in\mathcal{K}}, \mathcal{I})$ with C_{Ex}.

By the following quantity, which corresponds to an extension operator $E_{\mathcal{I}}$, we may give another more compact formulation of the solvability of the restricted PAPR problem:

$$\|E_{\mathcal{I}}\|_{l^2(\mathcal{I})\to L^\infty([0,1])} := \sup_{\|\mathbf{a}\|_{l^2(\mathcal{I})}\leq 1} \|E_{\mathcal{I}}\mathbf{a}\|_{L^\infty([0,1])}. \tag{6}$$

Thus by (5), we may say that the restricted PAPR reduction problem is solvable for $(\{\phi_n\}_{n\in\mathcal{K}}, \mathcal{I})$, with extension norm $C_{Ex} \geq 1$, if there exists an extension operator $E_{\mathcal{I}}$, fulfilling:

$$\|E_{\mathcal{I}}\|_{l^2(\mathcal{I})\to L^\infty([0,1])} \leq C_{Ex}.$$

For a reformulation of the general PAPR reduction problem, we define the following quantity:

$$\overline{\|E_{\mathcal{I}}\|}_{l^2(\mathcal{I})\to L^\infty([0,1])} := \sup_{\|\mathbf{a}\|_{l^2(\mathcal{I})}\neq 0} \frac{\|E_{\mathcal{I}}\mathbf{a}\|_{L^\infty([0,1])}}{\|\mathbf{a}\|_{l^2(\mathcal{I})}}. \tag{7}$$

As similar as done in the previous paragraph, we may give the following helpful formulation of the general PAPR reduction problem: We say that the PAPR reduction problem is solvable for $(\{\phi_n\}_{n\in\mathcal{K}}, \mathcal{I})$ with extension constant C_{Ex}, if there exists an extension operator E_I, for which $\overline{\|E_{\mathcal{I}}\|}_{l^2(\mathcal{I})\to L^\infty([0,1])} \leq C_{Ex}$ holds. In case that $E_{\mathcal{I}}$ is linear, (7) and (6) are basically the same, and is known as the operator norm of $E_{\mathcal{I}}$.

Back to our actual aim, we can define the optimal constant $\underline{C}_{Ex}$, which was already mentioned in the beginning of this subsection in the context of the solvability of the restricted PAPR reduction problem as follows:

$$\underline{C}_{Ex} := \inf \{C_{Ex} > 0 : \text{RPAPR prob. is solv. for } (\{\phi_n\}_{n \in \mathcal{K}}, \mathcal{I}) \text{ with } C_{Ex}\}. \tag{8}$$

By means of the extension operator, and $\|\cdot\|_{l^2(\mathcal{I}) \to L^\infty([0,1])}$, we can write the above expression by:

$$\underline{C}_{Ex} := \inf \left\{ \|E_{\mathcal{I}}\|_{l^2(\mathcal{I}) \to L^\infty([0,1])} : E_{\mathcal{I}} \text{ is an extension operator} \right\}. \tag{9}$$

Notice that solvability of the restricted PAPR reduction problem with the optimum extension constant $\underline{C}_{Ex}$, and accordingly/equivalently the existence of an extension operator giving a solution is not yet ensured. Further, in the interest of the general PAPR reduction problem, we may consider the following quantity:

$$C_{Ex}^* := \inf \{C_{Ex} > 0 : \text{PAPR prob. is solv. for } (\{\phi_n\}_{n \in K}, \mathcal{I}) \text{ with } C_{Ex}\},$$

which can be written by means of extension operators and $\overline{\|\cdot\|}_{l^2(\mathcal{I}) \to L^\infty([0,1])}$ as follows:

$$C_{Ex}^* := \inf \left\{ \overline{\|E_{\mathcal{I}}\|}_{l^2(\mathcal{I}) \to L^\infty([0,1])} : E_{\mathcal{I}} \text{ is an extension operator} \right\}. \tag{10}$$

The case $\underline{C}_{Ex} = \infty$ (resp. $C_{Ex}^* = \infty$), means that the restricted (resp. general) PAPR reduction problem is not solvable for $(\{\phi_n\}_{n \in \mathcal{K}}, \mathcal{I})$ (with any extension constant $C_{Ex} > 0$). Those cases can also clearly be formalized by means of the non-existence of an extension operator giving the solution. Without ensuring the solvability of the PAPR reduction problem in the optimum, we are able show that $\underline{C}_{Ex}$ and C_{Ex}^* are essentially the same:

LEMMA 7. *Let be* $\{\phi_n\}_{n \in \mathbb{N}}$, $\mathcal{K} \subset \mathbb{N}$, *and* $\mathcal{I} \subset \mathcal{K}$. *It holds for the quantities* (9) *and*(10):

$$\underline{C}_{Ex} = C_{Ex}^*.$$

Proof. In case that that the restricted (resp. general) PAPR reduction problem is not solvable for $(\{\phi_n\}_{n \in \mathcal{K}}, \mathcal{I})$, it follows immediately, that the general (resp. restricted) PAPR reduction problem $(\{\phi_n\}_{n \in \mathcal{K}}, \mathcal{I})$ is not solvable. In both cases, we have $\underline{C}_{Ex} = C_{Ex}^* = \infty$. For the remaining, assume that the general/restricted (by Remark 5, both are equivalent) PAPR reduction problem is solvable for $(\{\phi_n\}_{n \in \mathcal{K}}, \mathcal{I})$ with an extension constant $C_{Ex} \geq 1$.

By straightforward computation involving the definitions of C_{Ex}^* and $\underline{C}_{Ex}$, it is clear that $C_{Ex}^* \geq \underline{C}_{Ex}$. To show the reverse inequality, let $\epsilon > 0$ be arbitrary. By the property of the infimum, there exists an extension operator $E_{\mathcal{I}}^{(\epsilon)}$, with:

$$\underline{C}_{Ex} + \epsilon \geq \left\| E_{\mathcal{I}}^{(\epsilon)} \right\|_{l^2(\mathcal{I}) \to L^\infty([0,1])} \geq \underline{C}_{Ex}.$$

Now take an arbitrary $\mathbf{a} \in l^2(\mathcal{I})$, with $\|\mathbf{a}\|_{l^2(\mathcal{I})} \neq 0$. Subsequently, define the sequence $\mathbf{a}_1 := \mathbf{a}/\|\mathbf{a}\|$. Clearly $\|\mathbf{a}_1\|_{l^2(\mathcal{I})} = 1$. Thus we have:

$$\left\| E_{\mathcal{I}}^{(\epsilon)} \mathbf{a}_1 \right\| \leq \left\| E_{\mathcal{I}}^{(\epsilon)} \right\|_{l^2(\mathcal{I}) \to L^\infty([0,1])} \leq \underline{C}_{Ex} + \epsilon. \tag{11}$$

We can represent $E_{\mathcal{I}}^{(\epsilon)} \mathbf{a}_1$ as follows:

$$E_{\mathcal{I}}^{(\epsilon)} \mathbf{a}_1 = \sum_{k \in \mathcal{I}} \frac{a_k}{\|\mathbf{a}\|_{l^2(\mathcal{I})}} \phi_k + \sum_{k \in \mathcal{I}^c} b_k^{(\epsilon)}(\mathbf{a}) \phi_k, \tag{12}$$

where $\mathbf{b}^{(\epsilon)} := \left\{ b^{(\epsilon)} \right\}_{n \in \mathbb{N}} \in l^2(\mathcal{I}^c)$. Combining (11) and (12), we have:

$$\left\| \sum_{k \in \mathcal{I}} a_k \phi_k + \sum_{k \in \mathcal{I}^c} b_k^{(\epsilon)}(\mathbf{a}) \, \|\mathbf{a}\|_{l^2(\mathcal{I})} \, \phi_k \right\|_{L^\infty([0,1])} \leq (\underline{C}_{\mathrm{Ex}} + \epsilon) \, \|\mathbf{a}\|_{l^2(\mathcal{I})} .$$

Of course, the above relation holds for all $\mathbf{a} \in l^2(\mathcal{I})$, with $\|\mathbf{a}\|_{l^2(\mathcal{I})} \neq 0$. Thus we have an operator $\tilde{E}_{\mathcal{I}}^\epsilon : l^2(\mathcal{I}) \to L^\infty([0,1])$ given by:

$$\mathbf{a} \mapsto \sum_{k \in \mathcal{I}} a_k \phi_k + \sum_{k \in \mathcal{I}^c} b_k^{(\epsilon)}(\mathbf{a}) \, \|\mathbf{a}\|_{l^2(\mathcal{I})} \, \phi_k,$$

for which it holds:

$$\left\| \tilde{E}_{\mathcal{I}}^\epsilon \right\|_{l^2(\mathcal{I}) \to L^\infty([0,1])} \leq \underline{C}_{\mathrm{Ex}} + \epsilon.$$

Taking infimum over the left hand side of the above inequality, it yields $C_{\mathrm{Ex}}^* \leq \underline{C}_{\mathrm{Ex}} + \epsilon$, and since $\epsilon > 0$ is arbitrary, we have $C_{\mathrm{Ex}}^* \leq \underline{C}_{\mathrm{Ex}}$ as desired. □

Thus, to analyze the behaviour of the PAPR reduction problem in the optimal case, it is unnecessary to give a distinction between the restricted version and the general version of the problem.

4.5 Necessary and sufficient condition for solvability of the PAPR reduction problem

We aim in this section to give a necessary condition for the solvability of the PAPR reduction problem. In case, that $\{\phi_n\}_{n \in \mathbb{N}}$ is an orthonormal basis for $L^2([0,1])$, the condition given later is even sufficient. For ease of notations, let us first define the following subspaces of $L^1([0,1])$:

DEFINITION 8. *For an $\mathcal{I} \subset \mathbb{N}$, and an ONS $\{\phi_n\}_{n \in \mathbb{N}}$, we define the following subspaces of $L^1([0,1])$:*

$$\mathfrak{f}^1(\mathcal{I}) := \left\{ f \in L^1([0,1]) : f = \sum_{k \in \mathcal{I}} a_k \phi_k, \text{ for a } \{a_k\}_{k \in \mathcal{I}} \text{ in } \mathbb{C} \right\}$$

$$\mathfrak{f}_c^1(\mathcal{I}) := \left\{ f \in L^1([0,1]) : f = \sum_{k \in \mathcal{I}} a_k \phi_k, \text{ where } a_k \neq 0, \text{ for finitely many } k \in \mathcal{I} \right\}$$

Since, we are possibly dealing with infinite index set $\mathcal{I}$, the expression $f = \sum_{k \in \mathcal{I}} a_k \phi_k$ in the definition of $\mathfrak{f}^1(\mathcal{I})$ has to be understood w.r.t. the norm structure of $\mathfrak{f}^1(\mathcal{I})$, viz. inherited from $L^1([0,1])$. Specifically, $f = \sum_{k \in \mathcal{I}} a_k \phi_k$ means that the finite partial sum of $\sum_{k \in \mathcal{I}} a_k \phi_k$ converges to f, w.r.t. $\|\cdot\|_{L^1([0,1])}$. Notice that subspaces $\mathfrak{f}^1(\mathcal{I})$ and $\mathfrak{f}^1(\mathcal{I})_c$ depend in particular on the choice of the ONS. We will not emphasize the choice of the ONS in the notation for both mentioned subspaces, since it will be clear from the context. The following characterization of those subspaces is elementary to show:

LEMMA 9. *For a $\mathcal{I} \subset \mathbb{N}$, the following statements holds:*

1. *$\mathfrak{f}^1(\mathcal{I})$ is a closed subspace of $L^1(\mathcal{I})$*
2. *$\mathfrak{f}^1(\mathcal{I})$ is the closure of $\mathfrak{f}_c^1(\mathcal{I})$.*

To show the first statement, one can simply take a sequence $\{f_n\}_{n\in\mathbb{N}}$ in $\mathfrak{f}^1(\mathcal{I})$, which converges to an $f \in L^1([0,1])$. By involving the fact that $\phi_n \in L^\infty([0,1])$, $n \in \mathbb{N}$, one can show by simple application of the Hölder's inequality, that for arbitrary $k \in \mathcal{I}^c$, $\int_0^1 f(t)\phi_k(t)dt = \lim_{n\to\infty} \int_0^1 f_n(t)\phi_k(t)dt$ (in $\mathbb{C}$). Finally, by noticing $\int_0^1 f_n(t)\phi_k(t)dt = 0$, $\forall n$, the statement is shown. The second statement can be shown, simply by approximating each $f \in \mathfrak{f}^1(\mathcal{I})$, which is an (infinite) linear combination of some members of $\{\phi_n\}$ by its finite sum.

Now, we are ready to give a necessary condition for the solvability of the PAPR reduction problem:

THEOREM 10 (Theorem 5 in [6]). *Let* $\{\phi_n\}_{n\in\mathbb{N}}$ *be an ONS in* $L^2([0,1])$. *Given a subset* $\mathcal{I} \subset \mathbb{N}$ *and a constant* $C_{Ex} > 0$. *Assume that the PAPR reduction problem is solvable for* $(\{\phi_n\}_{n\in\mathbb{N}}, \mathcal{I})$ *with extension constant* C_{Ex}. *Then:*

$$\|f\|_{L^2([0,1])} \le C_{Ex}\|f\|_{L^1([0,1])}, \quad \forall f \in \mathfrak{f}^1(\mathcal{I}) \tag{13}$$

Proof. For $f \in \mathfrak{f}_c^1(\mathcal{I})$, (13) was shown in [7]. It remains to show (13) holds generally for $\mathfrak{f}^1(\mathcal{I})$. Let be $f \in \mathfrak{f}^1(\mathcal{I})$ arbitrary. Lemma 9 asserts, that there exists a sequence $\{f_n\}_{n\in\mathbb{N}}$ in $\mathfrak{f}^1(\mathcal{I})$, which converges to f w.r.t. $\|\cdot\|_{L^1([0,1])}$. Now we claim that $\{f_n\}_{n\in\mathbb{N}}$ converges to f w.r.t. $\|\cdot\|_{L^2([0,1])}$. To show this claim, notice that since (13) holds for functions in $\mathfrak{f}_c^1(\mathcal{I})$, $\{f_n\}$ is a Cauchy sequence in $L^2([0,1])$. Thus by completeness of $L^2([0,1])$, there exists $g \in L^2([0,1])$, for which $\{f_n\}_{n\in\mathbb{N}}$ converges to g, w.r.t $\|\cdot\|_{L^2([0,1])}$. It is well known that the convergence of sequence of functions in L^p-spaces implies the convergence of a subsequence of those almost everywhere (a.e.). Thus, there exists subsequences $\{n_k\} \subset \mathbb{N}$, and $\{\tilde{n}_k\}$ of $\mathbb{N}$, for which:

$$\lim_{k\to\infty} f_{n_k}(t) = f(t) \text{ and } \lim_{k\to\infty} f_{\tilde{n}_k}(t) = g(t), \quad \text{a.e. } t \in [0,1],$$

which gives $f(t) = g(t)$, a.e. $t \in [0,1]$, accordingly $f = g$, which gives the claim. So by previous claim, sequential continuity of norm, and the fact that (13) holds for functions in $\mathfrak{f}_c^1(\mathcal{I})$, we have:

$$\|f\|_{L^2([0,1])} = \lim_{n\to\infty}\|f_n\|_{L^2([0,1])} \le C_{ex}\lim_{n\to\infty}\|f_n\|_{L^1([0,1])} = \|f\|_{L^1([0,1])}.$$

$\square$

Before we continue, let us first give the following remarks:

REMARK 11. *By Remark 5, we can infer that it is also adequate in the above theorem only to require that the restricted instead of the general PAPR reduction problem is solvable.*

REMARK 12. *Now, suppose that there exists a constant* $C_{Ex} > 0$, *s.t. the following holds:*

$$\|f\|_{L^2([0,1])} \le C_{Ex}\|f\|_{L^1([0,1])}, \quad \forall f \in \mathfrak{f}^1(\mathcal{I}), \|f\|_{L^1([0,1])} \le 1. \tag{14}$$

It is obvious that (13) *holds. Indeed, to see this, take an arbitrary* $f \in \mathfrak{f}^1(\mathcal{I})$, *with* $\|f\|_{L^1([0,1])} \ne 0$. *By setting the function* $f/\|f\|_{L^1([0,1])}$ *in* (14), *and subsequent elementary computations, the claim holds. Thus to show that the PAPR reduction problem is not solvable with an extension constant* $C_{Ex} > 0$, *it is sufficient to show the "restricted" norm equivalence* (14). *Further, that* (13) *implies* (14), *is trivial. Summarily, we can infer that the condition* (14) *is equivalent with* (13).

In case that $\{\phi_n\}_{n\in\mathbb{N}}$ forms additionally an orthonormal basis, we have also the converse of Theorem 10:

PROPOSITION 13 (Theorem 5 in [6]). *Let $\{\phi_n\}_{n\in\mathbb{N}}$ be an orthonormal basis for $L^2([0,1])$, and let be $\mathcal{I}\subset\mathbb{N}$, and $C_{Ex} > 0$. If the following condition is fulfilled:*

$$\|f\|_{L^2([0,1])} \le C_{Ex}\|f\|_{L^1([0,1])}, \quad \forall f \in \mathfrak{f}^1(\mathcal{I}), \tag{15}$$

then the PAPR reduction problem is solvable for $(\{\phi_n\}_{n\in\mathbb{N}}, \mathcal{I})$ with extension constant C_{Ex}.

Proof. Let be $f \in \mathfrak{f}^1(\mathcal{I})$ arbitrary, having the representation $f = \sum_{k\in\mathcal{I}} c_k\phi_k$ w.r.t. $\|\cdot\|_{L^1([0,1])}$, i.e. $\sum_{k\in\mathcal{I}} c_k\phi_k$ converges to f w.r.t. $\|\cdot\|_{L^1([0,1])}$, for a sequence $\mathbf{c}$ in $\mathbb{C}$. Since (15) holds by assumption, $f = \sum_{k\in\mathcal{I}} c_k\phi_k$ holds also w.r.t. $\|\cdot\|_{L^2([0,1])}$. Now, take an arbitrary $\mathbf{a} \in l^2(\mathcal{I})$ and define the mapping $\Psi_{\mathbf{a}} : \mathfrak{f}^1(\mathcal{I}) \to \mathbb{C}$ by $\Psi_{\mathbf{a}}f := \sum_{k\in\mathcal{I}} c_k\overline{a_k}$. Linearity of $\Psi_{\mathbf{a}}$ is obvious. Further $\Psi_{\mathbf{a}}$ is bounded, since:

$$|\Psi_{\mathbf{a}}f| \le \|\mathbf{a}\|_{l^2(\mathcal{I})}\|\mathbf{c}\|_{l^2(\mathcal{I})} = \|\mathbf{a}\|_{l^2(\mathcal{I})}\|f\|_{L^2([0,1])} \le \|\mathbf{a}\|_{l^2(\mathcal{I})} C_{Ex}\|f\|_{L^1([0,1])} < \infty, \tag{16}$$

where the equality follows from the fact that $f = \sum_{k\in\mathcal{I}} c_k\phi_k$ w.r.t. $\|\cdot\|_{L^2([0,1])}$, and the assumption that $\{\phi_n\}_{n\in\mathbb{N}}$ is orthonormal, and the third inequality from (15). $\mathfrak{f}^1(\mathcal{I})$ is clearly a subspace of $L^1([0,1])$, and as we have already seen, $\Psi_{\mathbf{a}}$ is linear and bounded. Thus the Hahn–Banach theorem asserts the existence of a linear and bounded mapping $\tilde{\Psi} : L^1([0,1]) \to \mathbb{C}$, for which:

$$\tilde{\Psi}\tilde{f} = \Psi_{\mathbf{a}}\tilde{f}, \quad \forall f \in \mathfrak{f}^1(\mathcal{I}), \quad \text{and} \quad \left\|\tilde{\Psi}\right\|_{L^1([0,1])\to\mathbb{C}} = \|\Psi_{\mathbf{a}}\|_{\mathfrak{f}^1(\mathcal{I})\to\mathbb{C}}, \tag{17}$$

holds.

Further, since the dual space of $L^1([0,1])$ is $L^\infty([0,1])$, we can find a unique $g \in L^\infty([0,1])$, for which the following holds:

$$\tilde{\Psi}f = \int_0^1 f(t)\overline{g(t)}dt, \quad \forall f \in L^1([0,1]), \text{ and } \|g\|_{L^\infty([0,1])} = \left\|\tilde{\Psi}\right\|_{L^\infty([0,1])\to\mathbb{C}}. \tag{18}$$

As $L^\infty([0,1]) \subset L^2([0,1])$, it follows that g can be represented by means of the series $g = \sum_{k=1}^\infty d_k\phi_k$, for a sequence $\mathbf{d}$ in $l^2(\mathbb{N})$. By the first statement in (17), and the orthonormality of $\{\phi_n\}_{n\in\mathbb{N}}$, one can imply that $a_k = d_k$, for every $k \in \mathcal{I}$. Define a sequence $\mathbf{b} \in l^2(\mathcal{I}^c)$, by setting $b_k = d_k, \forall k \in \mathcal{I}^c$. Thus we have:

$$\left\|\sum_{k\in\mathcal{I}} a_k\phi_k + \sum_{k\in\mathcal{I}^c} b_k\phi_k\right\|_{L^\infty([0,1])} = \|g\|_{L^\infty([0,1])} = \left\|\tilde{\Psi}\right\|_{L^\infty([0,1])\to\mathbb{C}} = \|\Psi_{\mathbf{a}}\|_{L^\infty([0,1])\to\mathbb{C}}$$

$$\le C_{Ex}\|\mathbf{a}\|_{l^2(\mathcal{I})},$$

where the 2. equality follows from (16), the 3. from (17), and the inequality from (16), as desired. $\square$

REMARK 14. *By Remark 12, the condition (15) in the above theorem can clearly be softened by the condition (14).*

4.6 Further discussions on the optimal extension constant and the solvability of the PAPR reduction problem in the optimum

Proposition 13 sheds light on the discussions made in Section 4.4, about the optimal constant $\underline{C}_{\text{Ex}}$, which gives the lower bound of the extension constant, for which both the restricted – and the general PAPR reduction problem is solvable: Assume that $\{\phi_n\}_{n\in\mathbb{N}}$ forms an ONB for $L^2([0,1])$. Notice that $\underline{C}_{\text{Ex}}$ is basically the operator norm of the embedding Emb : $\mathfrak{F}^1(\mathcal{I}) \to L^2([0,1])$, $f \mapsto f$. Formally, we have:

$$\underline{C}_{\text{Ex}} := \inf\left\{c > 0: \|\text{Emb}f\|_{L^2([0,1])} \le c\,\|f\|_{L^1([0,1])}\,,\ \forall f \in \mathfrak{F}^1(\mathcal{I})\right\}.$$

In particular, if $\mathcal{I}$ is finite, $\underline{C}_{\text{Ex}}$ is always finite. By Theorem 10, the former case also occurs, if $\mathcal{I}$ is infinite, and PAPR reduction problem is solvable for $(\{\phi_n\},\mathcal{I})$ for some constant C_{Ex}.

In case that the considered ONS is complete, we can even ensure the solvability of the PAPR reduction problem in the optimal case:

PROPOSITION 15. *Let* $\{\phi_n\}_{n\in\mathbb{N}}$ *be an ONB. Assume that the PAPR reduction problem is solvable for* $(\{\phi_n\}_{n\in\mathbb{N}},\mathcal{I})$ *with a certain extension constant* $C_{\text{Ex}} > 0$, *then it is also solvable with the optimal extension constant* $\underline{C}_{\text{Ex}}$.

Proof. Let $\epsilon > 0$ be arbitrary. Then by the solvability assumption, and the definition of the infimum, there exists an extension operator $\text{E}_{\mathcal{I}}^{(\epsilon)}$, for which:

$$\left\|\text{E}_{\mathcal{I}}^{(\epsilon)}\right\|_{l^2(\mathcal{I})\to L^\infty([0,1])} \le \underline{C}_{\text{Ex}} + \epsilon.$$

Consequently, by Theorem 10, it follows that:

$$\|f\|_{L^2([0,1])} \le (\underline{C}_{\text{Ex}} + \epsilon)\,\|f\|_{L^1([0,1])}\,,\quad \forall \mathfrak{F}^1(\mathcal{I}).$$

The left hand side of the above inequality does not depend on ϵ. Thus it follows that $\|f\|_{L^2([0,1])} \le \underline{C}_{\text{Ex}}\|f\|_{L^1([0,1])}$. Finally, Proposition 13 asserts the solvability of PAPR reduction problem for $(\{\phi_n\}_{n\in\mathbb{N}},\mathcal{I})$ with $\underline{C}_{\text{Ex}}$, as desired. $\square$

4.7 On a weaker formulation of the PAPR reduction problem

An interesting and weaker formulation of the PAPR reduction problem can be given as follows:

Problem 16. Given $\mathcal{I} \subset \mathbb{N}$, and an ONS $\{\phi_n\}_{n\in\mathbb{N}}$. Let $\mathbf{a} \in l^2(\mathcal{I})$ be fixed, but arbitrary. Does there exists an $\mathbf{b} \in l^2(\mathcal{I}^c)$, for which

$$\left\|\sum_{k\in\mathcal{I}} a_k\phi_k + \sum_{k\in\mathcal{I}^c} b_k\phi_k\right\|_{L^\infty([0,1])} < \infty, \tag{19}$$

holds?

Notice that in the above formulation, we merely require the solvability of the PAPR reduction problem only for some coefficients of interests, rather than the solvability of the PAPR reduction problem for all sequences in $l^2(\mathcal{I})$, and the corresponding "uniform" control. With the weak formulation of the PAPR reduction problem, we identify the following optimal constant:

$$C_{\text{opt}}(\mathbf{a}) := \inf_{\mathbf{b}\in l^2(\mathcal{I}^c)}\left\|\sum_{k\in\mathcal{I}} a_k\phi_k + \sum_{k\in\mathcal{I}^c} b_k\phi_k\right\|_{L^\infty([0,1])}.$$

Notice that C_{opt} can be seen as a functional $C_{opt} : l^2(\mathcal{I}) \to \overline{\mathbb{R}}$, where $\overline{\mathbb{R}} = \mathbb{R} \cup \{-\infty, \infty\}$ denotes the extended real line. It is obvious that C_{opt} is non-negative. Thus C_{opt} can not take the value $-\infty$. In case that $C_{opt}(\mathbf{a}) = \infty$, for an $\mathbf{a} \in l^2(\mathcal{I})$ means, that one can not find $\mathbf{b} \in l^2(\mathcal{I}^c)$, for which (19) holds. Further, C_{opt} depends on the choice of information set $\mathcal{I}$, and the choice of orthonormal system $\{\phi_n\}_{n \in \mathbb{N}}$. The functional C_{opt} possesses some nice properties, which can be shown straightforwardly:

PROPOSITION 17. *Let $\mathcal{I} \subset \mathbb{N}$, and $\{\phi_n\}_{n \in \mathbb{N}}$ be an ONS. The corresponding functional C_{opt} is convex, in the sense that:*

1. $C_{opt}(\mathbf{a}) \geq 0$
2. *For $\lambda \in (-1, 1)$, $\lambda \neq 0$, and $\mathbf{a} \in l^2(\mathcal{I})$, it holds: $C_{opt}(\lambda \mathbf{a}) = |\lambda| C_{opt}(\mathbf{a})$.*
3. *For $\mathbf{a}^{(1)}, \mathbf{a}^{(2)} \in l^2(\mathcal{I})$. It holds: $C_{opt}(\mathbf{a}^{(1)} + \mathbf{a}^{(2)}) \leq C_{opt}(\mathbf{a}^{(1)}) + C_{opt}(\mathbf{a}^{(2)})$.*

An additional property for the functional C_{opt}, which we desire to have is that it is lower semi-continuous, in the sense that:

DEFINITION 18 (Lower Semi-continuity). *Given a normed space X, and a functional $p : X \to \overline{\mathbb{R}}$. Then p is said to be lower semi-continuous at the point $x \in X$, if $p(x_0) = -\infty$, or for each $h \in \mathbb{R}$, with $p(x_0) > h$, there exists $\delta > 0$, such that:*

$$p(x) > h, \quad \forall x \in \mathcal{B}_\delta(x_0),$$

where $\mathcal{B}_\delta(x_0)$ denotes the open ball around x_0, with radius δ, formally $\mathcal{B}_\delta(x_0) := \{y \in X : \|x_0 - y\|_X < \delta\}$. In case that p is lower semi-continuous at every point $x \in X$, then we say p is lower semi-continuous on X

Before we continue, let us first introduce the following notions: Let $\mathcal{B}$ be a Banach space, a set $\mathcal{M} \subseteq \mathcal{B}$ is said to be *nowhere dense* if int $\overline{\mathcal{M}} = \emptyset$, i.e., if the inner of the closure of $\mathcal{M}$ is empty. A set $\mathcal{M} \subseteq \mathcal{B}$ is said to be *of 1. category*, if it can be represented as a countable union of nowhere dense sets. In case that a set is of 1. category, then it is said to be *of 2. category*. The complement of a set of 1. category is defined as a *residual set*. Topologically, sets of 1. category can be seen as a small set, in the sense that they are negligible if compared to the whole space, sets of 2. category as sets, which are not small, and residual sets, each as a complementary set of a set of 1. category, can be seen as a large set. The Baire category theorem ensures that this categorization of sets of a Banach spaces is non-trivial, by stating that the whole Banach space $\mathcal{B}$ is not "small" in this sense, or can even not be "approximated" by such sets, i.e. it can not be written as a countable union of sets of 1. category, and that the residual sets are dense in $\mathcal{B}$, and closed under countable intersection. A property that holds for a residual subset of $\mathcal{B}$ is called a generic property. A generic property might not hold for all elements of $\mathcal{B}$, but for "typical" elements of $\mathcal{B}$. For more detailed treatment of the Baire category theorem we refer to standard textbooks such as [21, 24, 25].

As an application of Gelfand's theorem (see, e.g., Theorem 4 (1. VII) in [16]), we have the following characterization of the functional C_{opt}:

LEMMA 19. *Let $\{\phi_n\}_{n \in \mathbb{N}}$ be an ONS, and $\mathcal{I} \subset \mathbb{N}$. Further assume that the following holds:*

1. C_{opt} *is lower semi-continuous on $l^2(\mathcal{I})$*
2. *There exists a set of 2. category $\mathcal{M} \subset l^2(\mathcal{I})$, s.t. $C_{opt}(\mathbf{a}) < \infty$.*

Then there exists a constant, for which the following holds:

$$C_{opt}(\mathbf{a}) \leq C \|\mathbf{a}\|_{l^2(\mathcal{I})}, \quad \forall \mathbf{a} \in l^2(\mathcal{I}). \tag{20}$$

Clearly, the above lemma implies immediately the following statement, which gives a connection between C_{opt} and the optimal extension constant:

PROPOSITION 20. *Let be $\mathcal{I} \subset \mathbb{N}$, and $\{\phi_n\}_{n\in\mathbb{N}}$ an ONS in $L^2([0,1])$. If C_{opt} is lower-semi continuous and is finite on a set of 2. category in $l^2(\mathcal{I})$, then the infimum of the constant C, for which (20) hold is exactly $\underline{C}_{Ex}$, formally:*

$$\infty > \underbrace{\inf\left\{C > 0 : C_{opt}(\mathbf{a}) \leq C \|\mathbf{a}\|_{l^2(\mathcal{I})}, \forall \mathbf{a} \in l^2(\mathcal{I})\right\}}_{=:C_{opt}^*} = \underline{C}_{Ex}.$$

Proof. Since C_{opt} is lower semi-continuous, and finite on a set of second category, by Lemma 19, we can find a constant $C > 0$, for which it holds:

$$\inf_{\mathbf{b}\in l^2(\mathcal{I}^c)} \left\| \sum_{k\in\mathcal{I}} a_k\phi_k + \sum_{k\in\mathcal{I}^c} b_k\phi_k \right\|_{L^\infty([0,1])} = C_{opt}(\mathbf{a}) \leq C \|\mathbf{a}\|_{l^2(\mathcal{I})}, \quad \forall \mathbf{a} \in l^2(\mathcal{I}),$$

which shows the finiteness of C_{opt}^*. For an arbitrary $\epsilon > 0$, we find an $\mathbf{b}^{(\epsilon)} \in l^2(\mathcal{I})$, for which it holds:

$$\frac{\left\| \sum_{k\in\mathcal{I}} a_k\phi_k + \sum_{k\in\mathcal{I}^c} b_k^{(\epsilon)}\phi_k \right\|_{L^\infty([0,1])}}{\|\mathbf{a}\|_{l^2(\mathcal{I})}} \leq C + \epsilon, \quad \forall \mathbf{a} \in l^2(\mathcal{I}), \|\mathbf{a}\|_{l^2(\mathcal{I})} \neq 0,$$

which gives the observation, that there exists an extension operator $E_{\mathcal{I}}^{(\epsilon)}$, for which $\left\|E_{\mathcal{I}}^{(\epsilon)}\right\|_{l^2(\mathcal{I})\to L^\infty([0,1])} \leq C + \epsilon$, and the correspondingly the solvability of the PAPR reduction problem for $(\{\phi_n\}_{n\in\mathbb{N}}, \mathcal{I})$. Taking the infimum of $\|\cdot\|_{l^2(\mathcal{I})\to L^\infty([0,1])}$ over all extension operators, and subsequently noticing that $\epsilon > 0$ can be chosen arbitrarily (the infimum on the L.H.S. does not depend on ϵ!), we have $\underline{C}_{Ex} \leq C$. By taking the corresponding infimum over all constant C, we have as desired $\underline{C}_{Ex} \leq C_{opt}^*$, as desired.

To show the reverse inequality, notice that by argumentations made in the beginning of the proof, we have that $\underline{C}_{Ex}$ is finite. Thus for each $\epsilon > 0$, we find an extension operator $E_{\mathcal{I}}^{(\epsilon)}$, such that $\left\|E_{\mathcal{I}}^{(\epsilon)}\right\|_{l^2(\mathcal{I})\to L^\infty([0,1])} \leq \underline{C}_{Ex} + \epsilon$. The former asserts, that for a fixed $\mathbf{a} \in l^2(\mathcal{I})$, $\|\mathbf{a}\|_{l^2(\mathcal{I})} \neq 0$, we can find $\mathbf{b}^{(\epsilon)} \in l^2(\mathcal{I}^c)$, for which:

$$\frac{\left\| \sum_{k\in\mathcal{I}} a_k\phi_k + \sum_{k\in\mathcal{I}^c} b_k^{(\epsilon)}\phi_k \right\|_{L^\infty([0,1])}}{\|\mathbf{a}\|_{l^2(\mathcal{I})}} \leq \underline{C}_{Ex} + \epsilon.$$

Thus taking the infimum on the left hand side over all $\mathbf{b} \in l^2(\mathcal{I})$, and since $\epsilon > 0$ was arbitrarily chosen, and now the left hand side does not depend on ϵ, we have:

$$C_{opt}(\mathbf{a}) \leq \underline{C}_{Ex} \|\mathbf{a}\|_{l^2(\mathcal{I})},$$

which shows that $\underline{C}_{Ex} \geq C_{opt}^*$ (since $\mathbf{a} \in l^2(\mathcal{I})$ was arbitrarily chosen). $\qquad\square$

In case that the functional C_{opt} is lower semi-continuous, the finiteness of C_{opt} in a subset of $l^2(\mathcal{I})$, which is not too small (in particular, it is sufficient to have finiteness of C_{opt} on a ball in $l^2(\mathcal{I})$ with arbitrary small radius), implies already the solvability of Problem 16 for every sequences in $l^2(\mathcal{I})$

THEOREM 21. *Let be $\mathcal{I} \subset \mathbb{N}$, and $\{\phi_n\}_{n\in\mathbb{N}}$ an ONS in $L^2([0,1])$. Assume that C_{opt} is lower-semi continuous. If there exists a set of 2. category $\mathcal{M}$ in $l^2(\mathcal{I})$, for which $C_{opt}(\mathbf{a}) < \infty$, $\forall \mathbf{a} \in \mathcal{M}$, then the problem 16 is solvable for all $\mathbf{a} \in l^2(\mathcal{I})$.*

Proof. Since C_{opt} is lower semi-continuous, and finite on a set of second category, Proposition 20 asserts that $\underline{C}_{Ex}$ is finite. Thus, Proposition 15 asserts that the PAPR reduction problem is solvable for $(\{\phi_n\}_{n\in\mathbb{N}}, \mathcal{I})$, with $\underline{C}_{Ex}$, and a fortiori Problem 16 for every $\mathbf{a} \in l^2(\mathcal{I})$. □

As an argumentum e contrario of the above theorem, and by noticing that sets, which are not of 2. category, is of 1. category, we have the following statement:

COROLLARY 22. *Let be $\mathcal{I} \subset \mathbb{N}$, and $\{\phi_n\}_{n\in\mathbb{N}}$ an ONS in $L^2([0,1])$. Assume that C_{opt} is lower-semi continuous. If the problem 16 is not solvable for an $\mathbf{a} \in l^2(\mathcal{I})$, then the set $\mathcal{M}$, for which $C_{opt}(\mathbf{a}) < \infty$, $\forall \mathbf{a} \in \mathcal{M}$, is at most a set of 1. category in $l^2(\mathcal{I})$.*

The above corollary gives in some sense a strong statement: The inability of compensating the peak value of only a single waveform formed by a sequence $\mathbf{a} \in l^2(\mathcal{I})$ implies immediately the inability of compensating of "typical" waveforms forms by sequences in $l^2(\mathcal{I})$. The latter is an implication of the fact that in this case C_{opt} is finite only for sets of 1. category, thence it is infinite for residual sets.

5 Necessary condition for solvability of PAPR reduction problem for OFDM

Now we aim to analyze the PAPR reduction problem for OFDM systems. As already mentioned in the introduction, an OFDM transmission consists of superposition of sines weighted by information coefficients. The sines have the form:

$$e_n := e^{2\pi i(n-1)(\cdot)}, \quad n \in \mathbb{N}.$$

Clearly, $\{e_n\}_{n\in\mathbb{N}}$ forms an ONS in $L^2[0,1]$. Furthermore, it is well-known, that $\{e_n\}_{n\in\mathbb{N}}$ even forms an orthonormal basis for $L^2([0,1])$. Surprisingly, the PAPR reduction problem is connected to a deep result in mathematics, the so-called Szemerédi theorem [29], concerning to a certain subset of integers with additive structure, namely an arithmetic progression. The following subsection is devoted to that issue.

5.1 Deterministic approach to the PAPR problem

A problem in mathematics which has been raised interests in the last decades is the problem of finding or determining a so-called arithmetic progressions of a certain length in a given subset A of natural numbers. Let us first discuss about that object in the following:

Szemerédi theorem on arithmetic progressions.

DEFINITION 23 (Arithmetic progression). *Let be $m \in \mathbb{N}$. An arithmetic progression of length m is defined as a subset of $\mathbb{Z}$, which has the form:*

$$\{a, a + d, a + 2d, \ldots, a + (m - 1)d\},$$

for some integer a and some positive integer d.

For sum sets, i.e. sets with specific structures such as $A + A$, $A + A + A$, or $2A - 2A$, for an $A \subset \mathbb{N}$, there are some results concerning to the existence of arithmetic progressions within those sets. However, they require some insights into the structure of the subset A. For some detailed discussions concerning to this aspect, we refer to the excellent textbook [30].

We are mostly interested in the following subset:

DEFINITION 24 ((δ, m)-Szemerédi set). *Let $\mathcal{I}$ be a set of integers, $\delta \in (0,1)$, and $m \in \mathbb{N}$. The set $\mathcal{I}$ is said to be (δ, m)-Szemerédi, if every subset of $\mathcal{I}$ of cardinality at least $\delta |\mathcal{I}|$ contains an arithmetic progression of length m.*

The celebrated Szemerédi Theorem [29] gives a connection on the size of the set of consecutive numbers $[N]$, s.t. every subset $\mathcal{I}$ with a certain density relative to $[N]$, viz. $|\mathcal{I}|/N$, contains an arithmetic progression of a given length:

THEOREM 25 (Szemérédi Theorem [29]). *For any $m \in \mathbb{N}$, and any $\delta \in (0,1)$, there exists $N_{Sz} \in \mathbb{N}$, which depends on m and δ, s.t. for all $N \geq N_{Sz}$, $[N]$ is (δ, m)-Szemerédi.*

The cases $m = 1, 2$ are merely trivial. The case $m = 3$ was already proven earlier by Roth [26], for which he was awarded the Fields Medal in 1958. Szemerédi proved the result firstly for $m = 4$ in 1969, and recent result [28] finally in 1975. Finding the correct constant N_{Sz} is quiet challenging. Gowers showed that $N_{Sz}(\delta, m) \leq 2^{2^{\delta - c_m}}$, where $c_m = 2^{2^{m+9}}$. A lower bound of N_{Sz} is due to Rankin [23]. He has proven, that it holds $N_{Sz}(\delta, m) \geq \exp(C(\log(1/\delta))^{1 + \lfloor \log_2(m-1) \rfloor})$, for some constant $C > 0$. A better lower bound might be derived from [20].

For the asymptotic case, Szemérédi theorem is somehow unsatisfactory. It merely ensures the existence of arithmetic progressions of arbitrary length for subsets of $\mathbb{N}$ with positive upper density. Specifically, an equivalent statement of the Szemérédi theorem can be given as follows: Given a subset $\mathcal{A} \subset \mathbb{N}$, whose upper density is positive, i.e., $\limsup_{N \to \infty}(|\mathcal{A} \cap [N]|/N) > 0$. Then, there exists an arithmetic progression of length k, where k is an arbitrary natural number. A tightening of this statement is due to Green and Tao [11]. They showed, that the set of prime numbers $\mathcal{P}$ contains arithmetic progressions of arbitrary length. It is well known that the density of prime numbers in $[N]$, i.e. the quantity $|\mathcal{P} \cap [N]|/N$, $N \in \mathbb{N}$, is asymptotically $1/\log(N)$. Thus the density of the prime numbers in $\mathbb{N}$ is 0. A more general statement than the previous one was already conjectured by Erdős (see Conjecture 45), which still remains unsettled. We shall later give a discussion in Section 6.3, and show that this conjecture holds true for Walsh case.

Recently, it was shown by Conlon and Gowers [8], that for arbitrary $\delta \in (0,1)$ and $m \in \mathbb{N}$, one can asymptotically give a "sparse" (δ, m)-Szemerédi, by choosing randomly the elements from $[N]$ by some arbitrary small probability p (Call the corresponding set $[N]_p$):

THEOREM 26 (Conlon and Gowers [8]). *Given $\delta > 0$, and a natural number $m \in \mathbb{N}$. There exists a constant $C > 0$, s.t.:*

$$\lim_{N \to \infty} \mathbb{P}([N]_p \text{ is } (\delta, m)\text{-Szemerédi}) = 1, \quad \text{if } p > CN^{\frac{-1}{(m-1)}}.$$

Notice that the above theorem ensures the existence of a sequence $\{p_N\}$ in $(0, 1)$, tending to zero, for which:

$$\lim_{N \to \infty} \mathbb{P}([N]_{p_N} \text{ is } (\delta, m)\text{-Szemerédi}) = 1,$$

which justifies in particular the notion, that such a (δ, m)-Szemerédi is asymptotically "sparse" in $\mathbb{N}$, or specifically: has a density 0 a.s. in $\mathbb{N}$, i.e. $\left| [N]_p \right| / N = 0$, as $N \to \infty$.

A necessary condition for solvability of the PAPR reduction problem and arithmetic progressions. The existence of an arithmetic progression in a subset $\mathcal{I} \subset \mathbb{N}$ allows us to give a more specific necessary condition for the solvability of the PAPR reduction problem for OFDM systems than that given in Theorem 10:

LEMMA 27. *Let be $\mathcal{I} \subset \mathbb{N}$. Assume that there exists an arithmetic progression of length m in $\mathcal{I}$. Then, if the PAPR reduction problem is solvable for $(\{e_n\}_{n \in \mathbb{N}}, \mathcal{I})$ with a given $C_{Ex} > 0$, it follows:*

$$C_{Ex} > \frac{\sqrt{m}}{\frac{4}{\pi^2} \log\left(\frac{m}{2}\right) + C}, \tag{21}$$

for a fixed constant $C > 0$.

Proof. Consider the signal $f = \sum_{k=0}^{m-1} \frac{1}{\sqrt{m}} e_{a+dk}$. It is obvious, that $f \in \mathfrak{f}^1(\mathcal{I})$. Further, we have the following observation:

$$\|f\|_{L^2([0,1])} = 1, \quad \|f\|_{L^1([0,1])} < \frac{\frac{4}{\pi^2} \log\left(\frac{m}{2}\right) + C}{\sqrt{m}},$$

for some absolute constant $C > 0$. The equality above follows from the orthonormality of $\{e_n\}_{n \in \mathbb{N}}$, and the inequality follows from usual upper bound for Dirichlet kernel, respectively. Finally, by the assumption that PAPR reduction problem is solvable for $(\{e_n\}_{n \in \mathbb{N}}, \mathcal{I})$ with constant C_{Ex}, Theorem 10, and the fact $f \in \mathfrak{f}^1(\mathcal{I})$, we have:

$$1 = \|f\|_{L^2([0,1])} \le C_{Ex} \|f\|_{L^1([0,1])} < C_{Ex} \frac{\frac{4}{\pi^2} \log\left(\frac{m}{2}\right) + C}{\sqrt{m}},$$

as desired. □

In particular, the above lemma gives an insight into the structure of PAPR reduction problem: It asserts, that a necessary condition for the solvability of the PAPR reduction problem for $(\{e_n\}_{n \in \mathbb{N}}, \mathcal{I})$, with a certain constant C_{Ex}, is that $\mathcal{I}$ does not contain an arithmetic progression of arbitrary large length m, otherwise, the right hand side of the inequality (21) would dominate C_{Ex}. Further, to ensure that the above statement makes sense, we need to ensure the existence of an arithmetic progression of a given length. Szemerédi theorem gives the remaining arguments.

THEOREM 28. *Given $\delta \in (0, 1)$ and $m \in \mathbb{N}$, then there exists an $N_{Sz} \in \mathbb{N}$, depending on δ and m, s.t. for all $N \geq N_{Sz}$, the following holds:*

If the PAPR reduction problem is solvable for $(\{e_n\}_{n \in \mathbb{N}}, \mathcal{I})$ with $C_{Ex} > 0$, where $\mathcal{I} \subset [N]$, with $|\mathcal{I}| \geq \delta N$, then:

$$C_{Ex} > \frac{\sqrt{m}}{\frac{4}{\pi^2} \log\left(\frac{m}{2}\right) + C}, \tag{22}$$

for some $C > 0$.

Proof. By Theorem 25, we can fix the constant N_{Sz} depending on the choice of m and δ. Thus, in a subset $\mathcal{I} \subset [N]$, where $N \geq N_{Sz}$, for which $|\mathcal{I}| \geq \delta N$, we can find an arithmetic progression of length m. Correspondingly, Lem. 27 gives the remaining of the argument.   □

Given a desired $C_{Ex} > 0$. One may conclude from the above Theorem, that there is a restriction to the size of the information set such that the PAPR reduction problem is solvable. Notice that the statement given in the Theorem is somehow stronger: it gives a necessary condition for solvability of the PAPR reduction problem not only for a certain information set, but for all information set having density bigger than $\delta \in (0, 1)$ in $[N]$.

5.2 Probabilistic and asymptotic approach to the PAPR problem of the OFDM

Recently, there are some interests aroused in the tightened and probabilistic version of Theorem 28. We have already mentioned in the previous subsection that Szemerédi theorem is unsatisfactory in the asymptotic case, since it ensures the existence of an arithmetic progressions of arbitrary length only in sets of positive upper density in $\mathbb{N}$. Furthermore, Green and Tao [11] showed the existence of a set with density zero in $\mathbb{N}$, viz. the prime numbers, in which there are arithmetic progressions of arbitrary length.

As asserted by Conlon and Gowers [8], a set of density zero having arithmetic progressions of arbitrary length, can be constructed in a probabilistic way. By means of that result, we are able to give a negative statement about the solvability of the PAPR reduction problem with arbitrary extension constant in the asymptotic setting:

THEOREM 29. *Let be $m \in \mathbb{N}$, and $\delta \in (0, 1)$. Given a constant $C_{Ex} > 0$. Then, there is a constant C, s.t.:*

$$\lim_{N \to \infty} \mathbb{P}\left(A_{N,m,p}\right) = 1, \quad if \, p > \frac{C}{N^{\frac{1}{m-1}}},$$

where $A_{N,m,p}$ denotes the event: "The PAPR problem is not solvable for $(\{e_n\}_{n \in \mathbb{N}}, \mathcal{I})$ with

$$C_{Ex} \leq \frac{\sqrt{m}}{\frac{4}{\pi^2} \log\left(\frac{m}{2}\right) + C}, \tag{23}$$

where $C > 0$ is an absolute constant, for every subset $\mathcal{I} \subset [N]_p$ of size $|\mathcal{I}| \geq \delta N$."

Proof. Choose m sufficiently large, s.t. (22) does not hold. Further, choose $p \in (0, 1)$, s.t. $p > CN^{-\frac{1}{m-1}}$, with a suitable constant $C > 0$. Theorem 26 asserts that the set $[N]_p$ resulted by choosing elements of $[N]$ independently by probability p, is a (δ, m)-Szemerédi with probability tends to 1 as N tends to infinity. By the definition of (δ, m)-Szemerédi, the choice of m, and Lemma 27, the result follows immediately.   □

The point of the above theorem is that a set, for which the PAPR problem is not solvable for every subset having a relative density bigger than a given number, might be a large set, but still a sparse set in $\mathbb{N}$, since the probability p can be decreased toward 0 as N increases.

5.3 Further discussion and outlook

We have already seen that the PAPR reduction problem for $(\{e_n\}_{n\in\mathbb{N}}, \mathcal{I})$, for a fixed information set $\mathcal{I} \subset \mathbb{N}$, is not solvable with arbitrary (small) extension constant C_{Ex}, although infinite compensation set is allowed. This gives of course a restriction to the solvability of the PAPR reduction problem for fixed information set, finite compensation set, and with a given extension constant.

Also an interesting question is, how does the PAPR reduction problem for OFDM systems behaves with fixed threshold constant C_{Ex}, if the information set and the compensation set is finite. An answer was given in [6]. To discuss this, let us define the following quantity, which depends on $N \in \mathbb{N}$ and $C_{\text{Ex}} > 0$:

$$\mathcal{E}_N(C_{\text{Ex}}) := \max\left\{|\mathcal{I}| : \mathcal{I} \subset [N], \text{ PAPR prob. is solv. for } (\{e_n\}_{n=1}^N, \mathcal{I}) \text{ with } C_{\text{Ex}}\right\}.$$

Notice that in the above definition, we allow only finite compensation set. One result (Theorem 2) given in [6] is that the following limit holds:

$$\lim_{N\to\infty} \frac{\mathcal{E}_N(C_{\text{Ex}})}{N} = 0, \tag{24}$$

which says that if a given PAPR bound C_{Ex} is always satisfied, and if we allow only finite compensation set, then the proportion between the possible information set, for which the PAPR reduction problem is solvable, and the number of available tones, goes toward zero, as the latter goes toward infinity. Thus the size of that possible information set does not scale with N. For practical insight, (24) should be not strictly as an asymptotic statement. Rather, (24) has to be understand as restrictions on the existence of any arbitrarily large OFDM system of number N, for which solvability occurs for a certain information set $\mathcal{I}_N \subset [N]$ with density $|\mathcal{I}_N|/N$ in N, and for a given $C_{\text{Ex}} > 0$.

In case that the information set $\mathcal{I}$ is infinite, and the compensation set is the remaining $\mathbb{N} \setminus \mathcal{I}$, and the extension constant C_{Ex} is fixed, one can give in some sense a quantitative statement: If $\limsup_{N\to\infty}(|S(N)|/N) > 0$, where $S(N) := \mathcal{I} \cap [N]$, then the PAPR Problem is not solvable for $(\{e_n\}_{n\in\mathbb{N}}, \mathcal{I})$. A corresponding proof can be found in [6] (Theorem 6).

6 Solvability of PAPR problem for CDMA

We have already seen in the previous section (Lemma 27), that the solvability of the PAPR reduction problem in the OFDM case for an information set $\mathcal{I}$ is connected to the existence of a certain combinatorial object, viz. arithmetic progression of a certain length in $\mathcal{I}$. For DS-CDMA systems, whose carriers are the so-called Walsh functions, we shall see, that the derivation of an easy-to-handle necessary condition for the solvability of the PAPR reduction problem, based on a slightly different technique. In particular, it does not depend on the existence of an arithmetic progression in the considered information set $\mathcal{I}$, rather on the so called existence of the optimal Walsh sum of a given length in $\mathfrak{f}^1(\mathcal{I})$ (or with abuse of notations: the existence of the optimal Walsh sum of a given length in the information set $\mathcal{I}$). Before we go into detail, let us first define the Walsh functions, which serve as the carriers for CDMA systems.

The Rademacher functions r_n, $n \in \mathbb{N}$, on $[0, 1]$ are defined as the functions:

$$r_n(\cdot) := sign[\sin(\pi 2^n(\cdot))],$$

where sign denotes simply the signum function, with the convention $sign(0) = -1$. By means of the Rademacher functions, we can define the so called Walsh functions w_n, $n \in \mathbb{N}$, on $[0, 1]$ iteratively by:

$$w_{2^k+m} = r_k w_m, \quad k \in \mathbb{N}_0 \text{ and } m \in [2^k],$$

where w_1 is given by $w_1(t) = 1$, $t \in [0, 1]$.

Notice that the indexing of the Walsh functions used in this work differs slightly with the usual indexing, since it begins by the index 1 instead of 0. The Walsh functions form a multiplicative group with the identity w_1. Furthermore, the Walsh functions are each self-inverse, i.e. $w_k w_k = w_1$, for every $k \in \mathbb{N}$, and form an orthonormal basis for $L^2([0, 1])$. The orthonormality of Walsh functions asserts that for every $n \in \mathbb{N} \setminus \{1\}$, w_n (which can be written as $w_n w_1$), it holds:

$$\int_0^1 w_n(t)dt = 0. \tag{25}$$

A more detailed treatment concerning to those issues, and further properties of the Walsh functions can be read in [10, 12, 27]

6.1 Deterministic approach

A result concerning to the behavior of tone reservation scheme for Walsh systems is given in the following Theorem:

THEOREM 31. *Given $\delta \in (0, 1)$, and assume that $N := 2^n$, $n \in \mathbb{N}$ fulfills:*

$$N \geq \frac{3}{2} \left(\frac{2}{\delta}\right)^{2m} \quad \text{for some } m \in \mathbb{N}.$$

If the PAPR problem is solvable for $(\{w_n\}_{n \in \mathbb{N}}, \mathcal{I})$ with constant C_{Ex}, for a subset $\mathcal{I} \subset [N]$ having the density $|\mathcal{I}|/N \geq \delta$, then it holds:

$$C_{Ex} \geq 2^{\frac{m}{2}}.$$

We still have a long way to prove the above statement. But first, let us give an easy implication of the above theorem concerning the solvability of PAPR reduction problem.

COROLLARY 32. *Let be $N := 2^n$, $n \in \mathbb{N}$. Assume that:*

$$N \geq \frac{3}{2} \left(\frac{2}{\delta}\right)^{2m} \quad \text{for some } m \in \mathbb{N}, \text{ and } \delta \in (0, 1).$$

Given a desired $C_{Ex} > 0$. If $C_{Ex} < 2^{\frac{m}{2}}$, then the PAPR problem for $(\{w_n\}_{n \in \mathbb{N}}, \mathcal{I})$, where $|\mathcal{I}| \geq \delta N$ is not solvable with constant C_{Ex}.

REMARK 33. *Notice that, for a suitable N, which represents the number of the available carriers in a considered CDMA system, and a given maximum peak value C_{Ex}, the above corollary gives a restriction on the number m of the information bearing coefficients such that the PAPR reduction problem is solvable.*

The first step to give a proof of the above results is to give the following definitions:

DEFINITION 34. *Let be $1 \subset \mathbb{N}$ finite and $r \in \mathbb{N}$. The correlation between w_r and 1 is defined as the quantity:*

$$Corr(w_r, 1) = \int_0^1 w_r(t) \left| \sum_{k \in 1} w_k(t) \right|^2 dt.$$

Furthermore, for w_r, $r \neq 1$, and 1, we define the following sets:

- $\mathcal{M}(w_r, 1) := \left\{ k \in 1 : w_k w_{\tilde{k}} = w_r, \text{ for a } \tilde{k} \in 1 \right\}$
- $\underline{\mathcal{M}}(w_r, 1) := \left\{ k \in 1 : w_k w_{\tilde{k}} = w_r, \text{ for a } \tilde{k} \in 1, \text{ with } \tilde{k} > k \right\}$
- $\overline{\mathcal{M}}(w_r, 1) := \mathcal{M}(w_r, 1) \setminus \underline{\mathcal{M}}(w_r, 1)$

Notice that for each $k \in \mathcal{M}(w_r, 1)$, there exists exactly one[2] $\tilde{k} \in \mathcal{M}(w_r, 1)$, for which the requirement given in the definition of $\mathcal{M}(w_r, 1)$ holds. This observation gives immediately the facts, that in case $\mathcal{M}(w_r, 1) \neq \emptyset$, $\mathcal{M}(w_r, 1)$ is always of even cardinality, and $\underline{\mathcal{M}}(w_r, 1)$ and $\overline{\mathcal{M}}(w_r, 1)$ are non-empty. Some other nice properties regarding to the correlation function, can easily be given ([7]):

LEMMA 35. *Let be $N = 2^n$, $n \in \mathbb{N}$, $1 \subset [N]$, and $r \in [N]$. The following holds:*

1. $|\mathcal{M}(w_r, 1)| = 2 \left| \underline{\mathcal{M}}(w_r, 1) \right| = 2 \left| \overline{\mathcal{M}}(w_r, 1) \right|$
2. $Corr(w_r, 1) = |\mathcal{M}(w_r, 1)|$
3. $\sum_{r=1}^N Corr(w_r, 1) = |1|^2$
4. $\arg\max_{r \in [N] \setminus \{1\}} Corr(w_r, 1) \geq (|1|^2 - |1|)/N$

We have already show the first item in the above lemma. The proof of the second and third item can be found in [7] (Lemma 4.4 and Lemma 4.6). To show the last item, notice first that by the orthonormality of Walsh functions, $Corr(w_1, 1) = |1|$ holds. Further, we have the following computation, which gives the desired statement:

$$|1| + N \arg\max_{r \in [N] \setminus \{1\}} Corr(w_r, 1) = Corr(w_1, 1) + N \arg\max_{r \in [N] \setminus \{1\}} Corr(w_r, 1)$$

$$\geq \sum_{r=1}^N Corr(w_r, 1) = |1|^2$$

where the first equality follows from previous observation, and the second equality from item 3 in the above lemma.

---

2. Otherwise, suppose that there exists $\tilde{k}_1, \tilde{k}_2 \in \mathcal{M}(w_r, 1)$, $\tilde{k}_1 \neq \tilde{k}_2$, for which it holds:

$$w_k w_{\tilde{k}_1} = w_r = w_k w_{\tilde{k}_2}.$$

Multiplying the above equation by w_k, and involving the fact that Walsh functions are self inverse, we obtain the contradiction, $w_{\tilde{k}_1} = w_{\tilde{k}_2}$ as desired.

Now, we define the object that plays an important role (as important as the arithmetic progression for the proof of Lemma 27 and Theorem 28), for the proof of Theorem 31:

DEFINITION 36. *Let be* $1 \subset [N]$, *and* $m \in \mathbb{N}$. *We say that* $\mathfrak{f}^1(1)$ *contains an perfect Walsh sum (PWS) of the size* 2^m *if there exists* $f \in \mathfrak{f}^1(1)$ *that can be written in the form:*

$$f = w_{l_*} \prod_{n=1}^{m} (1 + w_{k_n}) = w_{l_*} \left(1 + \sum_{n=1}^{2^m - 1} w_{l_n}\right) \tag{26}$$

for some $l_* \in \mathbb{N}$, $l_1, \ldots, l_{2^m-1} \in \mathbb{N} \setminus \{1\}$ *mutually distinct, and* $k_n \in \mathbb{N}$, *for* $n \in [m]$.

We call also the function (26) a *perfect Walsh sum of the size* 2^m. With abuse of notation, we also say 1 is a PWS of size 2^m. Given a set $\tilde{1} \subset [N]$. We say $\tilde{1}$ contains a PWS of size 2^m if $\tilde{1}$ has a subset, which is also a PWS of size 2^m. The adjective "perfect" is due to the factorability of the PWS. Further, the norms of PWS can be computed explicitly:

LEMMA 37. *Let* $m \in \mathbb{N}$. *For an optimal Walsh sum* f *of the size* 2^m, *it holds:*

$$\|f\|_{L^1([0,1])} = 1 \quad and \quad \|f\|_{L^2([0,1])} = 2^{\frac{m}{2}}$$

Proof. Let f be an PWS of size 2^m, i.e., it has the representation (26). Then by computation, we have:

$$\|f\|_{L^1([0,1])} = \int_0^1 |f(t)| \, dt = \int_0^1 |w_{l_*}(t)| \left| \prod_{n=1}^{m} (1 + w_{l_n}(t)) \right| dt = \int_0^1 \prod_{n=1}^{m} (1 + w_{l_n}(t)) dt$$

$$= 1 + \sum_{k=1}^{2^m} \int_0^1 w_{n_k}(t) dt = 1.$$

The 3. equality follows from the fact, that Walsh functions are always of modulus 1 and non-negative if added by 1, since they take values between $\{-1, +1\}$. The 4. inequality follows from (25). The second statement is also not hard to established. Indeed, by setting $l_0 = 1$, since the Walsh functions are of modulus 1, and the orthonormality of Walsh functions, we have:

$$\|f\|_{L^2([0,1])}^2 = \int_0^1 |w_*(t)|^2 \left| \sum_{n=0}^{2^m - 1} w_{l_n} \right|^2 dt = \sum_{k,l=0}^{2^m - 1} \int_0^1 w_{n_k}(t) w_{n_l}(t) dt = 2^m.$$

$\square$

REMARK 38. *As the above Lemma asserts, the existence of PWS in an information set, allows one to give the (lowest) extension constant* C_{ex} *for which the following embedding inequality:*

$$\|f\|_{L^2([0,1])} \leq C_{Ex} \|f\|_{L^1([0,1])}, \quad \forall f \in \mathfrak{f}^1(1), \tag{27}$$

explicitly, in the following sense: Assume $1 \subset \mathbb{N}$ *is a PWS of size* 2^m, $m \in \mathbb{N}$. *It can be shown, that* $C_{Ex} \leq \sqrt{|1|}$ *(This inequality holds also, if the assumption that* 1 *is a PWS is dropped, and if another ONS is considered, instead of Walsh functions), and correspondingly* $C_{Ex} \leq 2^{\frac{m}{2}}$. *Further, by setting the corresponding PWS* $f \in \mathfrak{f}^1(1)$ *into (27), and by involving lemma 37, one obtains* $C_{Ex} \geq 2^{\frac{m}{2}}$. *Thus* $C_{Ex} = 2^{\frac{m}{2}}$. *In the subsequent work, we shall a slightly stronger statement: For a* $1 \subset [2^n]$, $n \in \mathbb{N}$, *it holds* 1 *is a PWS if and only if it holds for the minimum constant* C_{Ex} *in (27),* $C_{Ex} = \sqrt{|1|} = 2^{\frac{m}{2}}$, *for some* $m \in \mathbb{N}$.

The most important step for proving Theorem 31 is given in the following statement:

THEOREM 39. *Let $N = 2^n$, $n \in \mathbb{N}$, and $\delta \in (0,1)$. Then, for every subset $\mathcal{I} \subset [N]$ fulfilling:*

$$|\mathcal{I}| \geq \delta N \quad \text{and} \quad |\mathcal{I}| \geq 3 \left(\frac{2}{\delta} \right)^{2^m - 1},$$

for an $m \in \mathbb{N}$, $\mathfrak{f}^1(\mathcal{I})$ contains a perfect Walsh sum of size 2^m, or more explicitly, $\mathcal{I}$ contains a PWS.

The proof of the above theorem shall be given subsequently. In particular, it is based on Lemma 35 and basic properties of Walsh sums. A construction of such an object, provided that the assumption in the above theorem is fulfilled, can explicitly be given. Roughly, the above theorem says that if an information set $\mathcal{I}$ is large enough, then it has a subset that is a PWS. Now we are to give the desired result:

Proof of Theorem 31. Let δ, N, m be as required in the theorem. Take an $\mathcal{I} \subset [N]$ having density in $[N]$ bigger than δ. Assume that the PAPR problem is solvable for $(\{w_n\}_{n \in \mathbb{N}}, \mathcal{I})$, with a desired extension constant $C_{Ex} > 0$. By Theorem 10, we know that:

$$\|f\|_{L^2([0,1])} \leq C_{Ex} \|f\|_{L^1([0,1])}, \quad \forall f \in \mathfrak{f}^1(\mathcal{I}). \tag{28}$$

Further by the assumption that $|\mathcal{I}| \geq \delta N$, and $N \geq (3/2)(2/\delta)^{2^m}$, which assert that $|\mathcal{I}| \geq 3(2/\delta)^{2^m - 1}$, Theorem 39 asserts that $\mathcal{I}$ contains a PWS of size 2^m. We denote the corresponding sum by $f \in \mathfrak{f}^1(\mathcal{I})$. Noticing that f has the norm as given in Lemma 37 and setting those to (28), $2^{\frac{m}{2}} \leq C_{Ex}$ has to hold. □

6.2 Probabilistic approach

In spirit of Theorem 29, we aim to give some asymptotic statements regarding to the solvability of the PAPR reduction problem. The first step to establish such statements, is to give the following remark:

REMARK 40. *Let $m \in \mathbb{N}$, and $N \in \mathbb{N}$ be at first fixed. Notice that by Theorem 39, a sufficient condition for $\delta \in (0,1)$, s.t. $\mathcal{I}$ contains a PWS of size 2^m, where $\mathcal{I} \subset [N]$ is a subset, having the density $|\mathcal{I}|/N$ in $[N]$ bigger, i.e. $|\mathcal{I}| \geq \delta N$, is that:*

$$\delta N \geq 3 \left(\frac{2}{\delta} \right)^{2^m - 1}. \tag{29}$$

since the left side is monotone increasing, and the right side monotone decreasing with δ, it follows that there exists $\delta_N \in (0,1)$, s.t. we have equality in (29), with $\delta = \delta_N$. Some elementary computation yields:

$$\delta_N = 2 \left(\frac{3}{2N} \right)^{\frac{1}{2^m}}. \tag{30}$$

Of course, m and N has to be chosen appropriately, s.t. $\delta_N \in (0,1)$. In case that this issue has been considered, it is obvious that for all subsets $\mathcal{I} \subset [N]$ having density δ in $[N]$ bigger that δ_N, $\mathcal{I}$ contains a PWS of size 2^m.

Let $N \in \mathbb{N}$, and $p \in [0, 1]$. $[N]_p$ denotes again the random subset of $[N]$, in which each element is chosen independently from $[N]$ with probability p. By means of Theorem 39 and Remark 40, the following statement, which gives a probabilistic construction of a sparse set $\mathcal{I}$, with density zero (a.s.) in $\mathbb{N}$, for which $\mathcal{I}$ contains a PWS of an arbitrary size, can be established:

THEOREM 41. *Let be* $m \in \mathbb{N}$. *Then there is a sequence* $\{p_N\}$, *with* N *large enough, in* $(0, 1]$ *tending to zero, for which it holds:*

$$\lim_{N \to \infty} \mathbb{P}\left[[N]_{p_N} \text{ contains a PWS of size } 2^m\right] = 1$$

Proof. For $m \in \mathbb{N}$, and $\tau > 1$, choose $N \in \mathbb{N}$ large enough, s.t.:

$$p_N := \tau \delta_N \in (0, 1)$$

where δ_N is given by (30). Note that $\left|[N]_{p_N}\right|$ is binomial distributed. Correspondingly, we have $\mathbb{E}[\left|[N]_{p_N}\right|] = \tau \delta_N N$. We may give the estimate:

$$\mathbb{P}\left[\left|[N]_{p_N}\right| < \delta_N N\right] = \mathbb{P}\left[\left|[N]_{p_N}\right| < \frac{1}{\tau}\mathbb{E}[\left|[N]_{p_N}\right|]\right] \le \exp\left(-\frac{\left(\frac{\tau-1}{\tau}\right)^2 \mathbb{E}[\left|[N]_{p_N}\right|]}{2}\right)$$

$$= \exp\left(-CN^{\frac{2^m-1}{2^m}}\right) \xrightarrow[N \to \infty]{} 0, \quad C := \left(\frac{3}{2}\right)^{\frac{1}{2^m}} \frac{(\tau-1)^2}{\tau}, \tag{31}$$

where the estimation is based on the Chernoff bound. Thus, we have:

$$\mathbb{P}\left[[N]_{p_N} \text{ contains a PWS of size } 2^m\right] = \mathbb{P}\left[\left|[N]_{p_N}\right| \ge \delta_N N\right]$$
$$= 1 - \mathbb{P}\left[\left|[N]_{p_N}\right| < \delta_N N\right] \xrightarrow[n \to \infty]{} 1,$$

by (31). Clearly, p_N tends to zero as $N \to \infty$, as desired. $\square$

In analogy to Theorem 26, and as a tightening of the above theorem, we have the following statement:

THEOREM 42. *Let* $m \in \mathbb{N}$, *and* $\delta \in (0, 1)$. *Then there is a sequence* $\{p_N\}$, *with* N *large enough, in* $(0, 1]$, *tending to zero, for which it holds:*

$$\lim_{N \to \infty} \mathbb{P}[A_{N,m,\delta}] = 1,$$

where $A_{N,m,\delta}$ *denotes the event:*

$$A_{N,m,\delta} := \left\{\forall \mathcal{I} \subset [N]_{p_N}, \ |\mathcal{I}| \ge \delta \left|[N]_{p_N}\right| : \mathcal{I} \text{ contains a PWS of size } 2^m\right\}$$

Proof. For some $p_N \in [0, 1]$, define another event $\underline{A}_{N,m,\delta}$ by:

$$\underline{A}_{N,m,\delta} := \left\{[N]_{p_N} \ge \frac{\delta_N N}{\delta}\right\},$$

where δ_N is given by (30). Obviously, we have by Remark 40, $\underline{A}_{N,m,\delta} \subset A_{N,m,\delta}$, which gives:

$$\mathbb{P}[A_{N,m,\delta}] \ge \mathbb{P}\left[\underline{A}_{N,m,\delta}\right]. \tag{32}$$

Now, for certain $m \in \mathbb{N}$, $\delta > 0$, and arbitrary $\tau > 1$, choose $N \in \mathbb{N}$ sufficiently large, s.t.:

$$p_N := \frac{\tau \delta_N}{\delta} \in (0, 1)$$

$\left| [N]_{p_N} \right|$ is binomial distributed, with expectation $\tau(\delta_N/\delta)N$, and by computation similar to (31), we have:

$$\mathbb{P}\left[\underline{A}^c_{N,m,\delta} \right] \leq \exp\left(-CN^{\frac{2^m-1}{2^m}} \right) \underset{N \to \infty}{\longrightarrow} 0, \quad C := \left(\frac{3}{2} \right)^{\frac{1}{2m}} \frac{(\tau-1)^2}{\tau} \frac{1}{\delta}.$$

Together with (32), the desired statement is shown. □

Now, we are ready to give the desired applications of previous theorems to the solvability of PAPR reduction problems:

THEOREM 43. *Let be $m \in \mathbb{N}$. Given an extension constant $C_{Ex} > 0$, with $C_{Ex} < 2^{\frac{m}{2}}$. Then there exists a sequence $\{p_N\}$ in $(0, 1]$, with N large enough, tending to zero, s.t.:*

$$\lim_{N \to \infty} \mathbb{P}\left[\text{The PAPR problem is not solvable for } (\{w_n\}_{n \in \mathbb{N}}, [N]_{p_N}) \text{ with } C_{Ex} \right] = 1$$

Proof. Choose $m \in \mathbb{N}$. Let C_{Ex} be given, with $2^{\frac{m}{2}} > C_{Ex}$. Further, construct the sequence $\{p_N\}_{N \in \mathbb{N}}$ in $(0, 1]$ tending to zero s.t. the statement Theorem 41 holds. Thus, we know that the probability of $[N]_{p_N}$ containing a PWS of size 2^m tends to 1 as N goes to infinity. Notice that for each of such events, a corresponding PWS f has the norms $\|f\|_{L^1([0,1])} = 1$ and $\|f\|_{L^2([0,1])} = 2^{\frac{m}{2}}$. By the assumption on C_{Ex}, we have a function in $\mathfrak{F}^1([N]_{p_N})$, for which (13) does not hold. In this case, Theorem 10 asserts that the PAPR reduction problem is not solvable, as desired. □

THEOREM 44. *Let be $m \in \mathbb{N}$. Given an extension constant $C_{Ex} > 0$, with $C_{Ex} < 2^{\frac{m}{2}}$, and $\delta > 0$. Then there exists a sequence $\{p_N\}_{N \in \mathbb{N}}$ in $(0, 1]$ tending to 0, for which it holds:*

$$\lim_{N \to \infty} \mathbb{P}\left[B_{N,\delta} \right] = 1,$$

where $B_{N,\delta}$ denotes the event:
"The PAPR problem is not solvable for all $(\{w_n\}_{n \in \mathbb{N}}, \mathit{1})$ with C_{Ex}, where $\mathit{1} \subset [N]_{p_N}$, $|\mathit{1}| \geq \delta \left| [N]_{p_N} \right|$."

Proof. Choose $m \in \mathbb{N}$. Let C_{ex} be given, with $2^{\frac{m}{2}} > C_{Ex}$. Take the sequence $\{p_N\}_{N \in \mathbb{N}}$ in $(0, 1]$, for which the statement in Theorem 10 holds. For each $N \in \mathbb{N}$, consider the event $A_{N,m,\delta}$. For $[N]_{p_N}$ in this event, every subset $\mathit{1} \subset [N]_{p_N}$, having the density $|\mathit{1}| / \left| [N]_{p_N} \right|$, contains a PWS of size 2^m. By similar argument made in the previous theorem, and by the assumption made for m, we find a function $f \in \mathfrak{F}^1(\mathit{1})$ harming (13). Thus in this case, PAPR reduction is not solvable. As p_N tending to 0 and $\mathbb{P}[A_{N,m,\delta}]$ tending to 1, the desired statement is obtained. □

6.3   On the perfect Walsh sums and an Erdős conjecture

We have already seen in this section, and in Section 5, that the existence of a certain combinatorial object in the considered information set I allows us to turn Theorem 10, which gives a necessary condition for the solvability of PAPR reduction problem, into another theorem, which is easy to handle, such as Theorem 28 and Theorem 31. If the considered information set is infinite, we have already seen that, by a probabilistic method, it is possible to construct the corresponding combinatorial object, having the desired property, i.e. sparsity in the natural number, for both the Fourier and Walsh case (Theorem 26, Theorem 41, and Theorem 44). However to find such combinatorial objects with sparsity constraint deterministically is not easy. Concerning to arithmetic progressions, one of the famous conjecture of Erdős, says that a set of positive integer A, satisfying a certain constraint on its size, contains arbitrarily long arithmetic progressions:

CONJECTURE 45 (Erdős conjecture on arithmetic progressions). *Let $A \subset \mathbb{N}$ satisfy $\sum_{n \in A} n^{-1} = \infty$. Then A contains arbitrarily long arithmetic progressions. Specifically: For each $m \in \mathbb{N}$, there exists $n_0 \in \mathbb{N}$, such that $A \cap [n_0]$ contains an arithmetic progression of length m.*

Thus, if that conjecture holds true, a subset of $\mathbb{N}$ which is not too small but sparse contains arbitrarily long arithmetic progressions. The conjecture remains unsolved. However, it is due to Green and Tao that the set of prime numbers contains arbitrarily long arithmetic progressions (see Section 5.1). Further, it is clear that the set of prime numbers satisfies the condition given in the previously mentioned Erdős conjecture. For recent reports concerning the previously mentioned Erdős conjecture, we refer to [13].

For the combinatorial object, which is important for the Walsh case, i.e. the subsets of the natural numbers forming PWS, things turn out differently:

THEOREM 46 (Solution of Erdős problem for Walsh sums). *Let $I \subset \mathbb{N}$ satisfy:*

$$\sum_{k \in I} \frac{1}{k} = \infty. \tag{33}$$

Then I contains a PWS of arbitrary size. Specifically: For each $m \in \mathbb{N}$, there exists $n_0 \in \mathbb{N}$, such that $I \cap [2^{n_0}]$ contains a PWS of size m.

To prove the above theorem, we need the following statement:

LEMMA 47. *For every $m \in \mathbb{N}$, there exist some $n_0 \in \mathbb{N}$, such that for every dyadic number N, i.e. $N = 2^{\tilde{n}}$, for a $\tilde{n} \in \mathbb{N}$, fulfilling $N \geq 2^{n_0}$, it follows that for every subset $I \subset [N]$, having the cardinality:*

$$|I| \geq \frac{N}{(\log N)^2}, \tag{34}$$

I contains a PWS of the size 2^m.

Proof. Let $m \in \mathbb{N}$ be fixed, and n_0 the smallest natural number for which the following holds:

$$0 < 2 \left(\frac{3}{2 \cdot 2^{n_0}} \right)^{\frac{1}{2m}} < 1 \quad \text{and} \quad \frac{2^{n_0}}{(\log 2^{n_0})^2} \geq 2 \left(\frac{3}{2 \cdot 2^{n_0}} \right)^{\frac{1}{2m}} \frac{2^{n_0}}{(2^{n_0})^{\frac{1}{2m}}}.$$

Notice that N is always of the form $N = 2^{\tilde{n}}$, $\tilde{n} \in \mathbb{N}$. Since $N \geq 2^{n_0}$, it holds for $I \subset [N]$, fulfilling

(34):

$$|\mathcal{I}| \geq \frac{N}{(\log N)^2} \geq \frac{2^{n_0}}{(\log 2^{n_0})^2} \geq 2\left(\frac{3}{2 \cdot 2^{n_0}}\right)^{\frac{1}{2m}} \frac{2^{n_0}}{(2^{n_0})^{\frac{1}{2m}}}.$$

Accordingly by (30), $\mathcal{I}$ contains an PWS of size 2^m as desired. $\square$

So, now we are ready to give the corresponding proof:

Proof of Theorem 46. For $r \in \mathbb{N}$, define $\delta_{\mathcal{I}}(r)$ by:

$$\frac{|\mathcal{I} \cap [r]|}{r} =: \delta_{\mathcal{I}}(r),$$

and for $r = 0$, $\delta_{\mathcal{I}}(0) = 0$. Notice that by means of $\delta_{\mathcal{I}}$, we can write:

$$\sum_{r \in \mathcal{I}} \frac{1}{r} = \sum_{r=1}^{\infty} \frac{\delta_{\mathcal{I}}(r-1)}{r}. \tag{35}$$

We continue by the following computations:

$$\sum_{r=1}^{\infty} \frac{\delta_{\mathcal{I}}(r-1)}{r} = \sum_{r=2}^{\infty} \frac{\delta_{\mathcal{I}}(r)}{r+1} = \sum_{l=1}^{\infty} \sum_{r=2^l}^{2^{l+1}-1} \frac{\delta_{\mathcal{I}}(r)}{r+1} \leq \sum_{l=1}^{\infty} \frac{1}{2^l} \sum_{r=2^l}^{2^{l+1}-1} \delta_{\mathcal{I}}(r)$$

$$= \sum_{l=1}^{\infty} \frac{1}{2^l} \sum_{r=2^l}^{2^{l+1}-1} \frac{|\mathcal{I} \cap [r]|}{r} \leq \sum_{l=1}^{\infty} \frac{1}{2^l} \left|\mathcal{I} \cap [2^{l+1}]\right| \sum_{r=2^l}^{2^{l+1}-1} \frac{1}{r}, \tag{36}$$

where the second equality follows from the segmentation of the sum into parts. Using the estimation $(1/r) < \int_{r-1}^{r} (1/x) dx$, we can compute:

$$\sum_{r=2^l}^{2^{l+1}-1} \frac{1}{r} < \int_{2^l-1}^{2^{l+1}-1} \frac{dx}{x} = \log\left(\frac{2^{l+1}-1}{2^l-1}\right) = \log\left(\frac{2 - \frac{1}{2^l}}{1 - \frac{1}{2^l}}\right) \leq C_1,$$

where $C_1 > 0$ is a constant which does not depend on l. Substituting this in (36), we have:

$$\sum_{r=1}^{\infty} \frac{\delta_{\mathcal{I}}(r-1)}{r} \leq C_1 \sum_{l=1}^{\infty} \frac{\left|\mathcal{I} \cap [2^{l+1}]\right|}{2^l} = 2C_1 \sum_{l=1}^{\infty} \frac{\left|\mathcal{I} \cap [2^{l+1}]\right|}{2^{l+1}} = 2C_1 \sum_{l=1}^{\infty} \delta_{\mathcal{I}}(2^{l+1}).$$

Now, we claim that there exist infinitely many numbers l_t, $t \in \mathbb{N}$, satisfying:

$$\delta_{\mathcal{I}}(2^{l_t+1}) > \frac{1}{(\log(2^{l_t+1}))^2}. \tag{37}$$

Otherwise, we have $\delta_{\mathcal{I}}(2^{l+1}) > (1/(\log(2^{l+1}))^2)$, for all $l \in \mathbb{N}$, which gives the contradiction:

$$\infty = \sum_{r=1}^{\infty} \frac{\delta_{\mathcal{I}}(r-1)}{r} \leq 2C_1 \sum_{l=1}^{\infty} \frac{1}{(l+1)^2(\log 2)^2} < \infty,$$

where the equality follows from the assumption (33) and (35), and the last inequality follows from the finiteness of C_1 and the usual convergence of geometric series. Thus (37) holds.

For the last step of the proof, let $m \in \mathbb{N}$ be arbitrary. Let $n_0 = n_0(m) \in \mathbb{N}$ be the number, s.t. for every dyadic N with $N \geq 2^{n_0}$, every subset $\mathcal{A} \subset [N]$ having the cardinality $|\mathcal{A}| \geq N/(\log N)^2$ contains a PWS of size 2^m, as asserted by Lemma 47. By (37), we can find an $t_0 \in \mathbb{N}$, with $t_0 \geq n_0$, for which $|\mathcal{I} \cap [2^{t_0}]| > N/(\log N)^2$, which gives the remaining clue, that $\mathcal{I} \cap [2^{t_0}]$ contains a PWS of size 2^m, as desired.   □

As an immediate consequence of Theorem 46, we have the following statement, which is analogous to Green and Tao's theorem on the existence of arithmetic progressions of arbitrarily in the set of prime numbers:

COROLLARY 48. *Let $\mathcal{P} \subset \mathbb{N}$ denotes the set of prime numbers. Then, $\mathcal{P}$ contains an PWS of arbitrary length, i.e. for every $m \in \mathbb{N}$, there exists $n_0 \in \mathbb{N}$, s.t. $\mathcal{P} \cap [2^{n_0}]$ contains a PWS of size 2^m.*

Proof. Since $\sum_{k \in \mathcal{P}} k^{-1} = \infty$, Theorem 46 gives the remaining argument.   □

6.4 Further discussion and outlook

From Theorem 39, one can also answer the question, whether the PAPR reduction problem is solvable for $(\{w_n\}_{n \in \mathbb{N}}, \mathcal{I})$ with $C_{\text{Ex}} > 0$, where $\mathcal{I} \subset \mathbb{N}$ is infinite: Let $\{k_l\}_{l \in \mathbb{N}}$ be an enumeration of $\mathcal{I}$, $m \in \mathbb{N}$ be arbitrary but firstly fixed. Further choose $N \in \mathbb{N}$ large enough, s.t. δ_N given in (30) takes value in $(0, 1)$, and correspondingly compute the value $\lceil \delta_N N \rceil$. Now by Remark 40, we can find a PWS of size 2^m in $\{k_l\}_{l=1}^{\lceil \delta_N N \rceil}$, since $\left|\{k_l\}_{l=1}^{\lceil \delta_N N \rceil}\right| \geq \delta_N N$. Previous observation asserts immediately, that there exists a PWS of size 2^m in $\mathcal{I}$, since $\{k_l\}_{l=1}^{\lceil \delta_N N \rceil} \subset \mathcal{I}$. Now, for an $C_{\text{Ex}} > 0$, choose $m \in \mathbb{N}$ large enough, s.t. $2^m > C_{\text{Ex}}$, and observe that by the procedure given previously, there is a PWS of size 2^m in $\mathfrak{F}^1(\mathcal{I})$. By the choice of m, the existence of a PWS of size 2^m, Lemma 37, and Theorem 10, it follows immediately that the PAPR reduction is not solvable for $(\{w_n\}_{n \in \mathbb{N}}, \mathcal{I})$ for arbitrarily chosen C_{Ex}, and hence for any C_{Ex}.

In case that the information set $\mathcal{I}$ are dyadic integers, i.e. $\mathcal{I} = \left\{2^k\right\}_{k=1}^{K}$, for a $K \in \mathbb{N}$, one can expect in some sense a positive result: It was shown in [7] (Theorem 4.13), there is some constant C_{Ex}, s.t. the PAPR reduction problem is solvable for $(\{w_n\}_{n \in \mathbb{N}}, \left\{2^k\right\}_{k=1}^{K})$. In this case one can even find a finite compensation set, explicitly $[2^K] \setminus \left\{2^k\right\}_{k=1}^{K}$. A possible extension constant, is the constant B_1, for which the upper Khintchine's Inequality [35] (I.B.8) holds:

$$\left\|\sum_{k=1}^{K} a_k r_k\right\|_{L^2([0,1))} \leq B_1 \left\|\sum_{k=1}^{K} a_k r_k\right\|_{L^1([0,1))},$$

for all sequences $\{a_k\}_{k=1}^{K}$.

Acknowledgement. The authors thank Peter Jung, Gitta Kutyniok, and Philipp Walk for carefully reading the manuscript and providing helpful comments. H.B. thanks Arogyaswami Paulraj for the introduction into the PAPR Problem for OFDM in 1998 and the discussions since that time. We note that the PAPR problem for CDMA systems was posed by Bernd Haberland and Andreas Pascht, of Bell Labs, in 2000, and the first author thanks them for the discussions since that time. The work on this paper was accomplished at the Hausdorff Research Institute for Mathematics in Bonn. H.B. und E.T. thank the Hausdorff Research Institute for Mathematics for the support and the hospitality. The results of this work was presented at the Hausdorff Research Institute for Mathematics during the Hausdorff trimester program "Mathematics of Signal Processing". H.B. thanks also the DFG (German Research Foundation) for the support by the grant of Gottfried Wilhelm Leibniz Prize.

References

[1] AGILENT TECHNOLOGIES and RUMNEY, MORAY, LTE and the evolution to 4G wireless: Design and measurement challenges. 2.nd Ed. (2013), *Wiley Publishing*.

[2] J. BECK, Flat polynomials on the unit circle – note on a problem of Littlewood. *Bull. London Math. Soc.* **23** (1991), 269–277.

[3] B. BOCCALETTI, M. LÖFFLER and J. OPPENHEIM, How IT can cut carbon emissions. *McKinsey Quarterly*, October (2008).

[4] H. BOCHE and V. POHL, Signal representation and approximation – fundamental limits. *European Trans. Telecomm. (ETT)* **5** (2007), 445–456.

[5] H. BOCHE, B. FARELL, M. LEDOUX and M. WIESE, Expected supremum of a random linear combination of shifted kernels. *J. Fou. Ana. Appl.* (2011).

[6] H. BOCHE and B. FARELL, PAPR and the density of information bearing signals in OFDM. *EURASIP J. Adv. Sig. Proc.* **2011** (2011).

[7] H. BOCHE and B. FARELL, On the peak-to average power ratio reduction problem for orthogonal transmission schemes. *Internet Mathematics* **9** (2013), 265–296.

[8] D. CONLON and W. T. GOWERS, Combinatorial theorems in sparse random sets. arXiv:1011.4310, (2011).

[9] P. ERDŐS, some unsolved problems. *Michigan Math. J.* **4** (1957), 291–300.

[10] N. J. FINE, On the Walsh function. *Trans. Amer. Math. Soc.* **65** (1949), 372–414.

[11] B. GREEN and T. TAO, The primes contain arbitrarily long arithmetic progressions. *Annals of Mathematics* **167** (2008), 481—547.

[12] B. GOLUBOV, A. EFIMOV and V. SKVORTSOV, *Walsh series and transforms*. Math. and its App., Springer Netherlands, Netherlands, 1991.

[13] W. T. GOWERS, Erdős and arithmetic progressions, Erdős Centennial, Bolyai Society Mathematical Studies, **25**, L. Lovasz, I. Z. Ruzsa, V. T. Sos eds., Springer (2013), pp. 265–287.

[14] S. H. HAN and J.H. LEE, An overview of peak-to-average power ratio reduction techniques for multicarrier transmissions. *Trans. on Wireless Communications* **12** (2005), 56–65.

[15] J. P. KAHANE, Sur les polynomes a coefficient unimodulaires. *Bull. London Math. Soc.* **12** (1980), 321–342.

[16] L. W. KANTOROVITSCH and G. P. AKILOW, *Functional analysis*. Pergamon Pr. , 1982.

[17] S. LITSYN, *Peak power control in multicarrier communications*. Cambridge University Press, 2007.

[18] J. E. LITTLEWOOD, On polynomials $\sum \pm z^m$, $\sum \exp(\alpha_m i)$, $z = e^{i\theta}$. *J. London Math. Soc.* (2) **41** (1966), 367–376.

[19] F. L. NAZAROV, The Bang solution of the coefficient problem. *Algebra i Analiz* **9** (1997), no. 2, 272–287; English translation in St. Petersburg Math. J. **9**(1998), no. 2, 407–419.

[20] K. O'BRYANT, Sets of integers that do not contain long arithmetic progressions. *El. J. of Comb.* **18** (2011), R59.

[21] J. OXTOBY, *Measure and category: A survey of analogies between topological and measure spaces*. Springer-Verlag, New York, US, 1980.

[22] PARLIAMENTARY OFFICE OF SCIENCE AND TECHNOLOGY (UK), ICT and CO_2 emmissions. *Postnote*, December (2008).

[23] R. RANKIN, Sets of integers containing not more than a given number of terms in arithmetical progression. *Proc. Roy. Soc. Edinburgh Sect. A* **65** (1960/61), 332–344.

[24] W. RUDIN, *Real and complex analysis*. McGraw-Hill Book Comp., 1921, int. ed. 1987.

[25] W. RUDIN, *Functional analysis*. McGraw-Hill Book Comp., int. ed. 1991.

[26] K. F. ROTH, On certain sets of integers. *J. London Math. Soc.* **28** (1953), 245–252.

[27] F. SCHIPP, W. R. WADE and P. SIMON, *Walsh Series*. Akadémiai KiaDó, Budapest, Hungary, 1990.

[28] E. SZEMERÉDI, On sets of integers containing no 4 elements in arithmetic progression. *Acta Math. Ac. Sc. Hung.* **20** (1969), 89–104.

[29] E. SZEMERÉDI, On sets of integers containing no k elements in arithmetic progression. *Acta Arith.* **27** (1975), 199–245.

[30] T. TAO and V. VU, *Additive Combinatorics*. Princeton Math. Ser. 39, Cambridge University Press, Mass, USA, 2006.

[31] J. TELLADO, Peak to average power reduction for multicarrier modulation. Ph.D. Thesis Stanford University, 1999.

[32] J. TELLADO and J. M. CIOFFI, Peak to average power ratio reduction. U.S. patent application Ser. No. 09/062, 867, Apr. 20, 1998.

[33] J. TELLADO and J. M. CIOFFI, Efficient algorithms for reducing PAR in multicarrier systems. *Proc. IEEE ISIT* (1998), 191.

[34] P. WALK, H. BECKER and P. JUNG, OFDM channel estimation via phase retrieval. http://arxiv.org/abs/1512.04252 (2015).

[35] P. WOJTASZCZYK, *Banach Spaces for Analysts*. Cambridge Stud. in Adv. Math., Cambridge University Press, Cambridge, 1991.

[36] G. WUNDER, R. F. H. FISCHER, H. BOCHE, S. LITSYN and J. S. NO, The PAPR problem in OFDM transmission: New directions for a long-lasting problem. *IEEE Signal Processing Magazine* **30** (2013), 130–144.

Cryptology:
Methods, applications and challenges

Information processing by electronic devices leads to a multitude of security-relevant challenges. With the help of cryptography, many of these challenges can be solved and new applications can be made possible.

What methods are hereby used? On which mathematical foundations do they rest? How did the prevailing ideas and methods come about? What are the current developments, what challenges exist and which future challenges can be predicted?

1 Encrypted messages and more

Secret agents, online stores and pupils exchanging "secret messages" consisting of nonsensical symbols all use it: cryptography. The word "cryptography" is derived from the Greek words χρυπτός, 'hidden', and γράφειν, 'to write'; it is therefore about secret writing.

The sender *encrypts* the message and the receiver *decrypts* it with an agreed-upon secret. The endeavor to read encrypted messages without knowledge of the secret is called *cryptanalysis*. It is common to summarize both aspects under *cryptology*.

Due to the technical development in the field of electronics, the notions of cryptography and cryptology are nowadays used more broadly; the goals of cryptography now cover all aspects of security in processing, transmission and use of information in the presence of an adversary.

In this way, cryptographic methods have entered many different areas. One can use them to ensure *confidentiality* in any kind of electronic communication. They are used for *authentication* when unlocking a car or releasing an immobilizer, withdrawing money with a bank card or identifying oneself at a border with a passport, for example. Documents are nowadays often signed digitally with cryptographic methods, for example by a notary; like this the *non-repudiation* of agreements can be guaranteed. With digital signatures one can also guarantee the *integrity* of electronic data, that is, that the data has not been tampered with; this is for example used in passports.

What is the current state of cryptography in a world of electronic devices in which data acquisition and processing are continuously increasing? What are its foundations, what are its applications? How did the prevailing ideas and methods develop historically and what current developments and challenges are there?

2 Classic ideas

Until the end of the First World War, cryptology developed slowly and the schemes used were, mathematically speaking, elementary from today's point of view. Nonetheless, many notions and ideas of the past are still fundamental.

2.1 Approaches

To start with, some definitions: In using a cryptographic scheme, texts are encrypted, sent, received and then decrypted. The primary text is called *plain text* and the encrypted text is called *cipher text*; instead of *en-/decrypting* one can also speak of *en-/deciphering.*

Classically, there are two fundamentally different approaches to encrypting: First, one can apply cryptographic methods at the atomic level, that is, at the level of letters; second, one can start at the level of words.

The second approach is easier to describe because the method allows for less variation: With the help of a special *code book*, words of the standard language are replaced by *code words*. These code words can be other words of the standard language or arbitrary combinations of letters and numeric symbols.

For the first approach an enormous amount of schemes have been developed over the centuries. Two obvious ideas have thereby occurred over and over: *substitution* and *transposition.*

With the method of substitution, individual symbols (letters, the space character, numeric symbols, punctuation marks) are replaced by other symbols (or combinations of symbols). These symbols can be made up or they can be normal letters and numeric symbols. Even if may be appealing to use "mysterious" symbols, this does not make any difference from the point of view of security.

With the method of transposition, the order of the symbols in the text is changed according to a certain rule. A good example is: Write the text line by line in a table from the upper left corner to the lower right corner, then read off the columns of the text in a prearranged order.

Of course, one can combine these methods with each other. One can, for example, first use a code book and then a transposition and finally a substitution.

Classic cryptographic methods rely on common secrets between the sender and the receiver. Generally, the methods offer the possibility to encrypt texts with variable secrets. With the code book-method, it is the code book itself, with the substitution method it is the substitution table and with the transposition method as described above it is the number and order of the columns.

This secret is called the *key.* One can thus – at least in the examples discussed above – distinguish between the cryptographic scheme itself and the key. We shall see that this distinction is of particular relevance.

2.2 The race of cryptology

As nobody wants to use an insecure cryptographic scheme, potential schemes are tested for possible attacks in advance and schemes in use are continuously reviewed.

This establishes a race between designing and attacking schemes. Here we present the progression of this race exemplarily with the substitution method. The following presentation is idealized rather than historical, even though corresponding thoughts and developments did indeed occur in the course of centuries.

With the substitution method as described, one immediately notices that symbols like E or the space character occur in natural languages more often than others. In general, this also holds for a particular plain text. It is therefore reasonable to assume that, for example, the most frequent symbol in a cipher text corresponds to E or to the space character. A text of a few lines in a known language can, generally speaking, be recovered by considering the frequencies of the symbols. This encryption scheme can thus be considered broken. One might then ask: Is there a variant of the scheme for which the described attack is not feasible anymore?

Yes, there is one: One can ensure that in the cipher text all symbols appear with about the same frequency. This method is based on the idea that one symbol might be encrypted to various distinct symbols. Suppose that we use 1000 distinct symbols for writing cipher texts. If now the letter E appears with a probability of 12,7 % in a given language, 127 distinct symbols are assigned to it. When encrypting, for each occurrence of E one of the 127 symbols is chosen at random and used in its place. The other symbols of the plain text are encrypted in the same manner. In this way, the frequency analysis as described above fails. But how can one up with 1000 distinct symbols? Well, this is easier than one might first think: One starts with the ten numerical symbols from 0 to 9 and regards each string of three of these symbols as a symbol in itself. In concrete terms, the letter E is therefore represented by 127 chosen combinations of strings of three numerical symbols.

After the encryption scheme has been modified, it is natural to ask if the attack method can be modified as well. Yes, it can: One no longer considers the frequencies of symbols in a language and in the cipher text (in the concrete example, in the cipher text, a string of three numerical symbols is considered as one symbol) but of so-called *bigrams*, which are combinations of two symbols or of *trigrams*, that is, combinations of three symbols.

Again the scheme is broken, which raises the question if this encryption can again be improved or if one should maybe use a completely different method.

A natural answer to that question is that the concept of a substitution is so fundamental that it should be part of the method in any case. One might for example combine a substitution with a transposition.

So far, the attacks considered have not taken into account information about the sent messages themselves, but such information might also be used. For example, letters usually begin and end with a common salutation phrase and military messages are often highly standardized. If this leads nowhere, an attacker might also try to have the user of a cryptographic scheme encrypt a message foisted on him, such that the attacker has for a given plain text the corresponding cipher text. It is conceivable that the attacker can extract a sought-after information from a secret message with the help of such plain text – cipher text pairs even without finding the key.

3 New ideas

According to the technical development the history of cryptology can be divided into three periods:

1. The paper-and-pencil era until about the end of World War I.
2. The era of electric-mechanic cipher machines from about the end of World War I until about 1970.
3. The electronic era from about 1970.

As the name indicates, the first period was characterized by the fact that at most simple mechanical devices came to use for secret writing. For breaking schemes, beside ad hoc approaches mainly statistical methods as described in Section 2.2 were applied.

In the second period, for encryption next to schemes for writing by hand, electric-mechanic machines like the German Enigma were used. The increase of sophistication in the electronic-mechanic encryption machines was countered by cryptoanalytic methods which exceeded purely statistical techniques. Cryptanalysis was a driving force in the development and construction of the first electronic computing machines.

The electronic era started with the advent of data processing. As the methods developed in the beginning of the era are still being used or the current methods are direct successors of these methods, it can be seen as the present age of cryptology. This period is characterized not only by the used technology but also by its strive for scientific methods in cryptography, a strong connection to mathematics and a high innovation speed.

In this section we mainly want to retrace the development in this near past, whereby we focus on conceptual ideas.

3.1 Four seminal texts

We reduce the first two periods to four texts in which methods still of relevance today are developed. This condensed presentation does surely not do justice to the history of cryptology before the electronic era. A reader whose interest is aroused is invited to read the definitive book on the history of cryptology, David Kahn's *The Codebreakers*, first published in 1967 with a new edition from 1996 ([19]). For presentations of newer results on the history of cryptology we recommend the books *Codeknacker und Codemacher* by Klaus Schmeh ([31]) and *CryptoSchool* by Joachim von zur Gathen ([11]).

3.1.1   رسالة أبي يوسف يعقوب بن إسحاق الكنديّ في استخراج المعمّى إلى أبي العبّاس   *(Abū Yūsuf Ya'qūb ibn Ishāq al-Kindī: The missive on cryptanalysis to Abū l-'Abbās).*   By current knowledge, the first systematic presentation of cryptology originates from the Islamic middle ages. The author is the Aristotelian philosopher al-Kindī, who resided in Baghdad in the 9th century C.E.

This text was long considered lost, as were two other Arabic texts on cryptology from the 13th and 14th century. In the 1980s, for each of the three texts, one manuscript was found and subsequently edited and published ([25]).

In the manuscript presumably containing al-Kindī's treatise, different schemes based on substitution and transposition of letters are discussed; the schemes are even given in a tree diagram. The cryptanalysis is developed on the basis of frequency analysis and even bi- and digrams are considered.

By the way: Words of European languages like "cipher" (or "cypher"), "chiffre", "cifre", "cifra" or "Ziffer" come from the Arabic word for zero, صفر ("ṣifr") Over the course of time, these and similar words have meant zero, numerical symbol, scheme for en-/deciphering and cipher text.

3.1.2 *Leon Battista Alberti: De Componendis Cyfris (On the writing in ciphers).* The first passed down European text on cryptology was written in about 1466. Its author Leon Battista Alberti is regarded as the embodiment of a "universal Renaissance man"; he worked as an

Figure 1. The oldest known preserved cipher disk, a French disk from the time of Louis XIV (Source: Nicholas Gessler Collection)

architect and wrote remarkably many works about the most diverse subjects and also literary works.

Alberti advocates changing the substitution table during encryption. For this, he invented – how it seems – the *cipher disks*, which were popular devices for cryptography until the 19th century. These devices consist of two disks which are clinched in the middle, whereby the lower disk is larger than the upper one. On the boundaries of each of these disks an alphabet is written. For every position of the disks with respect to each other one therefore obtains a particular substitution.

For the scheme envisioned by Alberti, the order of one of the alphabets is permuted and the configuration of the two disks is changed after some words. In addition, Alberti proposed to use a code book for the most important words.

This concrete scheme does not seem to have been much in use, however. Rather, later a simpler but much less secure scheme became popular among laypersons of cryptology under the name *Vigenère scheme.* For this scheme, both alphabets are in identical (common) order and the position is changed after ever letter. The common secret is now a keyword (codeword). If this is for example DISK, then the A is first turned to the D, then to the I, then to the S and finally to the K, after which one starts anew. Professional cryptographers knew however that this scheme was rather weak and used more elaborate schemes involving code books. (For more information see Chapter 4 of [19].)

3.1.3 *Auguste Kerckhoffs: La Cryptographie Militaire.*   This work ([20]) from 1883 is arguably the one from the paper-and-pencil period which is most reffered to today. Kerckhoffs, a Dutch linguist and cryptographer residing in France, enunciates six principles for military cryptography, which were later called *Kerckhoffs' principles*. The first three are the most important ones. In slightly pointed and revised form, these are:

1. The scheme must be de facto, if not mathematically, unbreakable.
2. The usage of the encryption device must not require a secret, and it must be possible that such a device falls into the wrong hands without disadvantage.
3. The key must be transferable without written notes and kept in mind and exchanged at the will of the correspondents.

With respect to the first demand Kerckhoffs states: It is generally assumed that it is sufficient during war if a cipher system offers security for three or four hours. However, there is very well information which is important for more than a couple of hours. "Without enumerating all thinkable possibilities", Kerckhoffs mentions communication from a sieged city to the outside. That a good cryptographic scheme should offer security independently of all eventualities is an idea which has remained ever since.

The argumentation concerning the second demand is: Cryptography always demands secrets. Armies are now so large that one has to assume that the enemy knows all secrets which are known to a large number of soldiers. It is therefore imperative that only very few people must know of the secret. This means in particular that no extensive, hard to keep secret code books should be used. This demand is reinforced by the third principle.

These principles and their justification were seminal for the development of cryptography, particularly concerning the establishment of scientific methods. The first two principles are still considered to be fundamental for the design of cryptographic devices. Concerning the third principle, one still demands that the key can be exchanged easily but not that it can be kept in mind (cf. Section 3.2.7).

3.1.4 *Claude Shannon: Communication Theory of Secrecy Systems.*   Published in 1948 by the US-American Claude Shannon, this work ([32]) was of similar importance for the development of cryptology as the one by Kerckhoffs. Developing an abstract theory of en- and decryption, Shannon describes cryptology in mathematical terms like no one before him. For example, he describes encryption in a cipher system as a function in two variables, one for the key and one for the message (plain text). The claims are then formulated as mathematical theorems and proven accordingly.

Shannon asks how much information a cipher text gives about a plain text, given that the attacker already has some information. He uses stochastic and statistical methods as well as the so-called *information theory* developed by himself and introduces the notion of *perfect security* which captures that an attacker gains no information. Thereafter he proves that perfect security requires that the key written as a string of symbols is at least as long as the message to be transmitted securely. This illustrates that perfect security as defined by Shannon is incompatible with Kerckhoffs' principles.

Shannon also introduces two relevant principles for the construction of cipher schemes: *diffusion* and *confusion*. In both cases, it is the goal to impede the application of statistical methods for cryptanalysis. Diffusion means that small changes of the plain text affect large parts of the cipher text. Confusion as defined by Shannon means that the relationship between cipher text and key is difficult. This requirement was later extended by the demand that the relationship between cipher text and plain text shall also be complex.

3.2 The electronic era

During Word War II, a group of British scientists constructed one of the first vacuum-tube computers, called *Colossus*, to decipher the most high-level German military communication; their struggle was successful and one of the biggest successes of cryptanalysis ever.

For encryption and other aspects of cryptography computers were not relevant before the advent of electronic data processing at the end of the 1960s. The new technology soon went along with further changes: Handling and research with cryptography became more open, the notion of cryptography got a broader scope and in a completely new way mathematical concepts were employed. This development continues to the present day.

Computers store data as strings of bits; these bits are now the atoms that the letters were in the paper-and pencil period. It is natural to apply encryption at this level – what all schemes discussed in the following do. We recall that one can express natural numbers (which by definition shall also include the number 0) in the binary system, that is, by strings of bits. For example, the number 10 is represented by the string IOIO. The number of bits necessary to

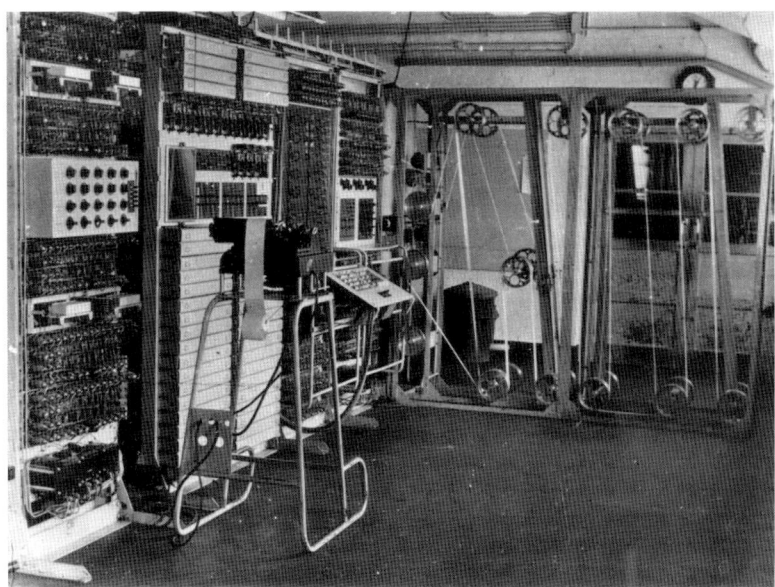

Figure 2. The Colossus computer (Source: British National Archive)

represent a number in this way is called the *bit-length* of the number; for a number $a \neq 0$ is about $\log_2(a)$. Moreover, by padding such strings with strings of 0's in front, one can identify bit-strings of a length exactly ℓ with the natural numbers smaller than 2^ℓ.

3.2.1 *From the Data Encryption Standard to the Advanced Encryption Standard.* In the year 1973 the predecessor of the US-American National Institute of Standards and Technology (NIST) made a public offer for a to-be standardized cipher algorithm (an *algorithm* being a method of computation). After no submitted algorithm was considered adequate, in 1977 a variant of an algorithm submitted by IBM was chosen as the *Data Encryption Standard (DES)*. Like all standards issued by NIST, also this standard has applied officially only to the US-government and its contractors. Nonetheless, it developed fast into a de facto industry standard – as was to be expected.

In the algorithm, blocks consisting of 64 bits each are encrypted. The algorithm consists of 16 rounds in which the same method is applied over and over again. At the beginning of each round, a part of the key and a part of the current intermediate result are bitwise merged via the *exclusive or (XOR)* operation. Thereafter the intermediate result is partitioned into small blocks of 6 bits each. With the help of fixed but "randomly looking" tables, each of the 6 bit long strings is changed to another sting. Finally, the blocks are transposed among each other.

One can say that the algorithm makes repeated uses of Shannon's ideas: Via the tables, one obtains confusion and via the transposition one obtains diffusion.

Up until now, no practically relevant attack on DES has been found that is faster than pure trial-and-error (which is also called the *brute force method*). A problem is however the key length of just 56 bits, which was right away criticized by the cryptographers Whitfield Diffie and Martin Hellman (who will play a crucial role in Section 3.2.4) as being too short. With the rapid development of computers the key length really got intolerably short. Because of this, in 1999 NIST initiated an open competition for a new encryption standard which should be known under the name of *Advanced Encryption Standard (AES)*. This time, there was a lively participation and a contribution from two Belgian cryptographers was chosen for the Advanced Encryption Standard.

The algorithm is similar to its predecessor: Again blocks are handled and the algorithm works in rounds, in each of which the key is fed in and confusion and diffusion is generated. In comparison with DES, this algorithm relies on mathematically clear and elegant constructions. With these constructions it can be proven that particular potential attacks are impossible.

As to be expected, no practically relevant attack has yet been found against AES – provided that an attacker does not have direct information about the computations in the algorithm. With a required key length of at least 128 bits the algorithm then seems to offer optimal security for many decades to come.

The assumption that an attacker does not have direct information about the computations seems to be innocent and clearly satisfied in practice. There are, however, surprising ways in which an attacker can obtain such information. Particularly, there are several practically relevant attacks on "straight-forward" implementations on AES which rely on an analysis of the running times (see [5]).

3.2.2 *Password encryption.* The development of multi-user computers led to a problem: If each user has a password, how can all these passwords be secured against espionage? In particular, how can it be ensured that a system administrator, who has access to the complete system, cannot read the passwords?

Let us assume that we have a function f which assigns to every password an "encrypted password" and which has the following properties: First, f can be computed in a fast manner and second, it is practically impossible to find for an encrypted password C a password P which is mapped to C via f, that is, with $f(P) = C$. Then one obtains the following authentication scheme:

Instead of storing the password P of a particular user, let us call her Ursa, one stores $C = f(P)$. If now a user claiming to be Ursa inputs a password P', one checks if $f(P) = f(P')$ holds. If this is the case, the user is authenticated as Ursa.

This idea was developed by Roger Needham at the University of Cambridge and called *one-way cipher* by his colleague Maurice Wilkies ([35]).

Building upon an encryption scheme like DES or AES, one can realize this idea as follows: One applies the encryption scheme with P as key and a constant plain text like $0\ldots0$; the result is then $f(P)$. If the enciphering scheme is secure, one obtains a function with the desired properties.

Besides the application itself, the idea is interesting for two reasons: First, it illustrates that there is a deep relationship between the classic goal of cryptography, confidentiality, and other goals like authentication. Second, the idea of a function f as described leads to a connection with *complexity theory*.

3.2.3 *Complexity theory.*

With the development of computers a new discipline emerged: computer science. Inside of computer science in turn developed *theoretical computer science* which comprises the theoretical study of algorithms in well-defined formal settings. From a scientific point of view, theoretical computer science is part of mathematics.

An important part of theoretical computer science is *complexity theory*. Here questions of the following kind are studied: Let a particular computational problem, for example the addition or the multiplication of natural numbers or the problem of factoring natural numbers, be given. How fast can then computations for larger and larger numbers be performed on an idealized elementary computing device? Thus, not computations for concrete inputs (also called *instances*) and also not inputs of a concrete order of magnitude are considered but all thinkable computations for all (infinitely many) inputs. One hereby imagines that the idealized computing device operates bit-wise, and one measures its running time accordingly. We exemplify this with the mentioned examples, starting with the addition of two natural numbers.

As remarked at the beginning of Section 3.2, the two natural numbers, say a and b, shall be given in binary representation. The running time shall be expressed with respect to the input length, which we denote by ℓ. (ℓ is about equal to the sum of the bit-lengths of the the the numbers, that is, to $\log_2(a) + \log_2(b)$ if $a, b \neq 0$.) With school-book addition one obtains a running time of at most $C \cdot \ell$ for some constant $C > 0$.

Analogously we consider the problem of multiplying two natural numbers a and b. With school-book multiplication one needs no more than ℓ additions of natural numbers of input length at most 2ℓ; one obtains thus a running time of at most $C' \cdot \ell^2$ for a constant $C' > 0$.

One expresses this as follows: The upper bound on the running time of the addition algorithm is *linear* in ℓ and the upper bound on the running time of the multiplication algorithm is *quadratic* in ℓ. There are also other methods for multiplication as the usual school-book method. For example, there is a method with which one can achieve a running time of at most $C'' \cdot \ell^{1,5}$ for a constant $C'' > 0$. This upper bound is from a certain size of ℓ onward better (that is, smaller) than the classic, quadratic one – independently of the constants C' and C''. In other words, the bound obtained via the alternative method is better than the classic one for all but finitely many inputs.

One says then that the bound obtained with the alternative method is *asymptotically* better. In complexity theory the focus is on such asymptotic statements, and from the point of view of complexity theory the bound obtained with the alternative method is considered to be better. Nevertheless, this does not say anything about which method may be faster for CONCRETE numbers a and b. Such statements are usually not addressed in complexity theory.

A running time which can be upper-bounded by $C \cdot \ell^k$ for some $C, k > 0$ is called *polynomial*. In complexity theory algorithms with polynomial running time are considered to be "fast" in a qualitative way and simply called "fast" or "efficient". To highlight the complexity theoretic approach, we use the term "qualitatively fast" and also modify other terms of complexity theory accordingly.

We see that the algorithms for the addition or for the multiplication two natural numbers have polynomial running time and are therefore qualitatively fast in the sense just defined.

Let us now consider the problem of integer factorization, where the integer is again variable. No method is known with which one can solve this problem qualitatively fast, that is, in polynomial time. This holds also if one allows that algorithms "throw dice" during a computation, an operation we allow from now on when we speak of an "algorithm". More specifically, no qualitatively fast algorithm is known that computes for a non-negligible portion of products pq of two prime numbers p, q of the same bit-length the factorization, whereby the notion "non-negligible" can also be defined in a precise manner. Even more, no qualitatively fast algorithm is known that computes for a non-negligible portion of products of two natural numbers of the same bit-length a factorization into two natural numbers of the same bit-length.

Following the ideas and notions of complexity theory, the situation just discussed can also be expressed as follows: Let f be the function which assigns to a tuple (m, n) of natural number m, n of the same bit-length its product mn, that is, $f(m, n) = mn$. This function is computable qualitatively fast. However, no algorithm is known with which one can compute for a non-negligible portion of values y of the function so-called *preimages* (which are here tuples (m, n) of natural numbers of the same bit-length with $mn = y$) in a qualitatively fast way. If there really is no such algorithm, the function considered is a so-called *one-way function*. With the notion of one-way functions the practical problem of "one-way encryption" as described in the preceding section is linked with complexity theory.

The now prevailing complexity theoretic point of view on cryptology is completely different from Shannon's. Whereas Shannon addressed the question *in how far* a cipher text determines the plain text or in how far one could compute information on plain texts from cipher texts if one had an *arbitrarily large* amount of computing power, the complexity theoretic point of view is: Can one in a certain sense compute plain texts from cipher texts in a *fast manner* or should this task considered to be infeasible?

Concerning the notion of "fast" one has however to act with caution: As already remarked, in complexity theory, concrete computations (like a concrete multiplication or factorization) are not considered. Rather, qualitative statements are made on the speed of solution methods for computational problems for ARBITRARY (and thus arbitrarily large) instances, that is, inputs. The statement that a particular function is a one-way function is therefore not a statement on the computational difficulty for concrete instances. There is thus a GAP between the complexity theoretic consideration and a possible practical application in cryptography like in password encryption. Closing this gap is not an easy task.

Also from a theoretical point of view there are problems concerning the notion of one-way function: Not a single function is proven to be indeed a one-way function – even if there are some good candidates for such functions, like the one just described.

The situation is even more tricky: If one can prove for a single function that it is one-way, one has solved the most famous open problem in theoretical computer science and one of the most prominent problems in mathematics, the *P versus NP problem*. This problem is considered to be one of the most difficult open questions of mathematics. It is one of the so-called *millennium problems* by the *Clay Mathematics Institute*; a solution is awarded with one million US dollars. A simple – however non-classic – formulation of this problem is as follows: Is there a qualitatively fast non-randomized algorithm which computes a solution to the following question, called the *subset sum problem*: Given arbitrarily many natural numbers $a_1, \ldots, a_k$ and a further natural number S, is the sum over SOME of these numbers equal to S? For example, for the four numbers $3; 7; 13; 21$ and $S = 31$ the answer is "yes" because $3 + 7 + 21 = 31$, whereas the answer is "no" if one changes S to 30 or to 32. Now the following is known: If there exists any one-way function, the *P versus NP* problem has a negative answer, that is, there is no such algorithm.

3.2.4 *Key exchange and cryptography with public keys.* According to Kerckhoffs' principles a cryptographic scheme should be distinguished from the key. The scheme should be generally known, whereas evidently the individual keys have to be kept secret. Two parties who wish to communicate with each other can publicly agree on a common scheme. But it seems to be clear, even self-evident, that the two parties cannot agree in public on a common key.

That this is nonetheless possible was shown by Whitfield Diffie and Martin Hellman in 1976 in a work with the seminal title "New directions in cryptography" ([7]). They presented a scheme which is now called the *Diffie–Hellman method* or *scheme.*

With the mentioned scheme, two people, which in cryptography are always called *Alice* and *Bob*, can agree on a common secret in public.

We briefly present the scheme. For this, we first recall the so-called *modulo computing*:

We choose a natural number $m \geq 2$, the so-called *modulus*. The most fundamental operation is now to take the remainder of an integer a with respect to division by m. The resulting integer, which is always between 0 and $m - 1$ (inclusively) is called the *remainder of a modulo m* and is denoted by a mod m.

Upon computation *modulo m* one computes in the set of integers $\{0, 1, \ldots, m - 1\}$ and after each operation the remainder modulo m is taken: If a and b are two natural numbers smaller than m, the result of *modulo addition* is $(a + b)$ mod m, the result of *modulo multiplication* is $(a \cdot b)$ mod m and for a natural number e the result of *modulo exponentiation* is a^e mod m.

For a prime number p and two natural numbers a, b between 1 and $p - 1$ (1 and $p - 1$ included) one has also $(a \cdot b)$ mod $p \neq 0$. This can be seen as an analogue to the fact that the product of two non-zero integers is also non-zero. One calls the domain $\{0, \ldots, p - 1\}$ with the given operations of computation the *finite prime field* with respect to the prime number p and denotes it by $\mathbb{F}_p$. These domains of computation have in various aspects analogous properties

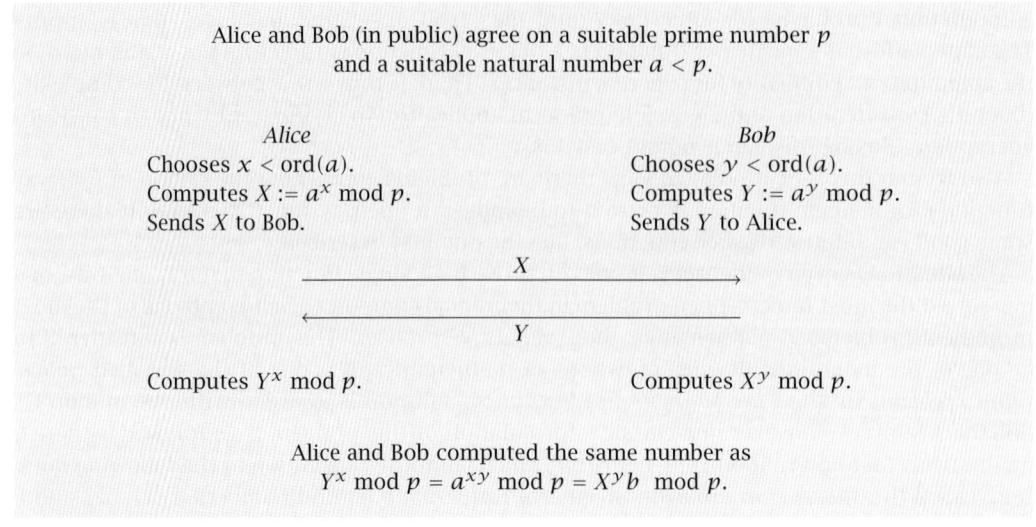

Alice and Bob (in public) agree on a suitable prime number p
and a suitable natural number $a < p$.

Alice	*Bob*
Chooses $x < \mathrm{ord}(a)$.	Chooses $y < \mathrm{ord}(a)$.
Computes $X := a^x \bmod p$.	Computes $Y := a^y \bmod p$.
Sends X to Bob.	Sends Y to Alice.

$$X \longrightarrow$$
$$\longleftarrow Y$$

Computes $Y^x \bmod p$.   Computes $X^y \bmod p$.

Alice and Bob computed the same number as
$$Y^x \bmod p = a^{xy} \bmod p = X^y b \;\bmod p.$$

Figure 3. The Diffie–Hellman protocol

to the domain of rational numbers, $\mathbb{Q}$; they can be seen as finite analogs of the infinitely large domain $\mathbb{Q}$.

An important aspect of modular computation with respect to the scheme by Diffie and Hellman is: If p, a and e are given, one can compute $a^e \bmod p$ qualitatively fast, that is, in polynomial time.

The scheme is now as follows: First, Alice and Bob agree on a (large) prime number p and a positive integer $a < p$. These two numbers can be known to everybody and might be shared by a large group of people. For this reason, we call p and a *public parameters*. Now, Alice chooses a natural number $x < p - 1$ and Bob chooses a natural number $y < p - 1$ at random. Both keep their numbers secret. Alice then computes $X := a^x \bmod p$ and sends this to Bob and he computes $Y := a^y \bmod p$ and sends this to Alice. Now, $X^y \bmod p = a^{xy} \bmod p = Y^x \bmod p$. This integer can be computed by both and shall be the common secret.

We now take the view of an eavesdropper who wants to compute the presumed secret. He receives p, a, X, Y and wants to compute $X^y \bmod p = Y^x \bmod p$. The problem to perform such a computation is now called the *Diffie–Hellman problem*.

A possible approach is to compute from p, a, X a natural number x with $a^x \bmod p = X$ because then the eavesdropper can easily also compute $Y^x \bmod p$. This computational problem is the so-called *discrete logarithm problem*.

A year after Diffie and Hellman's work, Ron Rivest, Adi Shamir and Leonard Adleman published an encryption scheme in which every user has a pair of keys consisting of a *private* and a *public* key ([29]). The idea of this so-called *RSA-scheme* is: Alice can distribute her public key and every person can use this key to send Alice an encrypted message, but only Alice can read the message with her private key. As one now uses two keys with distinctively different roles, one speaks here of an *asymmetric encryption scheme*, whereas the classical encryption schemes with a single common key or secret are called *symmetric encryption schemes*.

Also the RSA scheme is based on the difficulty of a computational problem, in this case of the problem of factoring the product of two (large) prime numbers; this problem was already discussed above. According to today's knowledge, the discrete logarithm problem (with random choice of p and a) and the factorization problem can be considered to be equally difficult at equal input length; more on the difficulty of the discrete logarithm problem in Sections 3.2.6 and 4.4.2.

In addition to encrypting texts, the RSA scheme can also be used for signing: A text can be signed by encrypting it with the private key. An alleged signature can then be checked by decryption with the public key and comparison with the original text. Cryptographic schemes relying on the use of public keys form what is now called the area of *public key cryptography*.

For their contributions to cryptography, both Rivest, Shamir and Adleman as well as Diffie and Hellman received the Turing Award, the most prestigious prize in computer science, named after one of the pioneers of computer science, Alan Turing: Rivest, Shamir and Adleman received the prize in 2002 and Diffie and Hellman in 2015.

3.2.5 *Protocols, active attackers and reductive security results.* Let us assume that Alice wants to communicate confidentially with Bob over the internet. She uses the Diffie–Hellman scheme with concrete parameters p and a to establish a common secret with Bob and then the cipher AES with the common secret as the key.

As just shown, an eavesdropper can break the Diffie–Hellman scheme (for concrete parameters p and a) if and only if he can solve the Diffie–Hellman problem for the same parameters. One could be tempted to conclude from this that Alice has chosen an adequate scheme for her goal if the Diffie–Hellman problem is unsolvable for the concrete parameters and no attack on AES is known.

This is however not the case. The reason is simple: From the moment of the key exchange on, an attacker could intercept her communication with Bob and masquerades as Bob with respect to Alice. Alice would then send confidential information to the attacker.

Maybe Alice would notice this after some time. An attacker can however proceed even more skillfully: He masquerades as Bob towards Alice and as Alice towards Bob and establishes a common key with each of them. He can then decrypt the messages from Alice to Bob and from Bob to Alice, read them, reencrypt them and send them off again. One speaks here of a so-called *man-in-the-middle attack.*

The demonstrated dysfunctionality of the scheme was due to the complete lack of authentication. The shown problems can however also be interpreted more abstractly: A cryptographic scheme can be INSECURE with respect to an ACTIVE ATTACKER even if it is SECURE AGAINST AN EAVESDROPPER, that is, a PASSIVE ATTACKER.

Whereas the obvious insecurity of the Diffie–Hellman scheme against active attacks highlights the importance to consider such attacks, such attacks were already relevant for classic cryptographic schemes. One idea for an active attack was already described in Section 2.2: One tricks a user of a cryptographic scheme to encrypt a given message and tries to obtain from the resulting pair of plain and cipher texts information about other received encrypted messages

Cryptographic schemes, in particular interactive cryptographic schemes, allow for a confusing magnitude of manipulation possibilities by active attackers. Some of these are evident, others in turn are not. To be convinced of the security of a scheme (for concrete parameters), one would like to have a strong argumentation that the scheme stays secure no matter which strategy an attacker chooses.

For just about every scheme in use today, it is thinkable that the security can be compromised even by passive attackers if for a particular basic algorithmic problem a new, surprisingly efficient algorithm is found. For example, the security of the Diffie–Hellman scheme with concrete parameters relies on the difficulty of the corresponding Diffie–Hellman problem with the same parameters. As such basic attacks cannot be avoided anyway, one tries to base the argumentation explicitly on the difficulty of some underlying problem. For a given task, like encryption, the goal is then to find an efficient scheme for which a large class of attacks can be ruled out if only a basic algorithmic problem is sufficiently difficult.

Such an approach requires adequate mathematically rigorous but also manageable definitions. Just the task to find such definitions is not an easy one.

The problems already start with the notion of "scheme". The usual mathematical definition is based on interacting algorithms whose inner computations are invisible for the attacker. One can also say that one abstracts from the inner computations and only considers the input-output-relationships. Such an abstracted scheme is called a *protocol* in cryptology. A protocol can be used for arbitrarily long inputs and comprises all necessary steps. For example, a protocol for the informally described Diffie–Hellman scheme begins with a setup phase. In this phase, after the input of some parameter size (like "1000 bits" or so), a suitable pair (p, a) is chosen. Such a protocol is to be distinguished from an *implementation* which consists of computer programs realizing the protocol. An *attacker* is then always an algorithm which interacts with the protocol.

An even greater challenge is finding adequate formal definitions for "secure" for different aspects of cryptography like establishing a common secret, encryption or decryption.

The established definitions mirror the complexity theoretic point of view described in Section 3.2.3. This means that not absolute statements on the security for concrete input lengths but QUALITATIVE statements for arbitrarily long input lengths are made. Following our general terminology introduced in Section 3.2.3, we emphasize this by using the term "qualitatively secure".

There are several, related formal definitions of *qualitatively secure* based on different attack scenarios. Most commonly, one nowadays bases the definitions on the idea of *games*. One hereby imagines that an "intelligent" attacker (also called *adversary*) plays a game against a simplistic *challenger*. One should hereby keep in mind that in fact the "attacker" is merely an algorithm.

To give an idea of this approach, we now give a slightly informal description of the strongest currently considered notion of qualitative security for symmetric encryption schemes, *Indistinguishability under adaptive chosen cipher text attack (IND-CCA2-security)*. It is based on the following game:

1. After the input of the key length, a secret key is chosen (in a randomized way) by a so-called setup algorithm and given to the challenger.
2. The attacker chooses some texts and sends them to the challenger with the request to either de- or encrypt them. The challenger sends the results back to the attacker.
3. The attacker chooses two different texts M_1 and M_2. He sends them both to the challenger. The challenger chooses one of the texts with equal probability, encrypts it to C and sends C back to the attacker.

4. The attacker again chooses some texts to be encrypted and some texts different from C to be decrypted. He sends them to the challenger who performs the desired operations and sends the results back to the attacker.

5. The attacker opts for M_1 or M_2.

During the whole game, the attacker can adapt his strategy according to previously obtained information. The attacker *wins* if he guesses correctly from which text he has received the cipher text. Nota bene: If the attacker merely guesses, he wins with a probability of $\frac{1}{2}$.

Let us call an attacker *qualitatively successful* if it is qualitatively fast and it wins with a probability which is non-negligibly larger than $\frac{1}{2}$. Then a scheme is called *IND-CCA2-secure* if there is no qualitatively successful attacker for the game just described.

Coming back to the game, let us note that in Step 3 the attacker may even choose M_1 or M_2 (or both) to be identical to a plain text chosen in Step 2. Similarly, in Step 4 the attacker may send the texts M_1 or M_2 to the challenger for encryption. The only request which is not allowed is to ask for the decryption of C. This implies that in order that an encryption scheme can be IND-CCA2-secure, it must be randomized.

To base definitions of security on games was a landmark idea. For this idea and related contributions to the mathematical foundations of cryptography, Shafi Goldwasser and Silvio Micali received the Turing Award in 2012.

On the basis of an attack scenario as the one given one can then try to establish a *reductive security result* for a given scheme. For this, one additionally fixes an underlying computational problem. A reductive security result, also called a *security reduction* or simply a *reduction*, is then a mathematical statement of the form: Every qualitatively successful attack on the protocol of the considered kind leads to a qualitatively fast algorithm for the computational problem. If such a result has been proven, one obtains: If there is no qualitatively fast algorithm for the computational problem, the protocol is qualitatively secure with respect to the considered kind of attacks.

Ideally, the security of suitable protocols (with respect to a wide range of attacks) for a multiplicity of cryptographic applications of modern times, like encryption, signature or authentication, would be reduced via reductive security results to a small number of algorithmic problems like to the factorization problem, the discrete logarithm problem or the Diffie–Hellman problem. These basic problem would then be studied exhaustively by the scientific community. This rigorous approach is particularly advocated in a two volume work by Oded Goldreich from 2001 and 2004 called *Foundations of Cryptography*, which can be seen as a first consolidation of the subject ([15]).

One should however note that, just as in Section 3.2.3, there is always a GAP between complexity theoretic considerations and praxis. Concerning the practical use of a protocol, there are several potential problems, even if a reductive security result for a broad attack scenario and a seemingly strong underlying problem has been proven:

- For a practically useful result it must be determined for which input lengths (or parameters) a protocol shall be used. If one takes the approach via reductive security results seriously, one must proceed as follows: One chooses a reductive security result with respect to a seemingly strong attack scenario and underlying computational problem. The result must not only qualitatively but explicitly and quantitatively relate the computational complexity of attacks and of the computational problem. One reasons for which input length this algorithmic problem is practically unsolvable. On the basis of these two statements, one computes how large the key length has to be in order that no practically relevant attack is possible if indeed the underlying problem is as difficult as assumed.

This is often not done, in particular because the key length would then be unmanageably large and/or the scheme too slow. Rather, often shorter key lengths are chosen or other scheme are considered which are inspired by the rigorously analyzed one, but nonetheless different.

- Self-evidently the implementation has to correspond to the description, it may therefore not contain any mistakes. To rule this out is difficult.
- Even on implementations which correspond to the specifications there are often attack possibilities. Notwithstanding that the attack scenarios considered are very broad, it is always assumed that an attacker does not know anything about the internal computations. However, concrete products often "radiate" in the literal and the figurative sense and this "radiation" might be be used for subtle attacks. An example are the attacks via running time already mentioned in Section 3.2.1.

3.2.6 *The influence of number theory.*   From a mathematical point of view, the discrete logarithm problem and the factorization problem fall not only in the realm of complexity theory but also in that of number theory. Thus the work by Diffie and Hellman created a connection between cryptology and this well-established field of pure mathematics. This connection is remarkable as mathematicians assumed just a few decades ago that especially number theory is immune against applications, in particular for military purposes. To this effect, the famous number theorist Godfrey Harold Hardy writes in his *A Mathematician's Apology* ([18]) from the war year 1940 that "real" (that is, deep) mathematics is "harmless and innocent" and concretely: "No one has yet discovered any warlike purpose to be served by the theory of numbers or relativity, and it seems very unlikely that anyone will do so for many years."

Cryptology is now unimaginable without number theory and related areas of mathematics. The importance of number theoretic methods gets particularly clear with the discrete logarithm problem, that is, the following algorithmic problem:

Given a prime number p, a positive integer $a < p$ and a further positive integer $b < p$ for which there is an x with $a^x \bmod p = b$, compute such an x.

The obvious first try is to solve the problem by brute force, that is, to test for given p, a, b consecutively for $x = 1, 2, 3, \ldots$ whether the equation $a^x \bmod p = b$ is satisfied.

To secure a system against this basic attack, the public parameters p, a must be chosen appropriately. But how can the running time be estimated for given p, a? Already for this basic question, elementary number theory is relevant:

As one might expect, the number of possible values $b = a^x \bmod p$ is crucial. This number is called the *order of a modulo p* and denoted $\mathrm{ord}(a)$. The values $a^x \bmod p$ lie between 1 and $p - 1$ (inclusively), therefore the order is at most $p - 1$. Furthermore, it holds, as can be shown: $a^{\mathrm{ord}(a)} \bmod p = 1$. If one multiplies this consecutively by a, one obtains $a^{\mathrm{ord}(a)+1} \bmod p = a$, $a^{\mathrm{ord}(a)+2} \bmod p = a^2 \bmod p$ and in full generality for every natural number e: $a^{\mathrm{ord}(a)+e} \bmod p = a^e \bmod p$. It follows that every possible value of $a^x \bmod p$ is taken for exactly one x between 0 and $\mathrm{ord}(a) - 1$ (inclusively). If one lets x run from 0 to $\mathrm{ord}(a) - 1$, there is thus exactly one x with $a^x \bmod p = b$. For fixed p and a and completely random b one needs in average $\frac{\mathrm{ord}(a)}{2}$ tries until one has found a solution.

Already Carl Friedrich Gauß proved that in his *Disquisitiones Arithmeticae*, published in 1801 ([12]), that for every prime number p there is an a of order $p - 1$. Let us consider such a pair, that is, let p be a prime and let $a < p$ be a positive integer of order $p - 1$.

For the brute force algorithm considered so far, one needs about a time which is given by p. By comparison: To compute for given p, a and an $x < p$ the value $a^x \bmod p$, one needs a running time which is about given by $\log_2(p)^3$. If for example p has 100 bits (about 30 positions in the decimal system), $\log_2(p)^3$ is about 300 whereas p is about 2^{100}, that is, about 10^{30}. The difference is enormous.

At the time of publication of the article by Diffie and Hellman it already was known that one cannot only obtain a running time of about p but of about $\sqrt{p}$. The idea of this can be described as follows: One computes numbers $a^c \bmod p$ and $a^d b \bmod p$ for arbitrary natural numbers c and d smaller than p and saves the results. Then one searches for a so-called collision $a^c \bmod p = a^d b \bmod p$. Such a collision leads to $a^{c-d} \bmod p = b$ if $c \geq d$ and to $a^{c-d+(p-1)} \bmod p = b$ if $c < d$. Maybe surprisingly one needs on average only about $\sqrt{p}$ results $a^c \bmod p$ and $a^d b \bmod p$ before one can find a collision as desired.

With the help of a classic number theoretic method, known under the name *Chinese Remainder Theorem*, this can be further improved. In total, one can obtain a running time of about $\sqrt{\ell}$, where ℓ is the largest prime divisor of $p - 1$.

Now, for a prime $p \geq 5$ the integer $p - 1$ is never prime as it is even and not 2. It is therefore of interest to consider primes p for which $\frac{p-1}{2}$ is also a prime. Such primes are called *Sophie-Germain primes*. Interestingly, it is not proven that there are infinitely many such primes, this is however conjectured.

If we consider Sophie-Germain primes, the running time of the best methods considered so far is again about $\sqrt{p}$.

There is however yet another method to solve the discrete logarithm problem, which can be called the *relation method*. This method is considerably more efficient than the collision method if the order of a is about that of p. It was already described by the mathematician Maurice Kraitchik in his *Théorie des Nombres* from 1922 ([23]) but fell into oblivion and was rediscovered after the publication of the work by Diffie and Hellman. We will not present the method here and solely remark that the relation method uses that one can factorize many natural numbers into products of substantially smaller (prime) numbers.

A natural question is now if there is a variant of the described discrete logarithm problem for which the mentioned algorithms do not work. For this one wants to substitute the domain of computation $\{1, \ldots, p - 1\}$ with modulo multiplication by another suitable domain with a completely different computing operation. It turns out that one cannot avoid the collision method under any circumstances. The reason is that the collision method relies directly on the computing operation itself. But is there a domain in which no better method is known, in which in particular the relation method does not work?

To put this idea into practice, Neal Koblitz and Victor Miller in 1981 independently proposed what is now called *elliptic curve cryptography*.

Elliptic curves are not ellipses, even if the name suggests this; the name relies on a "historical coincidence". However, one can explain what elliptic curves are by starting with the circle, which is a particular ellipse: The equation $x^2 + y^2 = 1$ describes the so-called *circle of unity* in the Cartesian coordinate system, that is, the circle of radius 1 around the origin. Every point P on the circle can be given by the angle α that it has to the y-axis; one then has $P = (\sin(\alpha), \cos(\alpha))$. If now such a point P and a further point $Q = (\sin(\beta), \cos(\beta))$ are given, one can add the angles and obtain in this way a new point $R = (\sin(\alpha + \beta), \cos(\alpha + \beta))$. Let us write $P \star Q$ for this point R, where the symbol "$\star$" is arbitrary and could be substituted by another symbol.

We obtain in this way a computing operation on the circle with radius 1, which might be called the "clock operation". This operation fulfills the usual rules of associativity and commutativity known from the addition or the multiplication of real numbers. Moreover, with $O := (0, 1)$ one has $P \star O = O \star P = P$ for every point P; the element $= O$ is therefore analogous to 0 for the addition of real numbers and to 1 for the multiplication of real numbers.

One does not need angles and trigonometric functions for the computing operation: Given points $P = (x_P, y_P)$ and $Q = (x_Q, y_Q)$ on the circle, the coordinates of the resulting point $R = P \star Q$ are given by the purely algebraic formulae

$$x_R = x_P y_Q + x_Q y_P \quad \text{and} \quad y_R = y_P y_Q - x_P x_Q . \tag{1}$$

If one now chooses a negative number d, the equation

$$x^2 + y^2 = 1 + d x^2 y^2 \tag{2}$$

describes an elliptic curve ([8], [2]). Interestingly, one can also define a computing operation on the curve. For points P and Q, the coordinates of the resulting point $R = P \star Q$ are now given by the formulae

$$x_R = \frac{x_P y_Q + x_Q y_P}{1 + d x_P x_Q y_P y_Q} \quad \text{and} \quad y_R = \frac{y_P y_Q - x_P x_Q}{1 - d x_P x_Q y_P y_Q} . \tag{3}$$

This operation is again associative and commutative and one again has $O \star P = P \star O = P$ for every point P.

One shall note that for $d = 0$ one would obtain again the circle with the computing operation (1), which is however not an elliptic curve. We also mention that the condition that d is negative ensures that the nominators in the formulae are always non-zero and remark – because we will use this shortly – that the condition on d is equivalent to d not being the square of another real number.

In cryptography, not solution sets of such equations over the real of the rational numbers but over finite computing domains are considered. Most often one uses the finite prime fields $\mathbb{F}_p = \{0, 1, \ldots p - 1\}$ described in Section 3.2.4, to which we want to restrict ourselves here.

So let p be a prime larger than 2 and $d \in \mathbb{F}_p$ with $d \neq 0$ and let us consider the solutions to the equation (2) in $\mathbb{F}_p$, that is, the tuples (x, y) with $x, y \in \mathbb{F}_p$ and $(x^2 + y^2) \bmod p = (1 + d x^2 y^2) \bmod p$. In order to obtain a computing operation on the solution set, one again has to ensure that the denominators in (3) are always non-zero. For this one uses the condition that d shall not be the square of another element, which is now also adapted to the computation in $\mathbb{F}_p$. This means that there shall be no $a \in \mathbb{F}_p$ with $a^2 \bmod p = d$; there are exactly $\frac{p-1}{2}$ such elements in $\mathbb{F}_p$.

The resulting domain of computation is usually denoted by $E(\mathbb{F}_p)$. The idea is now to substitute the domain $\{1, \ldots, p - 1\}$ with modular multiplication by such a domain $E(\mathbb{F}_p)$. Indeed, this is easily possible. One can then again speak of discrete logarithms and also of the *elliptic curve discrete logarithm problem* and one can adapt schemes as the one by Diffie and Hellman to this setting.

In the meantime, a large number of cryptographic schemes have been developed which rely on modular multiplication and which can be adapted to elliptic curves. Hereby in particular a

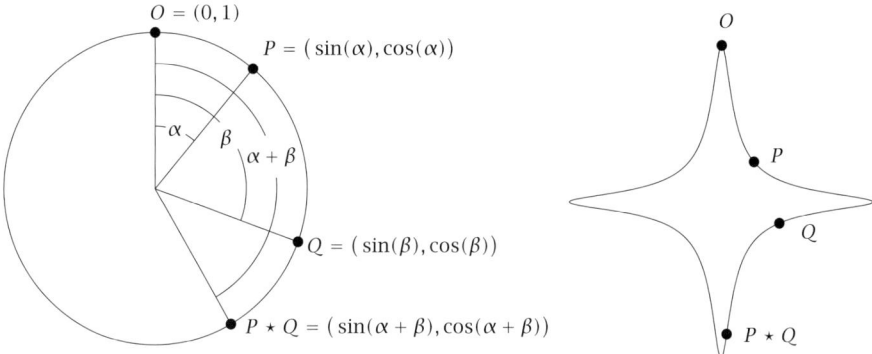

Figure 4. The unit circle (left) and the elliptic curve given by $x^2 + y^2 = 1 - 300x^2 y^2$ (right) with the computing operations

method by Taher ElGamal, which can be seen as an alternative to RSA, is worthwhile mentioning.

After 30 years of research, for most of the considered computing domains $E(\mathbb{F}_p)$ the collision method is still the most efficient method to solve the elliptic curve discrete logarithm problem. This means that for equal key length the variant of the protocol by Diffie and Hellman with elliptic curves leads to a much higher security level than the original protocol. The same holds true for other cryptographic protocols whose security rests on the discrete logarithm problem.

For example, it is recommended by the German Bundesamt für Sicherheit in der Informationstechnik (BSI) to use a key length of at least 250 bits if the discrete logarithm problem for elliptic curves is employed whereas for the classic problem and for the RSA-method a key length of at least 3000 bits is recommend ([4]).

A reader interested in public key cryptography and its number theoretic foundations might consider the book *Mathematics of Public Key Cryptography* by Steven Galbraith ([10]).

3.2.7 *Smart cards.* In the presentation of public key cryptography so far, a fundamental technical problem was put aside: How can the private key be secured from unwanted access? One can store the key by a variety of means. For example, in the 1980s the key could be saved on a floppy disk. To use the key it has however first to be loaded into the storage of a computer. This constitutes a security risk, in particular if the computer is connected to other computers, which is normally the case nowadays.

A clean solution is to create a small device containing its own miniature computer with storage and microprocessor. Then cryptographic applications like authentication and digital signature can be realized without the key ever leaving the device.

In the beginning of the 1980s miniaturization was advanced enough that the idea could be realized with very thin and about a square centimeter large microcomputers. Applied to plastic cards in credit card size, one obtains so the so-called *smart cards.*

The secret, the private key, shall not only not leave the card in the normal mode of operation, but under no circumstances shall it be possible to obtain usable information on the internal computations of the card, even if an attacker controls the environment of the card completely. This means for example that the card must not reveal information by radiation or by power consumption.

If the card falls into the wrong hands, it should be completely useless. For this, it is secured via a password, whereby the card locks itself by repeated faulty insertion. After this, even with physical manipulation or partial destruction of the card, it should be impossible to use the card or to obtain information on the key, which is after all stored on the card.

The magnitude of potential attacks poses a challenge for card manufacturers with which these however seem to cope well.

4 Current developments and challenges

4.1 Ubiquitous applications and interconnectedness

Whereas governments and companies have been using cryptography for data security already since the 1970s, since the spread of the World Wide Web, the everyday life of many, in industrialized countries arguably of most people cannot be imagined anymore without cryptography.

Public key cryptography is automatically used if one invokes a site with an address of the form `https://...`, which in turn is regularly the case if one transacts a payment. Here the user is informed about the use of cryptography, at other places this happens "transparently" for the user, that is, without him noticing anything.

Around the automobile alone, cryptography is employed multiple times. A classic application are electronic keys and immobilizers. Also in systems for road tolls, for example in the German Toll Collect system, cryptography is used. Since 2006, manipulations of the tachograph, mandatory for buses and trucks in the European Union, are prevented with cryptography. Completely new challenges are posed by communication from car to car, particularly if this influences the automatic behavior of cars. Additionally to using cryptography inside of applications, cryptography is also used to protect software in electronic devices against manipulation. As more and more electronic devices are used inside security relevant systems such as the brake or the steering system, this protection becomes ever more important.

Three interconnected trends are visible: More and more devices and products are equipped with microprocessors, these devices are more and more interconnected and thereby access to common storage outside of the devices is ever more important.

In both, business and private domains it becomes more and more common to not store data locally but in the "cloud" and to use services such as dropbox, google drive or icloud. All these services employ cryptography, it remains however the question in how far the data is really secure against access by employees or public authorities. If however the data is encrypted before sending it off, this problem is avoided. It then nonetheless remains the question if locally installed software does exactly what it is supposed to do or if it does not send off "a bit more".

With the interconnection, in the domestic domain too cryptographic challenges have developed or are going to develop in the near future. For example, many states of the European Union want to achieve that a major part of households are equipped with so-called "smart meters" by the year 2020. The thereby occurring data security problems shall be solved with cryptography. In the context of the German *Energiewende* (energy turn) and similar policies around the globe, "smart grids" with automatized turn on/off of devices by the grid administrator are promoted. Interesting cryptographic challenges are ahead here.

Even complete industrial complexes and critical infrastructures such as pipelines, electric grids or power stations are being connected to the internet, what gives rise to all kinds of horror scenarios. Here a radical solution against threat scenarios is obviously to separate the system control completely from the communication to the exterior. However, often one wants to avoid access from the exterior for security reasons while, yet at the same time allow external monitoring. As already the existence of a canal accessible from the exterior can be seen as a weakness, the establishment of a channel which physically makes data transfer to the interior impossible might be a solution. The company Waterfall Securities Solutions offers products based on a laser. With this technology systems can be created which offers a much higher level of security than electronic devices relying on cryptographic schemes (see Section 3.2.5).

4.2   The future of computers

The computing power of computers in relation to the amount of money to be spent has increased rapidly and without interruption since World War II. With this development, over and over, cryptographic systems once thought to be secure became attackable. What will hold the future and which impact will it have on cryptology?

Since the beginning of the 1970s, the integration density of microprocessors grew exponentially. It doubled about every two years, a fact which is known as *Moore's law*. Yet due to clear physical boundaries, this increase of integration density cannot continue arbitrarily. Concretely, currently in the most modern factories the so-called 14 nanometer technology is used, which is already remarkable if one considers that a silicon atom has a radius of about 0.1 nanometers. There are plans for a 5 nanometer technology for about the year 2020 for which one already has to fight with physical boundaries. Below 7 nanometers, one has to cope with the effect of quantum tunneling, which means that electrons can pass the logic gates unwantedly. One can therefore assume that the hitherto exponential grow will fade out relatively soon.

Yet it is thinkable that this development leads to a possibility to use quantum mechanical phenomena in a completely new way. The quantum world is a very strange world for humans. It surmounts human imagination over and over again, as can be seen with the different interpretations of quantum mechanics.

Already in the year 1981, the physicist Richard Feynman formulated the idea of a "quantum computer" and in the year 1994 the mathematician Peter Shor published a sketch of a potential quantum computer for the factorization and the discrete logarithm problem. He showed that in a mathematical model in which analogously to classic computational models quantum computers are described in an idealized way, these problems can be solved with polynomial cost, that is, qualitatively efficiently. For this breakthrough result Shor received the prestigious Rolf Nevanlinna Prize in mathematical aspects of information sciences by the International Mathematical Union in 1998.

The method for the computation of discrete logarithms on quantum computers is a generic method in the sense of Section 3.2.6. This means that just as the classic collision method it relies directly on the computing operation. The method is therefore not only applicable to the classic discrete logarithm problem modulo a prime number, but also to the problem in elliptic curves. Interestingly, the current main advantage of systems based on elliptic curves, the relatively short key length, could make such systems particularly vulnerable against quantum computers.

However, it is currently unclear if such a quantum computer will ever be built which can keep up with a classic computer. The hitherto tries in any case are slightly sobering: In the year 2001 with the help of a quantum computer based on Shor's ideas the number 15 was factored and the current record from the year 2012 is the number 21.[1]

In the potential quantum computers as envisioned by Shor, analogously to classic computers, the states are manipulated step by step in the course of time. There is also a different, more passive method to make quantum phenomena usable. With this method, Shor's ideas cannot be realized, but it is thinkable that for particular applications it leads to surprisingly fast computers. The company D-Wave develops computers which are based on this passive method. The computers seem to function as planned, but a prognosis on the capabilities of this technology currently seems to be hardly possible.

Interestingly, the American National Security Agency (NSA) seems to assume that quantum computers as envisioned by Shor can be realized in the coming decades. On August 19 2015, the agency announced that for publicly recommended cryptographic schemes, which includes schemes for communication with the US-government, it will "initiate a transition to quantum resistant algorithms in the not too distant future" ([27]).

4.3   Crypto currencies and crypto contracts

The crypto currency *bitcoin* is on everyone's lips. The development of its exchange rate is impressive. 10 000 bitcoins were offered by a programmer in the year 2010 for two pizzas, the prompt delivery of which led to the first payment in bitcoins. For the all time high up to now of 905 Euros per bitcoin in December 2013, one obtains, purely arithmetically, a solid price of about 9 million Euros. Also with the rate of around 400 Euros per bitcoin at the beginning of 2016, the pizzas were rather expensive.

Even though this development is impressive, at the moment there are no indications that bitcoins could indeed become relevant in day-to-day use.

The bitcoin payment system together with the anonymization software TOR is however one of the prominent technologies for anonymized trading platforms in the internet. The most well known of these markets was Silk Road, operating from 2011 to 2013. As this site was closed by the FBI in 2013, the agency declared that in about 1 million transactions a turnover of over 9 million bitcoins had been achieved. The physicist Ross Ulbricht was identified as the administrator of the site known under the pseudonym Dread Pirate Roberts and sentenced because of drug trafficking and other felonies to life long imprisonment without eligibility for parole. Ulbricht declared in the trail that he established the site because of idealism, "to empower people to be able to make choices in their life, for themselves in privacy and anonymity" and not being involved at a later time. One did not believe him.

Despite the drastic judgment, Silk Road will surely not be the last successful anonymous market place, by whatever motivation it will be established. A particularly interesting situation could occur if a state declared the operation of such a worldwide reachable marketplace legal.

---

1.   This is not a misprint; it is about the numbers 15 and 21, not about numbers with 15, respectively 21 bits.

Due to the outlook of a turnover in billions of Euros, particularly for smaller and poorer states there is an incentive to do so, if only the profit is taxed.

Even if the bitcoin system apparently does not prepare for a giant leap, this could be the case for an aspect of the bitcoin protocol: the *blockchain technology.*

Superficially, with the bitcoin protocol "electronic coins" are transferred. A first idea would be to administrate the rights on all coins in a single ledger. The administrator of the ledger would then be a kind of deposit bank (without lending). With the bitcoin protocol a more sophisticated decentralized scheme is realized. Public key cryptography on the basis of elliptic curves is used in about the following way: If Alice wants to transfer a coin to Bob, she signs with her private key a composition of the coin and Bob's public key. It must now be ensured that Alice indeed has the right to transfer the coin. For this, (essentially) all transactions since the beginning of the bitcoin system are stored. At first sight, this solution seems to contradict the desired anonymity of the bitcoin system. This is achieved by "Alice", "Bob" et cetera being only virtual concepts which are created anew over and which can act as fronts for arbitrary persons.

The ledger of transactions is not only stored once but in many different so-called knots. More concretely, new transactions are arranged in blocks and every 10 minutes in all knots the same new block is attached to the stored ledger. The resulting multiply stored ledger is called the *blockchain.*

By modifying the bitcoin protocol slightly, one can build systems for the decentralized storage and transfer of different categories of rights. To start with, one might use such a system for the management of bonds.

Bonds are administrated by so-called central security depositories and from a transaction to the settlement generally one to three days are passed during which the contracting partners have a mutual counterparty risk. With a scheme based on the blockchain technology, the central security depositories would be omitted and the settlement could occur a few minutes after the deal. This would be more efficient as well less risky.

Currently, there is a literal hype about the blockchain technology and it seems that in contrast to the anarchic bitcoin system this technology will indeed change the world of finance.

4.4   Edward Snowden and the NSA

The "Edward Snowden Story" is usually perceived by the public as a "real life thriller". The disclosures themselves have led to a vague feeling that "they surveil everything anyway".

With the disclosures indeed a substantial acquisitiveness became evident. On the other hand: That a secret service responsible for the surveillance and analysis of electronic communication surveils – presumably in line with the laws of its home country – exactly this communication and analyzes it with filter technologies should be obvious. It should be even more obvious that this technology is also used to surveil target subjects abroad, in particular after the corresponding country has suffered a massive terror attack.

It should have also been generally known that the sending of an email can be compared to the sending of a postcard, which everybody can read who gets it into his hand.

More interesting is the question concerning the abilities of the NSA regarding encrypted communication, a question on which speculations were made for years. The documents now in the public domain allow for the first time a look at the capabilities of the NSA in this area.

According to the documents, the NSA pursues a large-scale cryptanalytic program with the name *Bullrun*. In a presentation of the British partner service GCHQ it is written that Bullrun "covers the ability to defeat encryption used in specific network communications" and "includes multiple, extremely sensitive, sources and methods". It would offer "groundbreaking capabilities", would be "extremely fragile" and the addressees must "not ask about or speculate on sources or methods underpinning Bullrun successes" ([13]).

As however the further documents and known facts invite to such speculations, we will now do exactly this.

4.4.1 *Proactive approach.*   The NSA does not wait until schemes are established but rather attempts to steer the development into a direction favorable to the agency.

As per one of the revealed documents, in the years 2011 to 2013 an amount between 250 and 300 million US dollars was provided for the "SIGINT Enabling Project" ([28]). According to the project description, the "project actively engages the US and foreign IT industries to covertly influence and/or overtly leverage their commercial products' designs. These design changes make the systems in question exploitable through SIGINT collection [. . .] with foreknowledge of the modification. To the consumer and other adversaries, however, the systems' security remains intact." The resources of the project shall among others be used to "insert vulnerabilities" into systems and to "influence policies, standards and specifications for commercial public key technologies".

This apparently has been successful at least once, with the so-called *Dual-EC-Deterministic Random Bit Generator*, *Dual_EC_DRBG* for short.

Cryptographic protocols often need "randomness", which must first be generated in the computer. One can indeed generate "true randomness" by physical means, but this is rather time consuming. One therefore uses what is called a *pseudo random number generator* or a *deterministic random bit generator* to generate from a truly random bit-string a substantially longer bit-string. The essential requirements to such a generator are that it is very fast and that only with unrealistically large computing power the output can be distinguished from a truly random bit-string.

In the year 2006, the US-American standardization agency NIST published a "Recommendation for Random Number Generation Using Deterministic Random Bit Generators". As already remarked in Section 3.2.1, all standards and "recommendations" of NIST formally only apply to the US-government and its contractors, but often become de facto industry standards.

One class of generators in the document is the Dual_EC_DRBG. For this class of generators, also an exact specification with concrete parameters is contained in the document. This specification can be seen as the first standard with secret, but in hindsight obvious backdoor.

Already two years before the publication of the document of NIST the such specified generator was implemented by the company RSA Security as the default option in the widely used cryptographic software library BSAFE and only removed after the disclosures by Edward Snowden.

The "Dual_EC" stands for "dual elliptic curve" and indeed elliptic curves are of particular relevance for this class of generators. In contrast to other generators, the construction principle is particularly clear and mathematically elegant. This class of generators is a more efficient variant of the best known and most well studied class of generators in complexity oriented cryptography. For appropriate choices of parameters, the qualitative security of the generators can be reduced to a problem similar to the Diffie–Hellman problem.

In practice however, these generators have the disadvantage of being much slower than competing generators. Furthermore, as was already pointed out in 2005, the concrete generator specified by NIST can be distinguished from a truly random sequence, even if the derivation is small ([14]). Obviously, from the point of view of security, the parameters where chosen deliberately in the wrong way.

As argued in [3], the choice of parameters had its reasons: The generator is still reasonably secure for the user and at the same time it has an obvious backdoor for whoever chose the concrete parameters. This corresponds exactly to the the the goals in the document on "Bullrun" cited above. As the parameters for the concrete generator specified by NIST were officially computed by the NSA, one therefore cannot but assume that the NSA has a backdoor to this generator.

How could the NSA accomplish the masterpiece that a scheme with an obvious backdoor, which is in addition slow and has a security weakness independently of the backdoor, is built into a wildly used cryptographic software library as default option?

Obviously, no final answer is possible here, but there is some interesting information. According to Reuters, the documents brought into the public by Edward Snowden show that RSA Security received 10 million dollars by the NSA to make the standard the default method in the software library ([24]). RSA Security responded by stating that it "categorically" denies having "entered into a 'secret contract' with the NSA to incorporate a known flawed random number generator into its BSAFE encryption libraries". The company points out that it did indeed work with the NSA, which had "a trusted role in the community-wide effort to strengthen, not weaken, encryption" ([30]).

The NSA is indeed not a monolithic organization aimed at information gathering but also has a defensive arm, the Information Assurance Directorate. There is thus an internal conflict in the organization itself. This conflict was also addressed by a "Review Group on Intelligence and Communications Technologies" which was appointed by the American president Barack Obama after the disclosures by Edward Snowden ([6]). The committee asserted that the "NSA now has multiple missions and mandates, some of which are blurred, inherently conflicting, or both" and recommended to split off the Information Assurance Directorate from the NSA. However, not only were these plans not put into practice but somewhat ironically the NSA announced in February 2016 a reorganization, called NSA21, which goes in the opposite direction: the Information Assurance and the Signals Intelligence directorates shall be merged in a single directorate called Operations with employees working officially "on both sides" ([26]). It is not hard to predict that this merger will not help in regaining confidence by the industry in advice from the NSA.

4.4.2 Cryptanalysis.

Which "classic" cryptanalysis the NSA conducts exactly is not made clear by the documents. It would be a breakdown of security without comparison if an external staffer like Edward Snowden could get hold of such information. One can however get a general impression of applications of cryptanalysis and the general structure of systems.

Virtual private networks (VPNs) are cryptographically secured connections over the internet, which are for example used for external access to company nets. Often hereby the so-called *IPsec protocol* is used. According to the documents, the NSA has built a system for the analysis of data of such connections. Even if it is unknown how the system operates, in the article [1] the following is argued credibly:

Before the publication of the article, most connections could be started with a Diffie–Hellman key exchange modulo a prime p of 1024 bits. Now, already in the year 2005 researchers argued that it should be possible to build at a cost of one billion Euros a machine which can compute one instance of this size of the classic discrete logarithm in one year ([9]).[2]

By itself such a machine would be rather pointless for a secret service; after all, one not only wants to break one key exchange per year. Yet the problem to solve many discrete logarithms is in practice often much easier than one might think: According to information in [1], about two thirds of all key exchanges in VPNs using IPsec are conducted with respect to the same prime p. This means that one has to compute over and over again discrete logarithms modulo a single prime. Even if it is costly to start the computation of discrete logarithms modulo a fixed prime, after the initial phase it is surprisingly easy to compute arbitrarily many discrete logarithms.

Due to the falling of prices for computing power, the NSA could have built at the costs of some hundred million Euros a system with which it could surveil these VPNs and also the traffic with about one fifth of the most popular `https`-internet sites. This speculation is consistent with budget items for the NSA.

As this thought is now public, maybe exactly the case occurred against which it was warned in the presentation of the GCHQ mentioned in the beginning and the possibilities have been diminished already.

4.4.3 *What one should not forget.*   Since the disclosures by Edward Snowden, the NSA was very present in the media. In the crypto-scene there is a long tradition to call the top hacker "NSA". Also in this section capabilities of the NSA were discussed.

One should however not forget that besides the United States and its secret service partners Canada, United Kingdom, Australia and New Zealand, there are other countries with considerable means. And the five mentioned countries are after all states of the law, in which secret services face clear cut constraints, and lively democracies.

4.5 Specialization as a security threat

In all fields of science, the scientific progress leads to the necessity of more and more knowledge to comprehend or even to obtain a basic understanding of new works. There is therefore a general trend towards specialization. The formation of computer science and then of cryptology as a subdiscipline of computer science are aspects of this trend, a trend which now continues inside cryptology.

There are three aspects of the general trend towards specialization: First, works build on previous works; even if not directly results are used, a certain familiarity with definitions and techniques is necessary to read a work. This necessary background knowledge is constantly increasing. Second, works get more difficult to read even if one is familiar with the area. Third, not only are constantly new works added to the literature but the number of works published each year is also increasing.

---

2.   In [9] the factorization problem is treated. The discrete logarithm problem can be attacked with similar machines.

Nowadays, even for experts it is difficult to judge new works. To read a work without preparation, generally speaking it has to be very near to the personal research and even then one probably has to face several days of work. Sometimes, months or even a year of preparations are necessary before a work can be read and understood in detail.

This description fits to many areas of science, but for cryptography the implications are different than for pure mathematics, for example, because of the very goals of cryptography.

Why did the obvious backdoor in the Dual_EC_DRBG and in implementations like the one by RSA Securities not already cause a stir before the disclosures by Edward Snowden? Looking back, one notices that already in 2007, at the most prominent annual cryptologic conference, CRYPTO, two employees of Microsoft drew attention to the obvious backdoor in a short informal presentation ([33]). Despite this information, nobody seemed to have bothered to check whether the recommendation had been implemented. The author does not have an explanation of this, except maybe that he had never even heard about the NIST's "Recommendation" and the generator, even though he works in the very area of elliptic curve cryptography, and many of his colleagues had not either.

The real-life goal of the rigorous, mathematical approach to cryptographic schemes based on solid foundations, as described in Section 3.2.5, is to have an assurance against unexpected attacks. As discussed at the end of Section 3.2.5, there is always a gap between theory and praxis, which is not easy to close. Unfortunately, the factual carrying out of the rigorous approach to cryptographic schemes goes along with further problems:

The exact statements of scientific works in cryptography are often hard to understand and to interpret even for researchers working in the area and it is even harder to check whether the alleged proofs are correct. Not only this, but it does happen not too seldom that works with alleged reductive security results with respect to seemingly strong attack scenarios do not hold what they seem to promise. Sometimes, the exact contributions are misrepresented in the introduction; sometimes, the allegedly rigorous definitions on which the analysis relies are in fact unclear or not appropriate for the situation to be studied; sometimes, to obtain a proof for a particular attack scenario a scheme is designed which is then weak with respect to a straightforward attack outside of the attack scenario; sometimes, the underlying computational problem seems to be a contrived and artificial one which was only invented to obtain some kind of result; and last but not least, sometimes the alleged proofs are plainly wrong. Often such a problem has been discovered only after a scheme has been attacked successfully.

The fact that a number of protocols which were advertised as being "provably secure" (which actually means that some reductive security result had allegedly been established) have had unexpected real-life security holes has been critically assessed by Neal Koblitz, the coinventor of elliptic curve cryptography, and Alfred Menezes in an unusual article with the title "Another look at provable security" in the Journal of Cryptology and further articles with similar titles ([22]). According to Koblitz, "As with many other over-hyped ideas – fallout shelters in the 1950s, missile shields in the 1980s – 'proofs' of the security of a cryptographic protocol often give a false confidence that blinds people to the true dangers" ([21]).

The critique by Koblitz and Menezes has led to a fierce dispute in which the contrary position was particularly advocated by Oded Goldreich, the author of the *Foundations of Cryptography* ([16]). In his opinion, "Misconceptions about the meaning of results in cryptography are unfortunately all too common. But Koblitz and Menezes, besides pointing out some already known flaws in published purported proofs, only added to the confusion with an article which is full of such misconceptions as well." Flaws, misconceptions and misunderstandings would in any case only highlight the importance of a scientific approach to cryptography, an approach which is based on rigorous terminology and analyses ([17]).

Without further addressing the assessments by Koblitz and Menezes as well as by Goldreich, the author wants to emphasize that it is THE SCIENTIFIC PROGRESS ITSELF which goes along with increased specialization which in turn goes along with the careful reading and thinking through of many important works by a very small number of people. As humans have only limited time and limited intellectual capabilities and do err, there is no easy remedy against the security threat of misleading and wrong statements and publications or misconceptions by readers.

4.6   The picture at large

After existing in the penumbra and the influence of strongmen for centuries, cryptology has now stepped into the public.

There is an active research community with results in the public domain, there are established scientific principles, a never-ending stream of results, ideas for new applications as well as technical progress which makes new applications possible. Cryptographic schemes like crypto currencies or crypto contracts could have large consequences on the economy or even society as a whole.

But as in the past centuries so today there is the question if the used schemes and products in their daily use are really secure.

Like for other aspects of modern life, the layman has to rely here on specialists, who themselves only have a limited knowledge, can make mistakes or have other interests as the ones pretended. How can at least a partial remedy be found?

Well, a single specialist might make misleading statements or be misunderstood by a layman. Wrong statements by single persons are however rather irrelevant if there is a process in which worse ideas are refuted and better ideas can succeed.

Like for other areas it is also valid for cryptology: It is the right social institutions on which progress is built. Values like integrity, self critique, openness towards the new, a conduct based on transparency, factuality, cooperation and competition as well as the right formal institutions with clear goals free from conflicts of interest lead in a continuous improvement process to good ideas and products.

Acknowledgments.   I thank Marianne Diem, Werner Diem, Oded Goldreich, Wolfgang König, Alfred Menezes, Neal Koblitz, Eric Noeth and Sebastian Sterl for comments, discussions and help.

References

[1] DAVID ADRIAN, KARTHIKEYAN BHARGAVAN, ZAKIR DURUMERIC, PIERRICK GAUDRY, MATTHEW GREEN, J. ALEX HALDERMAN, NADIA HENINGER, DREW SPRINGALL, EMMANUEL THOMÉ, LUKE VA-LENTA, BENJAMIN VANDERSLOOT, ERIC WUSTROW, SANTIAGO ZANELLA-BÉGUELIN and PAUL ZIM-MERMANN, Imperfect forward secrecy: How Diffie–Hellman fails in practice, In *Procedings of the 22nd ACM SIGSAC Conference on Computer and Communications Security*, 2015.

[2] DANIEL BERNSTEIN and TANJA LANGE, Faster addition and doubling on elliptic curves. In *Advances in Cryptology – ASIACRYPT 2007*, Springer, Berlin, 2007, 29-50.

[3] DANIEL BERNSTEIN, TANJA LANGE and RUBEN NIEDERHAGEN, Dual EC: A Standardized Back Door. *Cryptology ePrint Archive*: Report 2015/767, 2015.

[4] BUNDESAMT FÜR SICHERHEIT IN DER INFORMATIONSTECHNIK, BSI Technische Richtlinie TR-02102-1, Kryptographische Verfahren: Empfehlungen und Schlüssellängen, 2015.

[5] ANNE CANTEAUT, Céderic Lauradoux and André Seznec, Understanding cache attacks. *INRIA Rapport de recherche* No. 5881, 2006.

[6] RICHARD CLARKE, MICHAEL MORELL, GEOFFREY STONE, CASS SUNSTEIN and PETER SWIRE, Liberty and Security in a Changing World, Report and Recommendations of The President's Review Group on Intelligence and Communications Technologies. The White House, 2013.

[7] WHITFIELD DIFFIE and MARTIN HELLMAN, New directions in cryptography. *IEEE Transactions on Information Theory* **2** (1976), 644-654.

[8] HAROLD EDWARDS, A normal form for elliptic curves. *Bulletin of the American Mathematical Society* **44** (2007), 393-422.

[9] JENS FRANKE, THORSTEN KLEINJUNG, CHRISTOF PAAR, JAN PELZL, CHRISTINE PRIPLATA and COLIN STAHLKE, SHARK. Presented at SHARCS – Special-purpose Hardware for Attacking Cryptographic Systems 2005. http://www.sharcs.org

[10] STEVEN GALBRAITH, *Mathematics of Public Key Cryptography*. Cambridge University Press, Cambridge, UK, 2012.

[11] JOACHIM VON ZUR GATHEN, *CryptoSchool*. Springer, Berlin, 2015

[12] CARL FRIEDRICH GAUSS, *Disquisitiones Arithmeticae*. Gerhard Fleischer, Leipzig, 1801.

[13] GCHQ, BULLRUN. Internal presentation. www.spiegel.de/media/media-35532.pdf

[14] KRISTIAN GJØSTEEN, Comments on the Dual-EC-DRBG/NIST SP 800-900, 2005. www.math.ntnu.no/~kristiag/drafts/dual-ec-drbg-comments.pdf

[15] ODED GOLDREICH, *Foundations of Cryptography I and II*. Cambridge University Press, Cambridge, UK, 2001 and 2004.

[16] ODED GOLDREICH, On post-modern cryptography. *Cryptology ePrint Archive*: Report 2006/461, 2006.

[17] ODED GOLDREICH, Conversation with the author, 2016.

[18] GODFREY HAROLD HARDY, *A Mathematician's Apology*. Cambridge University Press, Cambridge, UK, 1940.

[19] DAVID KAHN, *The Codebreakers – The Comprehensive History of Secret Communication from Ancient Times to the Internet*. Scribner, New York, 1996.

[20] AUGUSTE KERCKHOFFS, La cryptographie militaire. *Journal des sciences militaires* **9** (1883), 5-38 and 161-191.

[21] NEAL KOBLITZ, The uneasy relationship between mathematics and cryptography. *Notices of the AMS* **54** (2007), 972-979.

[22] NEAL KOBLITZ and ALFRED MENEZES, Another look at provable security. *Journal of Cryptology* **20** (2007), 3-37.

[23] MAURICE KRAITCHIK, *Théorie des Nombres*, Gauthier-Villars, Paris, 1922.

[24] JOSEPH MENN, Exclusive: Secret contract tied NSA and security industry pioneer. Reuters, Dec. 20, 2013.

[25] MOHAMMED MRAYATI, Y. MEER ALAM and M. H. TAYYAN, *Series on Arabic Origins of Cryptology* 1-3. KFCRIS & KACST, Riyadh, 2002-2003

[26] ELLEN NAKASHIMA, National Security Agency plans major reorganization. Washington Post, Feb. 2, 2016.

[27] NATIONAL SECURITY AGENCY, Suite B Cryptography, Aug. 19, 2015. www.nsa.gov/ia/programs/suiteb_cryptography/index.shtml

[28] NEW YORK TIMES, Secret documents reveal N.S.A. campaign against encryption, Sep. 5, 2013.

[29] RON RIVEST, ADI SHAMIR and LEONARD ADLEMAN, A method for obtaining digital signatures and public-key cryptosystems. *Communications of the ACM* **21** (1987), 120-126.

[30] RSA SECURITY, RSA response to media claims regarding NSA relationship, 2013. https://blogs.
     rsa.com/rsa-response

[31] KLAUS SCHMEH, *Codeknacker und Codemacher,* W3L, Bochum, 2014.

[32] CLAUDE SHANNON, Communication Theory of Secrecy Systems. *Bell System Technical Journal*
     28 (1948), 656 – 715.

[33] DAN SHUMOW AND NIELS FERGUSON, On the Possibility of a Back Door in the NIST SP800-90
     Dual EC Prng. CRYPTO Rump Session, 2007. http://rump2007.cr.yp.to/15-shumow.pdf

[34] PETER SHOR, Polynomial-time algorithms for prime factorization and discrete logarithms on a
     quantum computer. *SIAM Journal on Computing* **26** (1997), 1484–1509.

[35] MAURICE WILKES, *Time-Sharing Computer Systems.* American Elsevier, 1968.

A Mathematical view on voting and power

Werner Kirsch

In this article we describe some concepts, ideas and results from the mathematical theory of voting. We give a mathematical description of voting systems and introduce concepts to measure the power of a voter. We also describe and investigate two-tier voting systems, for example the Council of the European Union. In particular, we prove criteria which give the optimal voting weights in such systems.

1 Introduction

A voting system is characterized by a set V of voters and a collection of rules specifying the conditions under which a proposal is approved. Examples for the set V of voters comprise, e.g., the citizens of voting age in a country, the members of a parliament, the representatives of member states in a supranational organization or the colleagues in a hiring committee at a university. On any proposal to the voting system the voters may vote 'yes' or 'no'. Here and in the following we exclude the possibility of abstentions.[1]

Probably the most common voting rule is 'simple majority': A proposal is approved if more than half of the voters vote in favor of the proposal. This voting rule is implemented in most parliaments and committees. For some special proposals the voting rules may ask for more than half of the voters supporting the proposal. For example a two-third majority could be required for an amendment of the constitution. In such cases we speak of a 'qualified majority'.

In most countries with a federal structure (e.g., the USA, India, Germany, Switzerland, . . .) the legislative is composed of two chambers, one of which represents the states of the federation. In these cases the voters are the members of one of the chambers, a typical voting rule would require a simple majority in both chambers. However, the voting rules can be more complicated than this. For example in the USA both the President and the Vice-President are involved in the legislative process, the President by his right to veto a bill, the Vice-President as the President of the Senate (and tie-breaker in the Senate). In Germany the state chamber (called 'Bundesrat') can be overruled by a qualified majority of the Bundestag for certain types of laws, so called 'objection bills'.

In the Senate, the state chamber of the US legislative system, each state is represented by two senators. Since every senator has the same influence on the voting result ('one senator,

---

1. Voting systems with abstentions are considered in, for example, [11], [4] and [3].

one vote'), small states like Wyoming have the same influence in the Senate as large states like California. This is different in the Bundesrat, Germany's state chamber. Here the state governments have a number of seats (3 through 6) depending on the size of the state (in terms of population). The representatives of a state can cast their votes only as a block, i.e., votes of a state can't be split. This is a typical example of a weighted voting system: The voters (here: the states) have different weights, i.e., a specific number of votes.

Another example of a weighted voting system is the Board of Governors of the International Monetary Fund. Each member state represented by a Governor has a number of votes depending on the 'special drawing rights' of that country. For example the USA has a voting weight of 421962 (= 16.74% of the total weight) while Tuvalu has 756 votes, equivalent to 0.03%.

The Council of the European Union used to be a weighted voting system before the eastern extension of the European Union in 2004. Since then there is a more complicated voting system for the Council composed of two (or even three) weighted voting procedures.

In Section 2 of this paper we present a mathematical description of voting systems in general and specify our considerations to weighted voting systems in Section 3. In Section 4 we discuss the concept of voting power. Section 5 is devoted to a description of our most important example, the Council of the European Union. Section 6 presents a treatment of two-layer (or two-tier) voting systems. In such systems (e.g., the Council of the EU) representatives of states make decisions as members of a council. We raise and discuss the question of a fair representation of the voters in the countries when the population of the states is different in size. Finally, Section 7 presents a systematic probabilistic treatment of the same question.

The first four sections of this paper owe much to the excellent treatments of the subject in [12], [32] and [34]. In a similar way, Section 5 relies in part on [12].

2 A mathematical formalization

In real life voting systems are specified by a set a rules which fix conditions under which a proposal is approved or rejected. Here are a couple of examples.

Example 1. (1) The 'simple majority rule': A proposal is accepted if more than half of the voters vote 'yes'. More formally: If the voting body has N members then a proposal is approved if (and only if) the number Y of 'yes-votes' satisfies $Y \geq (N + 1)/2$. (Recall that we neglect the possibility of abstentions.)

(2) The 'qualified majority rule': A number of votes of at least rN is needed, N being the number of voters and r a number in the interval $(1/2, 1]$. Such a qualified majority is typically required for special laws, in particular for amendments to the constitution, for example with $r = 2/3$.

A simple majority rule is a special example of a qualified majority rule, with the choice $r = \frac{1}{2}(1 + \frac{1}{N})$.

(3) The 'unanimity rule': A proposal is approved only if *all* voters agree on it. This is a special case of (2), namely for $r = 1$.

(4) The 'dictator rule': A proposal is approved if a special voter, the dictator 'd', approves it.

(5) Many countries have a 'bicameral parliament', i.e., the parliament consists of two chambers, for example the 'House of Representatives' and the 'Senate' in the USA, the 'Bundestag' and the 'Bundesrat' in Germany.

One typical voting rule for a bicameral system is, that a bill needs a majority in both chambers to become law. This is, in deed, the case in Italy, where the chambers are called 'Camera dei Deputati' and 'Senato della Repubblica'. The corresponding voting rules in the USA and in Germany are more complicated and we are going to comment on such systems later.

(6) The UN Security Council has 5 permanent and 10 nonpermanent members. The permanent members are China, France, Russia, the United Kingdom and the USA. A resolution requires 9 affirmative votes, including all votes of the permanent members (veto powers).

The set of voters together with a set of rules constitute a 'voting system'. A convenient mathematical way to formalize the set of rules is to single out which sets of voters can force an affirmative decision by their positive votes.

DEFINITION 2. *A voting system is a pair $(V, \mathcal{V})$ consisting of a (finite) set V of voters and a subset $\mathcal{V} \subset \mathcal{P}(V)$ of the system of all subsets of V.*

Subsets of V are called coalitions, *the sets in $\mathcal{V}$ are called* winning coalitions, *all other coalitions are called* losing.

The set $\mathcal{V}$ consists of exactly those sets of voters that win a voting if they all agree with the proposal at hand.

Example 3. (1) In a parliament the set V of voters consists of all members of the parliament. Under simple majority rule, the winning coalitions are those which comprise more than half of the members of the parliament.

(2) If a body V decides according to the unanimity rule, the only winning coalition is V itself, thus $\mathcal{V} = \{V\}$.

(3) In a bicameral parliament consisting of chambers, say, H (for 'House') and S (for 'Senate'), the set of voters consists of the union $H \cup S$ of the two chambers. If a simple majority in both chambers is required a coalition M is winning if M contains more than half of the members of H *and* more than half of the members of S.

In more mathematical terms: $V = H \cup S$ (as a rule with $H \cap S = \emptyset$).

A coalition $A \subset V$ is winning, if

$$|A \cap H| > \frac{1}{2} |H| \quad \text{and} \quad |A \cap S| > \frac{1}{2} |S|$$

where $|M|$ denotes the number of elements in the set M.

(4) The voters in the UN Security Council are the permanent and the nonpermanent members. A coalition is winning if it comprises *all* of the permanent members *and* at least four nonpermanent members.

Later we'll discuss two rather complicated voting systems: the federal legislative system of the USA and the Council of the European Union.

In the following, we will *always* make the following assumption:

ASSUMPTION 4. *If $(V, \mathcal{V})$ is a voting system we assume that:*

(1) *The set of* all *voters is always winning, i.e., $V \in \mathcal{V}$.*
    'If all voters support a proposal, it is approved under the voting rules.'

(2) *The empty set $\emptyset$ is never winning, i.e., $\emptyset \notin \mathcal{V}$.*
    'If nobody supports a proposal it should be rejected.'

(3) *If a set A is winning (A ∈ 𝒱) and A is a subset of B, then B is also winning.*
  'A winning coalition stays winning if it is enlarged.'

REMARK 5. *If (V, 𝒱) is a voting system (satisfying the above assumptions) then we can recon-struct the set V from the set 𝒱, in fact V is the biggest set in 𝒱. Therefore, we will sometimes call 𝒱 a voting system without explicit reference to the underlying set of voters V.*

Some authors also require that if $A \in \mathcal{V}$ then $\complement A := V \setminus A \notin \mathcal{V}$. Such voting systems are called *proper*. As a rule, real world voting systems are proper. In the following, unless explicitly stated otherwise, we may allow improper voting systems as well.

Now, we introduce two methods to construct new voting systems from given ones. These concepts are implemented in many real world systems.

We start with the method of intersection. The construction of intersection of voting systems can be found in practice frequently. Many bicameral parliamentary voting systems are the intersections of the voting systems of the two constituting chambers.

DEFINITION 6. *We define the* intersection $(W, \mathcal{W})$ *of two voting systems $(V_1, \mathcal{V}_1)$ and $(V_2, \mathcal{V}_2)$ by*

$$W := V_1 \cup V_2$$
$$\mathcal{W} := \{M \subset W \mid M \cap V_1 \in \mathcal{V}_1 \quad \text{and} \quad M \cap V_2 \in \mathcal{V}_2\}. \tag{1}$$

We denote this voting system by $(V_1 \cup V_2, \mathcal{V}_1 \wedge \mathcal{V}_2)$ or simply by $\mathcal{V}_1 \wedge \mathcal{V}_2$.

Colloquially speaking: A coalition in $\mathcal{V}_1 \wedge \mathcal{V}_2$ is winning if 'it' is winning both in $\mathcal{V}_1$ and in $\mathcal{V}_2$. This construction can, of course, be done with more than two voting systems.

In an analogous way, we define the union of two voting systems.

DEFINITION 7. *We define the* union $(W, \mathcal{W})$ *of two voting systems $V_1, \mathcal{V}_1$ and $V_2, \mathcal{V}_2$ by*

$$W := V_1 \cup V_2$$
$$\mathcal{W} := \{M \subset W \mid M \cap V_1 \in \mathcal{V}_1 \quad \text{or} \quad M \cap V_2 \in \mathcal{V}_2\}. \tag{2}$$

We denote this voting system by $(V_1 \cup V_2, \mathcal{V}_1 \vee \mathcal{V}_2)$ or simply by $\mathcal{V}_1 \vee \mathcal{V}_2$.

We end this section with a formalization of the US federal legislative system.

Example 8 (US federal legislative system). We discuss the US federal system in more details. For a bill to pass Congress a simple majority in both houses (House of Representatives and Senate) is required. The Vice President of the USA acts as a tie-breaker in the senate. Thus a bill (at this stage) requires the votes of 218 out of the 435 representatives and 51 of the 100 senators or of 50 senators and the Vice President. If the President signs the bill it becomes law. However, if the President vetoes the bill, the presidential veto can be overruled by a two-third majority in both houses.

To formalize this voting system we define (with $|A|$ denoting the number of elements of the set A):

(1) The House of Representatives

Representatives	$R := \{R_1, \ldots, R_{435}\}$
Coalitions with simple majority	$\mathcal{R}_1 := \{A \subset R \mid |A| \geq 218\}$
Coalitions with two-third majority	$\mathcal{R}_2 := \{A \subset R \mid |A| \geq 290\}$

(2) The Senate

Senators	$S := \{S_1, \ldots, S_{100}\}$
Coalitions with at least half of the votes	$\mathcal{S}_0 := \{A \subset S \mid |A| \geq 50\}$
Coalitions with simple majority	$\mathcal{S}_1 := \{A \subset S \mid |A| \geq 51\}$
Coalitions with two-third majority	$\mathcal{S}_2 := \{A \subset R \mid |A| \geq 67\}$

(3) The President

$$V_p := \{P\} \qquad \mathcal{V}_p = \{V_p\}$$

(4) The Vice President

$$V_v := \{VP\} \qquad \mathcal{V}_v = \{V_v\}$$

Above the President (denoted by 'P') and the Vice President ('VP') constitute their own voting systems in which the only non empty coalitions are those containing the President and the Vice President respectively.

With these notations the federal legislative system of the USA is given by the set V of voters:

$$V = R \cup S \cup V_p \cup V_v$$

and the set $\mathcal{V}$ of winning coalitions:

$$\mathcal{V} = \left(\mathcal{R}_1 \wedge \mathcal{S}_1 \wedge \mathcal{V}_p\right) \vee \left(\mathcal{R}_1 \wedge \mathcal{S}_0 \wedge \mathcal{V}_v \wedge \mathcal{V}_p\right) \vee \left(\mathcal{R}_2 \wedge \mathcal{S}_2\right)$$

3 Weighted voting systems

DEFINITION 9. *A voting system $(V, \mathcal{V})$ is called a* weighted voting system *if there is a function $w : V \to [0, \infty)$, called the* voting weight, *and a number $q \in [0, \infty)$, called the* quota, *such that*

$$A \in \mathcal{V} \quad \Leftrightarrow \quad \sum_{v \in A} w(v) \geq q$$

Notation 10. For a coalition $A \subset V$ we set $w(A) := \sum_{v \in A} w(v)$, so a coalition A is winning if $w(A) \geq q$.

We also define the *relative quota* of a weighted voting system by $r := \frac{q}{w(V)}$. Consequently, A is winning if $w(A) \geq r\, w(V)$.

REMARK 11. *If $(V, \mathcal{V})$ is a voting system given by the weight function w and the quota q, then for any $\lambda > 0$ the weight function $w'(v) = \lambda\, w(v)$ together with the quota $q' = \lambda q$ define the same voting system.*

Examples 12. (1) A simple majority voting system is a weighted voting system (with trivial weights). The weight function can be chosen to be identically equal to 1 and the quota to be $\frac{N+1}{2}$ where N is the number of voters. The corresponding relative quota is $r = \frac{1}{2}(1 + \frac{1}{N})$.

(2) A voting system with unanimity is a weighted voting system. For example one may choose $w(v) = 1$ for all $v \in V$ and $q = |V|$ (or $r = 1$).

(3) In the German Bundesrat (the state chamber in the German legislative system) the states ('Länder' in German) have a number of votes depending on their population (in a sub-proportional way). Four states have 6 votes, one state 5, seven states have 4 votes and four states have 3 votes. Normally, the quota is 35, which is just more than half of the total weight,[2] for amendments to the constitution (as well as to veto certain types of propositions) a quota of 46 (two-third majority) is needed.

(4) The 'Council of the European Union' (also known as the 'Council of Ministers') is one of the legislative bodies of the EU (the other being the 'European Parliament'). In the Council of Ministers each member state of the EU is represented by one person, usually a Minister of the state's government. In the history of the EU, the voting system of the Council was changed a few times, typically in connection with an enlargement of the Union. Until 2003 the voting rules were given by weighted voting systems.

From 1995 to 2003 the EU consisted of 15 member states with the following voting weights:

Country	Votes	Country	Votes
France	10	Greece	5
Germany	10	Austria	4
Italy	10	Sweden	4
United Kingdom	10	Denmark	3
Spain	8	Finland	3
Belgium	5	Ireland	3
Netherlands	5	Luxembourg	2
Portugal	5		

The quota was given by $q = 62$, corresponding to a relative quota of 79%. Such quotas, well above 50%, were (and are) typical for the Council of the EU. This type of voting system is called a 'qualified majority' in the EU jargon.

After 2004, with the eastern extension of the EU, the voting system was defined in the Treaty of Nice, establishing a 'threefold majority'. This voting procedure consists of the intersection of three weighted voting system. After this the Treaty of Lisbon constituted a voting system known as the 'double majority'. There is a transition period between the latter two systems from 2014 through 2017. We discuss these voting systems in more detail in Section 5.

(5) The Board of Governors of the International Monetary Fund makes decisions according to a weighted voting system. The voting weights of the member countries are related to their economic importance, measured in terms of 'special drawing rights'. The quota depends on the kind of proposal under considerations. Many proposals require a relative quota of 70%. Proposals of special importance require a quota of even 85% which makes the USA a veto player in such cases (the USA holds more than 16% of the votes).

---

2. In a sense this is a simple majority of the weights. Observe, however, that in this paper we use term 'simple majority rule' only for systems with identical weight for all voters.

(6) The voting system of the UN Security Council does not seem to be a weighted one on first glance. In fact, the way it is formulated does not assign weights to the members. However, it turns out that one can find weights and a quota which give the same winning coalitions. Thus, according to Definition 9 it is a weighted voting system. For example, if we assign weight 1 to the non-permanent members and 7 to the permanent members and set the quota to be 39, we obtain a voting system which has the same winning coalition as the original one. Consequently these voting systems are the same.

The last example raises the questions: Can all voting systems be written as *weighted* voting systems? And, if not, how can we know, which ones can?

The first question can be answered in the negative by the following argument.

THEOREM 13. *Suppose $(V, \mathcal{V})$ is a weighted voting system and A_1 and A_2 are coalitions with $v_1, v_2 \notin A_1 \cup A_2$. If both $A_1 \cup \{v_1\} \in \mathcal{V}$ and $A_2 \cup \{v_2\} \in \mathcal{V}$ then $A_1 \cup \{v_2\} \in \mathcal{V}$ or $A_2 \cup \{v_1\} \in \mathcal{V}$ (or both).*

This property of a voting system is called 'swap robust' in [32]. There and in [34] the interested reader can find more about this and similar concepts.

Proof. Suppose w and q are a weight function and a quota for $(V, \mathcal{V})$.
By assumption

$$w(A_1) + w(v_1) \geq q \qquad \text{and} \qquad w(A_1) + w(v_1) \geq q$$

Thus

$$\big(w(A_1) + w(v_2)\big) + \big(w(A_2) + w(v_1)\big) \geq 2q$$

It follows that at least one of the summands has to be equal to q or bigger, hence

$$A_1 \cup \{v_2\} \in \mathcal{V} \qquad \text{or} \qquad A_2 \cup \{v_1\} \in \mathcal{V}$$

$\square$

From the theorem above we can easily see that, as a rule, bicameral are not weighted voting systems.

Suppose for example, the voting system consists of two disjoint chambers V_1 (the 'house') and V_2 (the 'senate') with $N_1 = 2n_1 + 1$ and $N_2 = 2n_2 + 1$ members, respectively, with $n_1, n_2 \geq 1$. Let us assume furthermore, that a proposal passes if there is a simple majority rule (by definition with equal voting weight) in both chambers.

So, a coalition of $n_1 + 1$ house members and $n_2 + 1$ senators is winning. Let C be a coalition of n_1 house members and n_2 senators and let h_1 and h_2 be two (different) house members not in C and, in a similar way, let s_1 and s_2 be two (different) senators not in C.

Set $A_1 = C \cup \{h_1\}$ and $A_2 = C \cup \{s_2\}$. Then $A_1 \cup \{s_1\}$ and $A_2 \cup \{h_2\}$ are winning coalitions, since they both contain $n_1 + 1$ house members and $n_2 + 1$ senators. However, $A_1 \cup \{h_2\}$ and $A_2 \cup \{s_1\}$ are both losing: The former coalition contains only n_2 senators, the latter only n_1 house members.

The above reasoning can be generalized easily (see, for example, Theorem 20).

We have seen that swap robustness is a necessary condition for weightedness. But, it turns out swap robustness is *not* sufficient for weightedness. A counterexample (amendment to the Canadian constitution) is given in [32].

There is a simple and surprising combinatorial criterion for weightedness of a voting system which is a generalization of swap robustness, found by Taylor and Zwicker [33], (see also [32] and [34]).

DEFINITION 14. *Let $A_1, \ldots, A_K$ be subsets of (a finite set) V.*
A sequence $B_1, \ldots, B_K$ of subsets of V is called a rearrangement *(or* trade*) of $A_1, \ldots, A_K$ if for every $v \in V$*

$$|\{k \mid v \in A_k\}| = |\{j \mid v \in B_j\}|$$

where $|M|$ denotes the number of elements of the set M.

In other words: From the voters in $A_1, \ldots, A_K$ we form new coalitions $B_1, \ldots, B_K$, such that a voter occurring r times in the sets A_k occurs the same number of times in the sets B_j.
   For example, the sequence $B_1 = \{1, 2, 3, 4\}, B_2 = \varnothing, B_3 = \{2, 3\}, B_4 = \{2\}$ is a rearrangement of $A_1 = \{1, 2\}, A_2 = \{2, 3\}, A_3 = \{3\}, A_4 = \{2, 4\}$.

DEFINITION 15. *A voting system $(V, \mathcal{V})$ is called* trade robust, *if the following property holds for any $K \in \mathbb{N}$:*
   If $A_1, \ldots, A_K$ is a sequence of winning coalition, i.e., $A_k \in \mathcal{V}$ for all k, and if $B_1, \ldots, B_K$ is a rearrangement of the $A_1, \ldots, A_K$ then at least one of the B_k is winning.
   $(V, \mathcal{V})$ is called M-trade robust, if the above conditions holds for all $K \leq M$.

THEOREM 16 (Taylor and Zwicker). *A voting system is weighted if and only if it is trade robust.*

It is straightforward to prove that any weighted voting system is trade robust. One can follow the idea of the proof of Theorem 13. The other direction of the assertion is more complicated, and more interesting. The proof can be found in [33] or in [34]. In fact, these authors show that every $2^{2^{|V|}}$-trade robust voting system is weighted.
   We have seen that there are voting system which can not be written as weighted system. However, it turns out that *any* voting system is an *intersection* of weighted voting systems.

THEOREM 17. *Any voting system $(V, \mathcal{V})$ is the intersection of weighted voting systems $(V, \mathcal{V}_1)$, $(V, \mathcal{V}_2), \ldots, (V, \mathcal{V}_M)$.*

Proof. For any losing coalition $L \subset V$ we define a weighted voting system (on V) by assigning the weight 1 to all voters *not* in L, the weight 0 to the voters *in* L and setting the quota to be 1. Denote the corresponding voting system by $(V, \mathcal{V}_L)$.
   Then the losing coalition in this voting systems are exactly L and its subsets. The winning coalitions are those sets K with

$$K \cap (V \setminus L) \neq \varnothing$$

Then

$$\mathcal{V} = \bigwedge_{L \subset V; L \notin \mathcal{V}} \mathcal{V}_L \tag{3}$$

In deed, if K is winning in *all* the $\mathcal{V}_L$, then K is not losing in $(V, \mathcal{V})$, so $K \in \mathcal{V}$. On the other hand, if $K \in \mathcal{V}$ it is not a subset of a losing coalition by monotonicity, hence $K \in \mathcal{V}_L$ for all losing coalitions L.   □

DEFINITION 18. *The dimension of a voting system $(V, \mathcal{V})$ is the smallest number M, such that $(V, \mathcal{V})$ can be written as an intersection of M weighted voting systems.*

It is usually not easy to compute the dimension of a given voting system. For example, the exact dimension of the voting system of the Council of the European Union according to the Lisbon Treaty is unknown. Kurz and Napel [21] prove that its dimension is at least 7.

In a situation of a system divided into 'chambers' we have the following results.

Example 19. Suppose $(V_i, \mathcal{V}_i), i = 1,\dots,M$ are voting systems with unanimity rule. Then the $(V_1 \cup \dots \cup V_M, \mathcal{V}_1 \wedge \dots \wedge \mathcal{V}_M)$ is a weighted voting system, i.e., the intersection of unanimity voting systems has dimension one. In deed, the composed system is a unanimous voting system as well and hence is weighted (see Example 12.2).

THEOREM 20. *Let* $(V_1, \mathcal{V}_1), (V_2, \mathcal{V}_2),\dots,(V_M, \mathcal{V}_M)$ *be simple majority voting systems with pairwise disjoint* V_i.

If for all i we have $|V_i| \geq 3$ *then the dimension of*

$$(V, \mathcal{V}) := (V_1 \cup \dots \cup V_M, \mathcal{V}_1 \wedge \dots \wedge \mathcal{V}_M)$$

is M.

Proof. First, we observe that for each i there is a losing coalition L_i and voters $\ell_i, \ell'_i \in V_i \setminus L_i$, such that $L_i \notin \mathcal{V}_i$, but $L_i \cup \{\ell_i\} \in \mathcal{V}_i$ and $L_i \cup \{\ell'_i\} \in \mathcal{V}_i$.

Suppose there were weighted voting systems $(U_1, \mathcal{U}_1),\dots,(U_K, \mathcal{U}_K)$ with $K < M$ such that their intersection is $(V, \mathcal{V})$. We consider the coalitions

$$K_i = V_1 \cup \dots \cup V_{i-1} \cup L_i \cup V_{i+1} \cup \dots \cup V_M$$

Then all K_i are losing in $(V, \mathcal{V})$, since L_i is losing in $\mathcal{V}_i$ by construction. Hence, for each K_i there is a j, such that K_i is losing in $(U_j, \mathcal{U}_j)$. Since $K < M$ there is a $j \leq K$ such that two different K_i are losing coalitions in $(U_j, \mathcal{U}_j)$, say K_p and K_q with $p \neq q$.

Now, we exchange two voters between K_p and K_q, more precisely we consider

$$K'_p := \left(K_p \setminus \{\ell_q\}\right) \cup \{\ell_p\}$$

$$\text{and} \quad K'_q := \left(K_q \setminus \{\ell_p\}\right) \cup \{\ell_q\}$$

By construction, both K'_p and K'_q are *winning* coalitions in $(V, \mathcal{V})$ and hence in $(U_j, \mathcal{U}_j)$. But this is impossible since K'_p and K'_q arise from two losing coalitions by a swap of two voters and $(U_j, \mathcal{U}_j)$ is weighted, hence swap robust. □

Example 21. The US federal legislative system (see Example 8) is not a weighted voting system due to two independent chambers (House and Senate). Moreover, it has two components (President and Vice President) with unanimity rule, and, as we defined it, it contains a *union* of voting systems. So, our previous results on dimension do not apply. It turns out, that it has dimension 2 (see [32]).

4 Voting power

Imagine two countries, say France and Germany, plan to cooperate more closely by building a council which decides upon certain questions previously decided by the two governments. The members of the council are the French President and the German Chancellor.

The German side suggests that the council members get a voting weight proportional to the population of the corresponding country. So, the French President would have a voting weight of 6, the German Chancellor a weight of 8, corresponding to a population of about 60 millions and 80 millions respectively. 'Of course', for a proposal to pass one would need more than half of the votes.

It is obvious that the French side would not agree to these rules. No matter how the French delegate will vote in this council, he or she will never ever affect the outcome of a voting! The French delegate is a 'dummy player' in this voting system.

DEFINITION 22. *Let* $(V, \mathcal{V})$ *be a voting system.*
A voter $v \in V$ *is called a* dummy player *(or dummy voter) if for any winning coalition A which contains v the coalition $A \setminus \{v\}$, i.e., the coalition A with v removed, is still winning.*

One might tend to believe that dummy players will not occur in real world examples. Surprisingly enough, they do.

Example 23 (Council of EEC). In 1957 the 'Treaty of Rome' established the European Economic Community, a predecessor of the EU, with Belgium, France, Germany, Italy, Luxembourg and the Netherlands as member states. In the Council of the EEC the member states had the following voting weights:

Country	Votes
Belgium	2
France	4
Germany	4
Italy	4
Luxembourg	1
Netherlands	2

The quota was 12.

In this voting system Luxembourg is a dummy player! Indeed, the minimal winning coalitions consist of either the three 'big' countries (France, Germany and Italy) or two of the big ones and the two medium sized countries (Belgium and the Netherlands). Whenever Luxembourg is a member of a winning coalition, the coalition is also winning if Luxembourg defects. This voting system was in use until 1973.

From these examples we learn that there is no *immediate* way to estimate the power of a voter from his or her voting weight. For instance, in the above example Belgium is certainly more than twice as powerful as Luxembourg. Whatever 'voting power' may mean in detail, a dummy player will certainly have *no* voting power.

In the following we will try to give the term 'voting power' an exact meaning. There is no doubt that in a mathematical description only *certain aspects* of power can be modelled. For example, aspects like the art of persuasion, the power of the better argument or external threats will not be included in those mathematical concepts.

In this section we introduce a method to measure power which goes back to Penrose [27] and Banzhaf [2]. It is based on the definition of power as the ability of a voter to change the outcome of a voting by his or her vote. Whether my vote 'counts' depends on the behavior of the other voters. We'll say that a voter v is 'decisive' for a *losing* coalition A if A becomes winning if v joins the coalition, we call v 'decisive' for a *winning* coalition if it becomes losing if v leaves this coalition. More precisely:

DEFINITION 24. *Suppose* $(V, \mathcal{V})$ *is a voting system. Let* $A \subset V$ *be a coalition and* $v \in V$ *a voter.*

(1) *We call* v winning decisive *for* A *if* $v \notin A$, $A \notin \mathcal{V}$ *and* $A \cup \{v\} \in \mathcal{V}$.
    We denote the set of all coalitions for which v *is winning decisive by*

$$\mathcal{D}^+(v) := \{A \subset V \mid A \notin \mathcal{V};\; v \notin A;\; A \cup \{v\} \in \mathcal{V}\} \tag{4}$$

(2) *We call* v losing decisive *for* A *if* $v \in A$, $A \in \mathcal{V}$ *and* $A \setminus \{v\} \notin \mathcal{V}$.
    We denote the set of all coalitions for which v *is losing decisive by*

$$\mathcal{D}^-(v) := \{A \subset V \mid A \in \mathcal{V};\; v \in A;\; A \setminus \{v\} \in \mathcal{V}\} \tag{5}$$

(3) *We call* v decisive *for* A *if* v *is winning decisive or losing decisive for* A.
    We denote the set of all coalitions for which v *is decisive by*

$$\mathcal{D}(v) := \mathcal{D}^+(v) \cup \mathcal{D}^-(v). \tag{6}$$

The 'Penrose–Banzhaf Power' for a voter v is defined as the portion of coalitions for which v is decisive. Note that for a voting system with N voters there are 2^N (possible) coalitions.

DEFINITION 25. *Suppose* $(V, \mathcal{V})$ *is a voting system,* $N = |V|$ *and* $v \in V$. *We define the* Penrose–Banzhaf power $PB(v)$ *of* v *to be*

$$PB(v) = \frac{|\mathcal{D}(v)|}{2^N}$$

REMARK 26. *The Penrose-Banzhaf power associates to each voter* v *a number* $PB(v)$ *between* 0 *and* 1, *in other words* PB *is a function* $PB : V \to [0, 1]$. *It associates with each voter the fraction of coalitions for which the voter is decisive.*

If we associate to each coalition the probability $\frac{1}{2^N}$, *thus considering all coalitions as equally likely, then* $PB(v)$ *is just the probability of the set* $\mathcal{D}(v)$. *Of course, one might consider other probability measure* $\mathbb{P}$ *on the set of all coalitions and define a corresponding power index by* $\mathbb{P}(\mathcal{D}(v))$. *We will discuss this issue later.*

If a coalition A is in $\mathcal{D}^-(v)$ then $A \cup \{v\}$ is in $\mathcal{D}^+(v)$ and if A is in $\mathcal{D}^+(v)$ then $A \setminus \{v\}$ is in $\mathcal{D}^-(v)$. This establishes a one-to-one mapping between $\mathcal{D}^+(v)$ and $\mathcal{D}^-(v)$. It follows that

$$|\mathcal{D}^+(v)| = |\mathcal{D}^-(v)| = \frac{1}{2}\,|\mathcal{D}(v)|. \tag{7}$$

This proves:

PROPOSITION 27. *If* $(V, \mathcal{V})$ *is a voting system with* N *voters and* $v \in V$ *then*

$$PB(v) = \frac{|\mathcal{D}^+(v)|}{2^{N-1}} = \frac{|\mathcal{D}^-(v)|}{2^{N-1}} \tag{8}$$

We also define a normalized version of the Penrose–Banzhaf power.

DEFINITION 28. *If* $(V, \mathcal{V})$ *is a voting system with Penrose-Banzhaf power* $PB : V \to [0, 1]$ *then we call the function* $NPB : V \to [0, 1]$ *defined by*

$$NPB(v) := \frac{PB(v)}{\sum_{w \in V} PB(w)}$$

the Penrose-Banzhaf index *or the* normalized Penrose-Banzhaf power.

The Penrose-Banzhaf index quantifies the share of power a voter has in a voting system.

PROPOSITION 29. *Let* $(V, \mathcal{V})$ *be a voting system with Penrose-Banzhaf power* PB *and Penrose-Banzhaf index* NPB.

(1) *For all* $v \in V : 0 \le PB(v) \le 1$ *and* $0 \le NPB(v) \le 1$. *Moreover,*

$$\sum_{v \in V} NPB(v) = 1. \tag{9}$$

(2) *A voter* v *is a dummy player if and only if* $PB(v) = 0$ ($\Leftrightarrow NPB(v) = 0$).
(3) *A voter* v *is a dictator if and only if* $NPB(v) = 1$.

As an example we compute the Penrose-Banzhaf power and the Penrose-Banzhaf index for the Council of the EEC.

Country	Votes	PB	NPB
Belgium	2	3/16	3/21
France	4	5/16	5/21
Germany	4	5/16	5/21
Italy	4	5/16	5/21
Luxembourg	1	0	0
Netherlands	2	3/16	3/21

For small voting bodies (as for the above example) it is possible to compute the power indices with pencil and paper, but for bigger systems one needs a computer to do the calculations. For example, the programm IOP 2.0 (see [5]) is an excellent tool for this purpose.

For a parliament with N members and equal voting weight (and any quota) it is clear that the Penrose-Banzhaf *index* $NPB(v)$ is $\frac{1}{N}$ for any voter v. This follows from symmetry and formula (9).

It is instructive (and useful later on) to compute the Penrose-Banzhaf *power* in this case.

THEOREM 30. *Suppose* $(V, \mathcal{V})$ *is a voting system with* N *voters, voting weight one and simple majority rule.*

Then the Penrose-Banzhaf power $PB(v)$ *is independent of the voter* v *and*

$$PB(v) \approx \frac{2}{\sqrt{2\pi}} \frac{1}{\sqrt{N}} \qquad as \ N \to \infty. \tag{10}$$

REMARK 31. *By* $a(N) \approx b(N)$ *as* $N \to \infty$ *we mean that* $\lim_{N \to \infty} \frac{a(N)}{b(N)} = 1$

Theorem 30 asserts that the Penrose-Banzhaf power in a body with simple majority rule is roughly inverse proportional to the *square-root* of the number N of voters and not to N itself as one might guess at a first glance. So, in a system with four times as much voters, the Penrose-Banzhaf power of a voter is one half $(= \frac{1}{\sqrt{4}})$ of the power of a voter in the smaller system. The reason is that there are much more coalitions of medium size (with about $N/2$ participants) than coalitions of small or large size. This fact will be important later on!

The proof is somewhat technical and can be omitted by readers who are willing to accept the theorem without proof.

Proof. We treat the case of odd N, the other case being similar. So suppose $N = 2n + 1$. A voter v is decisive for a losing coalition A if and only if A contains exactly n voters (but not v). There are $\binom{2n}{n}$ such coalitions. Hence, by (10)

$$BP(v) = \frac{1}{2^{2n}} \binom{2n}{n}. \tag{11}$$

Now, we use Stirling's formula to estimate $\binom{2n}{n}$. Stirling's formula asserts that

$$n! \approx n^n \, e^{-n} \sqrt{2\pi n} \qquad \text{as } n \to \infty.$$

Thus, as $N \to \infty$ we have:

$$\binom{2n}{n} \approx \frac{(2n)^{2n} \, e^{-2n} \, 2 \sqrt{\pi n}}{n^{2n} \, e^{-2n} \, 2 \pi n}$$

$$= \frac{2^{2n}}{\sqrt{\pi} \sqrt{n}}$$

so

$$BP(v) \approx \frac{2}{\sqrt{2\pi}} \frac{1}{\sqrt{N}}$$

$\square$

Instead of using *decisiveness* as a basis to measure power one could use the voter's *success*. A procedure to do so is completely analogous to the considerations above: We count the number of 'times' a voter agrees with the result of the voting ('is successful').

DEFINITION 32. *Suppose $(V, \mathcal{V})$ is a voting system. For a voter $v \in V$ we define*

the set of positive success

$$S_+(v) = \{A \in \mathcal{V} \mid v \in A\} \tag{12}$$

the set of negative success *by*

$$S_-(v) = \{A \notin \mathcal{V} \mid v \notin A\} \tag{13}$$

and the set of success

$$S(v) = S_+(v) \cup S_-(v) \tag{14}$$

REMARK 33. *If A is the coalition of voters agreeing with a proposal, then $A \in S_+(v)$ means, the proposal is approved with the consent of v, similarly $A \in S_-(v)$ means, the proposal is rejected with the consent of v.*

DEFINITION 34. *The* Penrose–Banzhaf rate of success $Bs(v)$ *is defined as the portion of coalitions such that v agrees with the voting result, more precisely:*

$$Bs(v) = \frac{|S(v)|}{2^N}$$

where N is the number of voters in V.

REMARK 35. *For all v we have $Bs(v) \geq 1/2$, in particular, a dummy player v has $Bs(v) = 1/2$.*

There is a close connection between the Penrose–Banzhaf power and the Penrose–Banzhaf rate of success.

THEOREM 36. *For any voting system $(V, \mathcal{V})$ and any voter $v \in V$*

$$Bs(v) = \frac{1}{2} + \frac{1}{2} PB(v) \tag{15}$$

This is a version of a theorem by Dubey and Shapley [8]. It follows that the success probability of a voter among N voters in a body with simple majority rule is approximately $\frac{1}{2} + \frac{1}{\sqrt{2\pi}} \frac{1}{\sqrt{N}}$.

The above results makes it essentially equivalent to define voting power via decisiveness or via success. However, equation (15) is peculiar for the special way we count coalitions here. If we don't regard all coalitions as equally likely, (15) is not true in general (see [23]).

5 The council of the European Union: A case study

In many supranational institutions the member states are represented by a delegate, for example a member of the country's government. Examples of such institutions are the International Monetary Fund, the UN Security Council and the German Bundesrat (for details see Examples 12). Our main example, which we are going to explain in more detail, is the Council of the European Union ('Council of Ministers'). The European Parliament and the Council of Ministers are the two legislative institutions of the European Union.

In the Council of the European Union each state is represented by one delegate (usually a Minister). Depending on the agenda the Council meets in different 'configurations', for example in the 'Agrifish' configuration the agriculture and fishery ministers of the member states meet to discuss questions in their field. In each configuration, every one of the 28 member countries is represented by one member of the country's government. The voting rule in the Council has changed a number of times during the history of the EU (and its predecessors). Until the year 2003 the voting rule was a weighted one. It was common sense that the voting weight of a state should increase with the state's size in terms of population. The exact weights were not determined by a formula or an algorithm but were rather the result of negotiations among the governments. The weights during the period 1958-1973 are given in Example 23, those during the period 1995-2003 are discussed in Example 12 (4). The three, later four, big states, France, Germany, Italy, and the United Kingdom, used to have the same number of votes corresponding to a similar size of their population, namely around 60 millions. After German unification, the German population suddenly increased by about one third.

Table 1. Penrose–Banzhaf power indices for the Nice treaty (Population: in % of the EU population. Square root: Ideal power according to Penrose. Deviation: Difference between actual and ideal power in % of the square root power)

Country	Population	Weight	Power	Square root	Deviation
Germany	15.9	29	7.6	9.1	−16.5
France	13	29	7.6	8.2	−7.3
United Kingdom	12.7	29	7.6	8.1	−6.2
Italy	12	29	7.6	7.9	−3.8
Spain	9.2	27	7.2	6.9	4.3
Poland	7.5	27	7.2	6.2	16.1
Romania	3.9	14	4.2	4.5	−6.7
Netherlands	3.3	13	3.9	4.2	−7.1
Belgium	2.2	12	3.6	3.4	5.9
Greece	2.2	7	2.1	3.3	−36.4
Czech Republic	2.1	12	3.6	3.3	9.1
Portugal	2.1	12	3.6	3.3	9.1
Hungary	1.9	12	3.6	3.2	12.5
Sweden	1.9	10	3	3.1	−3.2
Austria	1.7	10	3	3	0
Bulgaria	1.4	10	3	2.7	11.1
Denmark	1.1	7	2.1	2.4	−12.5
Finland	1.1	7	2.1	2.4	−12.5
Slovakia	1.1	7	2.1	2.4	−12.5
Ireland	0.9	12	3.6	2.2	63.3
Croatia	0.8	7	2.1	2.1	0
Lithuania	0.6	7	2.1	1.7	23.5
Slovenia	0.4	4	1.2	1.5	−20
Latvia	0.4	4	1.2	1.4	−14.3
Estonia	0.3	4	1.2	1.2	0
Cyprus	0.2	4	1.2	0.9	33.3
Luxembourg	0.1	4	1.2	0.8	50
Malta	0.1	3	0.9	0.7	28.6

This fact together with the planned eastern accession of the EU were the main issues at the European Summit in Nice in December 2000. The other big states disliked the idea to increase the voting weight of Germany beyond their own one while the German government pushed for a bigger voting weight for the country. The compromise found after nightlong negotiations 'in smoky back-rooms' was the 'Treaty of Nice'. In mathematical terms, the voting system of Nice is the intersection of three (!) weighted voting systems, each system with the same set of voters (the Ministers), but with different voting rules. In the first system a simple majority of the member states is required. The second system is a weighted voting system the weights of which are the result of negotiations (see Table 1). In particular the four biggest states obtained 29 votes each, the next biggest states (Spain and Poland) got 27 votes. In the Treaty of Nice two inconsistent quotas are stipulated for the EU with 27 members: At one place in the treaty the quota is set to 255, in another section it is fixed at 258 of the 345 total weight![3] With the accession of Croatia in 2013 the quota was set to 260 of a total weight of 352. In the third voting system, certainly meant as a concession towards Germany, the voting weight is given by the population of the respective country. The quota is set to 64%. With these rules the Nice procedure is presumably one of the most complicated voting systems ever implemented in practice.

It is hopeless to analyze this system 'with bare hands'. For example, it is not at all obvious to which extend Germany gets more power through the third voting system, the only one from which Germany can take advantage of its bigger population compared to France, Italy and the UK. One can figure out that the Penrose–Banzhaf power index of Germany is only negligibly bigger than that of the other big states, the difference in Penrose–Banzhaf index between Germany and France (or Italy or the UK) is about 0.000001. If instead of the voting according to population Germany had been given a voting weight of 30 instead of 29, this difference would be more than 1600 times as big! (for more details see Table 1 and the essay [16]).

In 2002 and 2003 the 'European Convention', established by the the European Council and presided by former French President Valéry Giscard d'Estaing, developed a 'European Consti-tution' which proposed a new voting system for the Council of the EU, the 'double majority'. The double majority system is the intersection of two weighted voting systems, one in which each member state has just one vote, the other with the population of the state as its voting weight. This seems to resemble the US bicameral system ('The Connecticut Compromise'): The House with proportional representation of the states and the Senate with equal votes for all states.

Presumably the reasoning behind the double majority rule is close to the following: On one hand, the European Union is a union of *citizens*. A fair representation of citizens, so the reasoning, would require that each state has a voting weight proportional to its population. On the other hand, the EU is a union of independent *states*, in this respect it would be just to give each state the same weight. The double majority seems to be a reasonable compromise between these two views.

The European Constitution was not ratified by the member states after its rejection in ref-erenda in France and the Netherlands, but the idea of the double majority was adopted in the 'Treaty of Lisbon'. The voting system in the Council, according to the Treaty of Lisbon is 'essentially' the intersection of two weighted voting systems.

In the first voting system ($\mathcal{V}_1$) each representative has one vote (i.e., voting weight = 1), the relative quota is 55%. In the second system ($\mathcal{V}_2$) the voting weight is given by the population of the respective state, the relative quota being 65%.

Actually, a third voting system ($\mathcal{V}_3$) is involved, in which each state has voting weight 1 again, but with a quota of 25 (more precisely three less than the number of member states). The voting system $\mathcal{V}$ of the Council is given by:

$$\mathcal{V} := (\mathcal{V}_1 \wedge \mathcal{V}_2) \vee \mathcal{V}_3$$

In other words: A proposal requires either the consent of 55% of the states which also repre-sent at least 65% of the EU population or the approval by 25 states.

---

3.   The self-contradictory Treaty of Nice was signed and ratified by 27 states.

The third voting system does not play a big role in practice, but is merely important psychologically as it eliminates the possibility that three big states alone can block a proposal. This rule actually adds 10 winning coalitions to the more than 30 million winning coalitions if only the two first rules were applied.

Compared to the Treaty of Nice the big states like Germany and France gain power by the Lisbon system, others in particular Spain and Poland lose considerably. Not surprisingly, the Polish government under Premier Minister Jarosław Kaczyński objected heavily to the new voting system. They proposed a rule called the 'square root system', under which each state gets a voting weight proportional to the square root of its population. In fact, the slogan of the Polish government was 'Square Root or Death'.

The perception of this concept in the media as well as among politicians was anything but positive. For example, in a column of the Financial Times [28], one reads: 'Their [the poles'] slogan for the summit – "the square root or death" – neatly combines obscurity, absurdity and vehemence' and: 'Almost nobody else wants the baffling square root system . . .'. In terms of the Penrose–Banzhaf indices, the square root system is to a large extend between the Nice and the Lisbon system. The square root system was finally rejected by the European summit.

With three rather different systems under discussion and two of them implemented the question arises: What is a just system? This, of course, is not a mathematical question. But, once the concept of 'justice' is clarified, mathematics may help to determine the best possible system.

One way to approach this question is to consider the influence citizens of the EU member states have on decisions of the Council. Of course, this influence is rather indirect by the citizens ability to vote for or against their current government. A reasonable criterion for a just system would be that every voter has the same influence on the Council's decisions regardless of the country whose citizens he or she is. This approach will be formalized and investigated in the next section.

6 Two-tier voting systems

In a direct democracy the voters in each country would instruct their delegate in the Council by public vote how to behave in the Council.[4] Thus the voters in the Union would decide in a two-step procedure. The first step is a public vote in each member state, the result of which would determine the votes of the delegates in the Council and hence the final decision. In fact, such a system is (in essence) implemented in the election of the President of the USA through the Electoral College.[5]

Modern democracies are -almost without exceptions- representative democracies. According to the idea of representative democracy, the delegate in the Council of Ministers will act on behalf of the country's people and is -in principle- responsible to them. Consequently, we will assume idealistically (or naively?) that the delegate in the Council knows the opinion of the voters in her or his country and acts in the Council accordingly. If this is the case we can again regard the decisions of the Council as a two-step voting procedure in which the first step -the

---

4. Of course, the voters in the Union could also decide directly then, we'll talk about this in the next section.

5. As a rule, the winner of the public vote in a state appoints *all* electors of that state. This is different only in Nebraska and Maine.

public vote- is invisible, but its result is known or at least guessed with some precision by the government and moreover is obeyed by the delegate.

In such a 'two-tier' voting system we may speak about the (indirect) influence a voter in one of the member states has on the voting in the Council. Now, we define these notions formally.

DEFINITION 37. *Let $(S_1, S_1), \ldots, (S_M, S_M)$ be voting systems ('M states') with $S := \bigcup_{i=1}^{M} S_i$ ('the union'). Suppose furthermore that $C = \{c_1, \ldots, c_M\}$ ('Council with delegates of the states') and that (C, C) is a voting system.*

For a coalition $A \subset S$ define

$$\Phi(A) = \{ c_i \mid A \cap S_i \in S_i \} \tag{16}$$

and

$$S = \{A \subset S \mid \Phi(A) \in C\} \tag{17}$$

The voting system (S, S) is called the two-tier voting system *composed of the lower tier voting systems $(S_1, S_1), \ldots, (S_M, S_M)$ and the upper tier voting system (C, C). We denote it by $S = \mathcal{T}(S_1, \ldots, S_M; C)$.*

Example 38. The Council of the EU can be regarded as typical two-tier voting system. We imagine that the voters in each member state decide upon proposals by simple majority vote (e.g. through opinion polls) and the Ministers in the Council vote according to the decision of the voters in the respective country.

We call systems as in the above example 'simple two-tier voting systems', more precisely:

DEFINITION 39. *Suppose $(S_1, S_1), \ldots, (S_M, S_M)$ and (C, C) with $C = \{c_1, \ldots, c_M\}$ are voting systems. The corresponding two-tier voting system (S, S) with $S = \mathcal{T}(S_1, \ldots, S_M; C)$ is called a* simple two-tier voting system *if the set S_i are pairwise disjoint and the S_i are simple majority voting systems.*

We are interested in the voting power exercised indirectly by a voter in one of the states S_i. For the (realistic) case of simple majority voting in the states and arbitrary decision rules in the Council we have the following result. In its original form this result goes back to Penrose [27].

THEOREM 40. *Let (S, S) be a simple two-tier voting system composed of $(S_1, S_1), \ldots, (S_M, S_M)$ and (C, C) with $C = \{c_1, \ldots, c_M\}$. Set $N_i = |S_i|$, $N = \sum_{i=1}^{M} N_i$ and $N_{\min} = \min_{1 \leq i \leq M} N_i$.*

If PB_i is the Penrose–Banzhaf power of c_i in C, then the Penrose–Banzhaf power $PB(v)$ of a voter $v \in S_k$ in the two-tier voting system (S, S) is asymptotically given by:

$$PB(v) \approx \frac{2}{\sqrt{2\pi N_k}} \, PB_k \qquad \text{as } N_{\min} \to \infty \tag{18}$$

Proof. To simplify the notation (and the proof) we assume that all N_i are odd, say $N_i = 2n_i + 1$. The case of even N_i requires an additional estimate but is similar otherwise.

A voter $v \in S_k$ is critical in S for a losing coalition A if and only if v is critical for the losing coalition $A \cap S_k$ in S_k *and* the delegate c_k of S_k is critical in C for the losing coalition $\Phi(A)$.

Under our assumption, for any coalition B in S_i either $B \in S_i$ or $S_i \setminus B \in S_i$. [6] So, for each i

---

6.   For even N_i this is only approximately true. Therefore, the case of odd N_i is somewhat easier.

there are exactly 2^{N_i-1} winning coalitions in S_i and the same number of losing coalitions. For each coalition $K \subset C$ there are consequently 2^{N-M} different coalitions A in S with $\Phi(A) = K$.

According to Theorem 30 there are $\approx \frac{2}{\sqrt{2\pi N_k}} 2^{N_k-1}$ losing coalition B in S_k for which v is critical. So, for each losing coalition $K \subset C$ there are approximately

$$\frac{2}{\sqrt{2\pi N_k}} 2^{N-M}$$

losing coalitions for which v is critical in S_k.

There are $2^{M-1} PB_k$ losing coalitions in C for which c_k is critical, hence there are

$$2^{N-1} PB_k \frac{2}{\sqrt{2\pi N_k}}$$

losing coalitions in S for which $v \in S_k$ is critical. Thus

$$PB(v) \approx \frac{1}{2^{N-1}} 2^{N-1} PB_k \frac{2}{\sqrt{2\pi N_k}} = \frac{2}{\sqrt{2\pi N_k}} PB_k$$

$\square$

There is an important – and perhaps surprising – consequence of Theorem 40. In a two-tier voting system as in the theorem it is certainly desirable that all voters in the union have the *same* influence on decisions of the Council regardless of their home country.

COROLLARY 41 (Square Root Law by Penrose). *If (S, S) is a simple two-tier voting system composed of $(S_1, S_1), \ldots, (S_M, S_M)$ with $N_i = |S_i|$ and (C, C) with $C = \{c_1, \ldots, c_M\}$. Then for large N_i we have:*

The Penrose–Banzhaf power $PB(v)$ in S for a voter $v \in S_k$ is independent of k if and only if the Penrose–Banzhaf power PB_i of c_i is given by $C\sqrt{N_i}$ for all i with some constant C.

Thus, the optimal system (in our sense) is (at least very close to) the 'baffling' system proposed by the Polish government! Making the voting weights proportional to the square root of the population does not give automatically power indices proportional to that square root. However, Wojciech Słomczyński and Karol Życzkowski [30] from the Jagiellonian University Kraków found that in a weighted voting system for the Council in which the *weights* are given by the square root of the population and the relative quota is set at (about) 62%, the resulting Penrose–Banzhaf index follows the square root law very accurately. This voting system is now known as the *Jagiellonian Compromise*. Despite the support of many scientists (see, e.g., [17,24,25]), this system was ignored by the vast majority of politicians.

Table 1 shows the Penrose–Banzhaf power indices for the Nice system and compares it to the square root law, which is the ideal system according to Penrose. There is a pretty high relative deviation from the square root law. Some states, like Greece and Germany, for example, get much less power than they should, others, like Poland, Ireland and the smaller states gain too much influence. All in all there seems to be no systematic deviation.

The same is done in Table 2 for the Lisbon rules. Under this system, Germany and the small states gain too much power while all medium size states do not get their due share. Assigning a weight proportional to the population is over-representing the big states according to the Square Root Law. In a similar manner, giving all states the same weight is over-representing the

Table 2. Penrose–Banzhaf power indices for the Lisbon treaty (Population: in % of the EU population. Square root: Ideal power according to Penrose. Deviation: Difference between actual and ideal power in % of the square root power)

Country	Population	Power	Square root	Deviation
Germany	15.9	10.2	9.1	12.2
France	13	8.4	8.2	2.5
United Kingdom	12.7	8.3	8.1	1.6
Italy	12	7.9	7.9	0
Spain	9.2	6.2	6.9	−9.8
Poland	7.5	5.1	6.2	−18.5
Romania	3.9	3.8	4.5	−15.9
Netherlands	3.3	3.5	4.2	−16.4
Belgium	2.2	2.9	3.4	−14.6
Greece	2.2	2.9	3.3	−14.3
Czech Republic	2.1	2.8	3.3	−14
Portugal	2.1	2.8	3.3	−15.6
Hungary	1.9	2.8	3.2	−11.5
Sweden	1.9	2.7	3.1	−13
Austria	1.7	2.6	3	−11.3
Bulgaria	1.4	2.5	2.7	−8.6
Denmark	1.1	2.3	2.4	−3.2
Finland	1.1	2.3	2.4	−2.6
Slovakia	1.1	2.3	2.4	−2.1
Ireland	0.9	2.2	2.2	2.2
Croatia	0.8	2.2	2.1	4.6
Lithuania	0.6	2	17	12.3
Slovenia	0.4	2	1.5	40.9
Latvia	0.4	2	1.4	36.7
Estonia	0.3	1.9	1.2	61.9
Cyprus	0.2	1.8	0.9	95.5
Luxembourg	0.1	1.8	0.8	140
Malta	0.1	1.8	0.7	170.8

small states, if equal representation of all citizens is aimed at. One might hope that the Lisbon rules compensate these two errors. But this is not the case. The Lisbon rules over-represent both very big and very small states, but under-represents all others.

One might hope that the Nice or the Lisbon system may observe the square root law at least approximately if the quota are arranged properly. This is not the case, see [19].

The indices in Table 1 and Table 2 were computed using the powerful program IOP 2.0 by Bräuninger and König [5].

7 A probabilistic approach

In this section, we sketch an alternative approach to voting, in particular to the question of optimal weights in two-tier voting systems, namely a probabilistic approach. This section is mathematically more involved than the previous part of this paper, but it also gives, we believe, more insight to the question of a fair voting system.

7.1 Voting measures and first examples

We regard a voting system $(V, \mathcal{V})$ as a system that produces output ('yes' or 'no') to a random stream of proposals. We assume that these proposals are totally random, in particular a proposal and its opposite are equally likely. The proposal generates an answer by the voters, i.e., it determines a coalition A of voters that support it. If A is a winning coalition the voting system's output is 'yes', if A is losing, the output is 'no'.

It is convenient to assume (without loss of generality) that $V = \{1, 2, \ldots, N\}$. We denote the voting behavior the voter i by $X_i \in \{-1, +1\}$. X_i depends, of course, on the proposal ω under consideration. $X_i(\omega) = 1$ means voter i agrees with the proposal, $X_i(\omega) = -1$ means i rejects ω. The voting result of all voters is a vector $X = (X_1, \ldots, X_N) \in \{-1, +1\}^N$. The random input generates a probability distribution $\mathbb{P}$ on $\{-1, +1\}^N$ and thus makes the X_i random variables.

The voting rules associate to each voting vector X a voting outcome: 'Yes' or 'No'.

We assume that the voters act rationally, at least in the sense that they either agree with a proposal or with its opposite, but never with both. Since we regard a proposal and its counter-proposal as equally likely rationality implies that the probability $\mathbb{P}$ is invariant under changing all voters' decisions, in the sense of the following definition.

DEFINITION 42. *A probability measure $\mathbb{P}$ on $\{-1, +1\}^N$ is called a* voting measure *if*

$$\mathbb{P}(X_1 = x_1, \ldots, X_N = x_N) = \mathbb{P}(X_1 = -x_1, \ldots, X_N = -x_N) \tag{19}$$

for all $x_1, x_2, \ldots, x_N \in \{-1, 1\}$.

REMARK 43. *Definition 42 implies in particular that*

$$\mathbb{P}(X_i = 1) = \mathbb{P}(X_i = -1) = \frac{1}{2} \tag{20}$$

Thus the distribution of any single X_i is already fixed. Note, however that there is still a great deal of freedom to choose the measure $\mathbb{P}$. It is the correlation structure that makes voting measures differ from one another. This correlation structure describes how voters may be influenced by each other or by some other factors like common believes or values, a state ideology, a dominant religious group or other opinion makers.

Notation 44. (1) Given a voting measure $\mathbb{P}$ on $\{-1, +1\}^N$ we use the same letter $\mathbb{P}$ to denote an associated measure on the set of all coalitions given by:

$$\mathbb{P}(A) := \mathbb{P}(X_i = 1 \text{ for } i \in A \text{ and } X_i = -1 \text{ for } i \notin A) \tag{21}$$

(2) To shorten notation we use the short hand

$$\mathbb{P}_B(x_1, x_2, \ldots, x_N) := \mathbb{P}(X_1 = x_1, X_2 = x_2, \ldots, X_N = x_N)$$

Example 45. (1) If we assume the voters cast their votes independently of each other we obtain the probability measure we already encountered in Remark 26:

$$\mathbb{P}_B(x_1, \ldots, x_N) = \frac{1}{2^N} \tag{22}$$

for all $(x_1, \ldots, x_N) \in \{+1, -1\}^N$. We call this voting measure the *independence measure* or the *Penrose-Banzhaf measure*, since it leads to the Penrose-Banzhaf power index.

(2) Another measure which occurs in connection with the Shapley-Shubik power index ([29, 31]) is the measure

$$\mathbb{P}_S(x_1, \ldots, x_N) = \frac{1}{2^{N+1}} \int_{-1}^{1} (1 + p)^k (1 - p)^{N-k} \, dp \tag{23}$$

where $k = |\{i \mid x_i = 1\}|$.

We call this voting measure the *Shapley-Shubik measure*. We will discuss it and its generalization, the 'common believe measure', below.

(3) Extreme agreement between the voters may be modelled by the voting measure $\mathbb{P}_{\pm 1}$ ('unanimity measure') which is concentrated on $(-1, -1, \ldots, -1)$ and $(1, 1, \ldots, 1)$, i.e.,

$$\mathbb{P}_{\pm 1}(x_1, x_2, \ldots, x_N) = \begin{cases} \frac{1}{2}, & \text{if } x_i = 1 \text{ for all } i; \\ \frac{1}{2}, & \text{if } x_i = -1 \text{ for all } i; \\ 0, & \text{otherwise.} \end{cases} \tag{24}$$

7.2 Basic examples

Now we introduce two classes of voting measures which we will discuss below in connection with two-tier voting systems.

7.2.1 The Common Believe Measure.

The 'common believe' voting measure is a generalization of the Shapley-Shubik measure (see [14]). In this model there is a common believe in a group (e.g., a state). For example, there might be a dominant religion inside the state with a strong influence on the people in certain questions of ethics.

This 'believe' associates to a proposal a probability with which voters inside the group will agree with this proposal. The common believe is a random variables Z with values in the interval $[-1, 1]$ and distribution (=measure) μ, i.e., $\mu(I) = \mathbb{P}(Z \in I)$ for any interval I. $Z > 0$ models a collective tendency in favor of the proposal at hand. This tendency increases with increasing Z, analogously $Z < 0$ means a tendency against the proposal. More precisely, if $Z = \zeta$ then the voters still decide independent of each other, but with a probability

$$p_\zeta = \frac{1}{2} (1 + \zeta) \qquad \text{for 'yes'} \tag{25}$$

and

$$1 - p_\zeta = \frac{1}{2} (1 - \zeta) \qquad \text{for 'no'.} \tag{26}$$

DEFINITION 46. *If μ is a probability measure on $[-1, 1]$ with $\mu([a, b]) = \mu([-b, -a])$ then we call the voting measure $\mathbb{P}_\mu$ on $\{-1, +1\}^N$ defined by*

$$\mathbb{P}_\mu(A) := \int_{-1}^{1} \mathbb{P}_\zeta(A) \, d\mu(\zeta) \tag{27}$$

the common believe voting measure *with collective measure μ.*

Here

$$P_\zeta(A) := p_\zeta^{|A|} (1 - p_\zeta)^{N-|A|} \tag{28}$$

is a product measure.

REMARK 47. (1) *The probability p_ζ defined in (25) is chosen such that the expectation is given by ζ.*
(2) *The condition $\mu([a,b]) = \mu([-b,-a])$ ensures that $\mathbb{P}_\mu$ is a voting measure.*
(3) *For $\mu = \delta_0$ we obtain the Penrose-Banzhaf measure $\mathbb{P}_B$.*
(4) *If μ is the uniform distribution on $[-1,1]$ we recover the Shapley-Shubik measure $\mathbb{P}_S$ of (23).*

7.2.2 *The Curie-Weiss model.* Finally, we introduce a model which originates in the statistical physics of magnetism. In this original context the meaning of the random variables X_i is the state of an elementary magnet pointing 'up' (for $X_i = 1$) or 'down' ($X_i = -1$). In statistical physics one is interested in describing collective phenomena, in particular alignment, of the magnets. This collective behavior depends on an external parameter, the temperature T. It is common in physics to introduce the 'inverse temperature' $\beta = \frac{1}{T}$. At low β (high temperature) one expects rather random (i.e., almost independent) behavior of the magnets, while for high β one expects that most of the magnets point into the same direction. So, β measures the strength of the (positive) correlation between the magnets, in our case between the voters.

One of the easiest models in physics which actually shows such a behavior is the 'Curie-Weiss' model. We will describe and use this model in the context of voting. While the Common Believe model describes values (or prejudices), the Curie-Weiss measure models the tendency of voters to agree with one another.

DEFINITION 48. *For given $\beta \geq 0$ the Curie-Weiss measure $\mathbb{P}_\beta$ is the probability measure*

$$\mathbb{P}_\beta(x_1, x_2, \ldots, x_N) := Z^{-1} e^{\frac{\beta}{2N} (\sum_{i=1}^N x_i)^2} \tag{29}$$

where $Z = Z_\beta$ is the normalization which makes $\mathbb{P}_\beta$ a probability measure, i.e.,

$$Z = \sum_{x_1, x_2, \ldots, x_N \in \{-1, +1\}} e^{\frac{\beta}{2N} (\sum_{i=1}^N x_i)^2} \tag{30}$$

REMARK 49. *We are mainly interested in the behavior of random variables such as $\sum_{i=1}^N X_i$ for large N. Hence we actually consider sequences $\mathbb{P}^{(N)}$ of voting measures on $\{-1, +1\}^N$ and $N = 1, 2, \ldots$. For the independence measure as well as for the common believe measure $\mathbb{P}^{(N)}$ is just the restriction of the corresponding measure on the infinite dimensional space $\{+1, -1\}^{\mathbb{N}}$ to $\{-1, +1\}^N$. Note, however, that the Curie-Weiss measures depend explicitly on the parameter N and are not the restrictions of a measure on the infinite dimensional space.*

The behavior of random variables distributed to $\mathbb{P}_\beta = \mathbb{P}_\beta^{(N)}$ change drastically at the inverse temperature $\beta = 1$. In physical jargon such a phenomenon is called a 'phase transition'.

THEOREM 50. *Suppose the random variables $X_1, \cdots, X_N$ are $\mathbb{P}_\beta^{(N)}$-distributed*

Curie-Weiss random variables and set $m_N = \frac{1}{N} \sum_{i=1}^N X_i$ then

(1) *If $\beta \le 1$ then*

$$m_N \overset{D}{\Rightarrow} \delta_0 \tag{31}$$

(2) *If $\beta > 1$ then*

$$m_N \overset{D}{\Rightarrow} \frac{1}{2}(\delta_{-m(\beta)} + \delta_{m(\beta)}) \tag{32}$$

where $m(\beta)$ is the unique (strictly) positive solution of

$$\tanh(\beta t) = t \tag{33}$$

Above $\overset{D}{\Rightarrow}$ denotes convergence in distribution. and δ_a denotes the Dirac measure at the point $a \in \mathbb{R}$, defined by

$$\delta_a(M) := \begin{cases} 1, & \text{if } a \in M; \\ 0, & \text{otherwise.} \end{cases}$$

A proof of the above theorem as well as additional information on the Curie–Weiss model can be found in [9], [35], and [15]

The above classes of voting measures introduced above give rise to power indices and success measures (see [18]). Instead of exploring them in this direction we will discuss their use in the description of two-tier voting systems in the next section.

7.3 Two-tier systems and public vote

In this section we apply the probabilistic approach to get more insight into two-tier voting systems. In this approach we try to minimize the discrepancy between the voting result in the two-tier system and the voting result in a general public vote.

Typical examples, we try to model, are the Council of the EU or the Electoral College of the USA. We will assume in the following that voters from different states (e.g. member states of the EU) vote independently of each other. However, inside the states we allow correlations between voters.

Throughout this section we assume the following situation: (S, S) is a simple two-tier voting system composed of $(S_1, S_1), \ldots, (S_M, S_M)$ and (C, C) with $C = \{c_1, \ldots, c_M\}$. We set $N_\nu = |S_\nu|$, $N = \sum_{\nu=1}^{M} N_\nu$. The vote of the ith voter in state S_ν will be denoted by $X_{\nu i} \in \{+1, -1\}$.

A public vote in the state S_ν is given by:

$$P_\nu = \sum_{i=1}^{N_\nu} X_{\nu i}. \tag{34}$$

Since we assume the voting system in S_ν is simple majority, a proposal is approved within S_ν if $P_\nu > 0$ and rejected otherwise.

A public vote in S is given by:

$$P = \sum_{\nu=1}^{M} P_\nu = \sum_{\nu=1}^{M} \sum_{i=1}^{N_\nu} X_{\nu i} \tag{35}$$

Suppose now, that the voting system in the Council C is given by weights w_ν and a quota q. Define the function χ by:

$$\chi(x) = \begin{cases} 1, & \text{if } x > 0; \\ -1, & \text{otherwise.} \end{cases} \tag{36}$$

As above, we assume that the delegates in the Council vote according to the public vote in their respective country, i.e., c_v will vote 1 if $\sum_{i=1}^{N_v} X_{vi} > 0$. Thus the vote in the Council will be:

$$C = \sum_{v=1}^{M} w_v \, \chi \Big(\sum_{i=1}^{N_v} X_{vi} \Big) \tag{37}$$

We remark that both P and C depend on the proposal in question.

Our goal is to choose the weights w_v as good as possible. It would be desirable to have $P = C$, however a moment's reflection shows that there is no choice of the weights which give $P = C$ for *all* proposals ω (i.e., for all possible distributions of 'yes' and 'no' among the voters). So, the best we can hope for is to choose w_v such that the discrepancy between P and C is minimal *in average*.

DEFINITION 51. *Let $\mathbb{P}$ be a voting measure and denote the expectation with respect to $\mathbb{P}$ by $\mathbb{E}$. We call the number*

$$\Delta(w_1, \ldots, w_M) = \Delta_{\mathbb{P}}(w_1, \ldots, w_M) := \mathbb{E}\left(|P - C|^2 \right) \tag{38}$$

the democracy deficit *of the two-tier voting system with respect to the voting measure $\mathbb{P}$. We call weights $w_1, \ldots, w_M$ optimal (with respect to $\mathbb{P}$), if they minimize the function $\Delta_{\mathbb{P}}$.*

Just as the power index and the success rate the democracy deficit depends on the choice of a voting measure. The choice of a good voting measure depends on the particular situation as well as the specific goal of our consideration. We'll comment on this point later on.

In the following, we will consider only such voting measures for which voters from different states are independent. The case of correlations of voters across state borders is more complicated. First results in this direction can be found in [22] and [20].

THEOREM 52. *Suppose that the voters X_{vi} and $X_{\rho j}$ in different states ($v \neq \rho$) are independent under the voting measure $\mathbb{P}$. Then the democracy deficit $\Delta_{\mathbb{P}}(w_1, \ldots, w_M)$ is minimal if the weights are given by:*

$$w_v = \mathbb{E}\left(\Big| \sum_{i=1}^{N_v} X_{vi} \Big| \right) \tag{39}$$

For the proof see [14].

The quantity $M_v = \big| \sum_{i=1}^{N_v} X_{vi} \big|$ is the margin with which the voters in state S_v decide, in other words, it is the difference in votes between the winning and the losing part of the voters. So, the representative of S_v in the Council is actually backed by M_v voters, not by *all* voters in S_v. The optimal weight w_v according to Theorem 52 is thus the expected margin of a decision of the voters in S_v. We regard this result as rather intuitive.

Theorem 52 tells us that the optimal weight depends on the correlation structure within the states S_v. The simplest case are uncorrelated (actually independent) voters within the states.

This is modelled by the independence measure, i.e., the Penrose–Banzhaf measure $\mathbb{P}_B$. Since the random variables X_{vi} are independent (even within the states), we know by the central limit theorem that the random variables $\frac{1}{\sqrt{N}} \sum_{i=1}^{N_v} kirsch$ converge in law to a standard normal distribution. Thus, we infer (as $N_v \to \infty$):

$$w_v = \mathbb{E}\left(\Big| \sum_{i=1}^{N_v} X_{vi} \Big| \right) \approx C \sqrt{N_v}. \tag{40}$$

Hence we proved:

THEOREM 53. *The optimal weights w_v for independent voters are proportional to $\sqrt{N_v}$ for large N_v.*

This result is close in spirit to the square root law by Penrose. In fact, Felsenthal and Machover [12] call it the second square-root rule.

Let us now consider the voting measure $\mathbb{P}_\mu$ inside the states and suppose that $\mu \neq \delta_0$. In this case the voters inside a state are not independent. In fact, it is not hard to see that for $i \neq j$

$$\mathbb{E}(X_{vi} X_{vj}) = \int_{-1}^{1} t^2 \, d\mu(t) > 0. \tag{41}$$

We have (see [14])

THEOREM 54. *Suppose the collective measure μ is not concentrated in the single point 0, then (as $N_v \to \infty$)*

$$\mathbb{E}_\mu \left(\left| \sum_{i=1}^{N_v} X_{vi} \right| \right) \approx C N_v. \tag{42}$$

So, the optimal weights for the Common Believe Measure are proportional to N_v.

There is a generalization of this theorem when the collective measure $\mu = \mu_N$ may depend on N. With an appropriate choice of μ_N one can get $w_v \sim N^\alpha$ for any $1/2 \leq \alpha \leq 1$ (for a proof of both the theorem and its generalization see [14]).

These examples suggest that positive correlation between voters and thus collective behavior leads to a higher optimal voting weight in the Council. This conjecture is supported by the following two results.

THEOREM 55. *Let $\mathbb{P}_\beta$ be the Curie-Weiss voting measure. Then the optimal weights for the Council are given by*

$$w_v = \mathbb{E}_\mu \left(\left| \sum_{i=1}^{N_v} X_{vi} \right| \right) \approx \begin{cases} C_\beta \sqrt{N}, & \text{for } \beta < 1; \\ C_1 N^{3/4}, & \text{for } \beta = 1; \\ C_\beta N, & \text{for } \beta > 1. \end{cases} \tag{43}$$

For a proof of this theorem see [14] (in combination with [9] or [15]).

There is a common pattern behind the above result which we summarize now.

THEOREM 56. *Suppose $\mathbb{P}_\mu$ is a voting measure on the simple two-tier voting system $S = \mathcal{T}(S_1, \ldots, S_M; C)$ under which voters in different states are independent and set $\Sigma_N^{(v)} = \sum_{i=1}^{N_v} X_{vi}$.*
(1) *If*

$$\frac{1}{N} \Sigma_N^{(v)} \overset{\mathcal{D}}{\Longrightarrow} \mu$$

with $\mu \neq \delta_0$, then the optimal weight w_v is asymptotically given by

$$w_v \approx \int |t| \, d\mu(t) \, N$$

(2) *If*

$$\frac{1}{N} \Sigma_N^{(v)} \overset{\mathcal{D}}{\Rightarrow} \delta_0 \qquad and \qquad \frac{1}{N^\alpha} \Sigma_N^{(v)} \overset{\mathcal{D}}{\Rightarrow} \rho$$

with $\rho \neq \delta_0$, then the optimal weight w_v is asymptotically given by

$$w_v \approx \int |t| \, d\rho(t) \, N^\alpha$$

The above considerations raise the question which voting measure to choose in a particular situation. As usual, the answer is: It depends!

If one wants to describe a particular political situation at a specific time one should try to infer the voting measure from statistical data. Such an approach may be of use, for example, for opinion polls and election forecasts.

A quite different situation occurs if one wants to design a constitution for a political union. Such a concept should be independent of the current political constellation in the states, which is subject to fluctuate, and even the particular states under consideration, as the union might be enlarged in the future. In this case it seems that the best guess is to choose the independence measure.

This particular voting measure has a tendency to perhaps give big states less power than they might deserve under a more realistic voting measure. However, it seems to the author that this 'mistake' is less severe than to give big states more power than they ought to have.

8 A short outlook

In this paper we have considered various situations when mathematics has something to say about politics. We have shown that real world voting systems, such as the system for the Council of the EU or the Electoral College, are so complex that 'common sense' is simply not enough to understand the system. In fact, mathematical tools are necessary to analyze them. Sometimes, common sense is even misleading. For example, most people tend to believe that a voting weight proportional to the population would be fair for the Council of the EU.

Every day's experience shows that politicians are reluctant to ask scientists (and especially mathematicians) how to do, what they believe is their job, for example how to design a voting system. This empirical fact has certainly various roots. One is, we believe, that mathematicians have only very occasionally made clear that they work on problems with relations to politics. Another reason is that there are considerable cultural differences between the world of mathematics and the world of politics. Whatever the reasons may be, we believe a discussion between politicians and voting theorists would be beneficial to both sides, and to normal voters.

In particular for the voting system in the Council of the EU it would be helpful to contact voting theorists before the next reform. It is striking that politicians agreed at different times on two voting systems with almost opposite defects, while the reasonable system proposed by the Polish government was neglected (if not ridiculed). There was a petition signed by more than 50 scientists from various European countries sent to all governments of the EU member states, which explained the benefits of the square root voting system for the Council of Ministers. Only one of the (then) 25 governments reacted.

We believe that a voting system which can be considered as unjust on a scientific basis will certainly not promote the idea of a unified Europe.

There is also a positive example of interaction between politicians and mathematicians, namely in the fields of fair allocation of seats in a parliament. This is the biproportional apportionment which was implemented by the mathematician Friedrich Pukelsheim for the Swiss Canton Zürich and subsequently for other Swiss Cantons. This system allows a representation in a parliament which is both proportional with respect to parties and with respect to regions. For the biproportional apportionment method in theory and practice see Pukelsheim [26] and references therein.

References

[1] AIGNER, M., *Discrete Mathematics*. AMS, 2007

[2] BANZHAF, J., Weighted voting doesn't work: a mathematical analysis. *Rutgers Law Review* **19**, 317-343, 1965.

[3] BIRKMEIER, O., *Machtindizes und Fairness-Kriterien in gewichteten Abstimmungssystemen mit Enthaltungen*. PhD-thesis (in German), Augsburg, 2011.

[4] BIRKMEIER, O., KÄUFL, A. and PUKELSHEIM, F., Abstentions in the German Bundesrat and ternary decision rules in weighted voting systems. *Statist. Decisions* **28** 1, 1-16. 2011.

[5] BRÄUNINGER, T. and KÖNIG, T., Indices of power (IOP 2.0). *Software*, downloaded from http://www.tbraeuninger.de on January 7, 2016.

[6] CICHOCKI, M. and ŻYCZKOWSKI, K. (Eds.), *Institutional Design and Voting Power in the European Union*. Ashgate, 2010.

[7] COLEMAN, J. S., Control of collectivities and the power of a collectivity to act. In: Lieberman B (ed) *Social choice*. Gordon and Breach, London, 1971.

[8] DUBEY, P. and SHAPLEY, L. S., Mathematical properties of the Banzhaf power index. *Mathematics of Operations Research* **4**, 99- 131, 1979.

[9] ELLIS, R., *Entropy, large deviations, and statistical mechanics*. Springer, 1985.

[10] FARA, R., LEECH, D. and SALLES, M. (Eds.), *Voting Power and Procedures*. Springer, 2014.

[11] FELSENTHAL, D. and MACHOVER, M., Ternary voting games. *International Journal of Game Theory* **26**, 335-351, 1997.

[12] FELSENTHAL, D. and MACHOVER, M., *The measurement of voting power: Theory and practice, problems and paradoxes*. Cheltenham: Elgar, 1998.

[13] HOLLER, M. and NURMI, H. (Eds.), *Power, Voting and Voting Power: 30 Years After*. Springer, 2013.

[14] KIRSCH, W., On Penrose's square-root law and beyond. *Homo Oeconomicus* **24**, 357-380, 2007.

[15] KIRSCH, W., A survey on the method of moments. Preprint, book in preparation.

[16] KIRSCH, W., Die Formeln der Macht (in German). *Die Zeit*, 15. 3. 2001.

[17] KIRSCH, W., Europa, nachgerechnet (in German). *Die Zeit*, 9. 6. 2004.

[18] KIRSCH, W., Efficiency, Dicisivness and Success in Weighted voting Systems, Preprint.

[19] KIRSCH, W. and LANGNER, J., Invariably suboptimal: An attempt to improve the voting rules of the treaties of Nice and Lisbon. *Journal of Common Market Studies* **49.6** 1317-1338, 2011.

[20] KIRSCH, W., LANGNER, J., The Fate of the Square Root Law for Correlated Voting. In: [10], 147-158.

[21] KURZ, S. and NAPEL, S., Dimension of the Lisbon voting rules in the EU Council: A challenge and new world record. Preprint Universität Bayreuth.

[22] LANGNER, J., *Fairness, Efficiency and Democracy Deficit*. PhD-thesis, FernUniversität Hagen, 2012.

[23] LARUELLE, A. and VALENCIANO, F., Assessing success and decisiveness in voting situations. *Soc. Choice Welfare* 24 (1), 171-197, 2005.

[24] LUEF, W., Das Geheimnis der Quadratwurzel (in German). *Die Zeit (Austrian Edition)*, 21. 6. 2007.

[25] PUKELSHEIM, F., Der Jagiellonische Kompromiss (in German). *Neue Zürcher Zeitung*, 20. 6. 2007.

[26] PUKELSHEIM, F., *Proportional Representation*, Springer, 2014.

[27] PENROSE, L. S., The Elementary Statistics of Majority Voting. *J. Roy. Statist. Soc.* **109** (1), 53-57, 1946.

[28] RACHMAN, G., Square root of the EU's problems, The Financial Times, June 11, 2007 cited from http://www.ft.com (visited on March 23, 2016).

[29] SHAPLEY, L. S. and SHUBIK, M., A method for evaluating the distribution of power in a committee system. *American Political Science Review* **48**, 787-792, 1954.

[30] SŁOMCZYŃSKI, W.; ŻYCZKOWSKI, K., Penrose voting system and optimal quota. *Acta Physica Polonica* **B37** 11, 3133-43, 2006.

[31] STRAFFIN, P. D., Power indices in politics. In: Brams, S. J., Lucas, W. F. and Straffin, P. D. (Eds.), *Political and related models*. Springer, 1982.

[32] TAYLOR, ALAN D. and PACELLI, ALLISON M., *Mathematics and politics*. Springer, 2008.

[33] TAYLOR, ALAN D. and ZWICKER, WILLIAM S., A Characterization of weighted Voting. *Proc. Am. Math. Soc.* **115** (4), 1089-1094, 1992.

[34] TAYLOR, ALAN D. and ZWICKER, WILLIAM S., *Simple Games*. Princeton University Press, 1999.

[35] THOMPSON, C., *Mathematical statistical mechanics*. Princeton University Press, 1972.

Numerical methods and scientific computing for climate and geosciences

Jörn Behrens

Studying the climate, weather or other geoscientific phenomena is strongly related to simulation based knowledge gain, since the climate system, for example, is not assessable by laboratory experiments. In these simulations, mathematical models as well as numerical methods play a crucial role in many aspects of the knowledge work-flow. We will describe the general set-up of geoscientific models, and explore some of the applied mathematical methods involved in solving such models. One of the paramount problems of geoscientific simulation applications is the large span of scales that interact.

1 Introduction

In the beginning of the 20th century a rigorous mathematical description of weather phenomena was proposed by Norwegian scientist Vilhelm Bjerknes [7]. Only then, by formulating weather as a fluid dynamics problem, it became feasible to derive equation sets and solve those equations with the aim to forecast future weather conditions. Lewis Fry Richardson then dreamt of "Weather Prediction by Numerical Process" [23]. While he failed initially due to unstable numerical methods, insufficiently accurate initial conditions and unsuitable computing power, his dream eventually became true with the invention of programmable computers in 1950. The brightest heads of their time, John von Neumann and co-authors, integrated numerically the barotropic vorticity equation in order to demonstrate the earliest numerical weather prediction [9].

Since then, amazing progress has been made in accurately forecasting the weather. Today, it is taken for granted that a five-day forecast is accurate and reliable. International agreements on the level or United Nations organizations grant access to data and simulation results for national forecasting centers. Air traffic, agriculture, renewable energies, commodity markets rely vitally on forecasts of weather and climatic conditions.

Similarly, other geoscientific application fields developed very dramatically by introducing and utilizing mathematical methods [22]. One driving element for all these approaches was simulation based knowledge gain. Computers and computational modeling gave the opportunity to experiment in disciplines where experimentation in physical space is hardly ever possible – we cannot conduct large-scale climate experiments in the field.

One of the great challenges in geoscientific modeling is the wide range of spatial and temporal scales, determining the behavior of the systems involved. The famous "butterfly effect", attributed to the influential work on nonlinear dynamical systems by Lorenz [19], explains the large scale effect a small scale perturbation can have on the system. And on the other hand, large scale behavior strongly – and non-linearly – affects local phenomena. For example the local inundation behavior in the wadden area of the North Sea is driven by (global) tidal elevation. It is this interaction of scales that cannot be described to complete satisfaction yet.

In this manuscript, we will explore the basic mathematical formulations of current geophysical fluid dynamics problems (Section 2). We will look into the range of scales and develop an understanding of scale interaction (Section 3). While the range of scales poses one problem to mathematical and numerical modeling of geosciences problems, uncertainty in data and modeling is another important topic described in Section 4. One of the methods tackling uncertainty is adjoint modeling, which in itself is a viable strategy to answer inverse problems. An example is given in Section 5. One important application of geoscientific modeling with large impact on societies is the simulation of geo-hazards. This will be described in Section 6, before concluding remarks are given.

2 Modeling geosciences – the basic equation sets

Many geoscientific processes can be described by general fluid dynamics models: The atmospheric or oceanic circulation is a fluid dynamics process [13]. When chemical reactions, or thermal processes are considered, a thermodynamical model can be applied. The movement of glaciers can be modeled as a thermodynamical or an elasto-plastic process, with corresponding model equations. Mantle convection and volcanic magma plumes can be simulated by fluid dynamics processes with complicated material properties. Earthquake rupture processes are often modeled by elasto-plastic equation sets with seismic processes described by wave equations.

In particular, starting from first principles like mass, momentum, and energy conservation, one can derive three coupled sets of equations for the velocity components of a fluid, the density and the temperature. Further equations describe the basic property of the fluid (equation of state), moisture in the atmosphere, chemical tracers, radiative transport, rheological properties (for example of ice), salinity in the ocean, etc.

Often these complicated equation sets are simplified taking into consideration the "thin layer" property of the earth's diverse layers (mantle, oceans, atmosphere). More generally, diverse simplifications of Navier-Stokes equations coupled with advection-diffusion-reaction type and wave equations form the basic building blocks for modeling phenomena in the geosciences. These are often highly complicated non-linear partial differential equations to be solved numerically.

In order to solve such equation sets initial and boundary conditions need to be given. Specifying these is a difficult problem in itself. While initial conditions can in principle be determined by measurements, it is a challenging problem to provide these values. In many cases it is not possible or very difficult to gather enough measurement values in order to provide information for uniquely solving the equations. Furthermore, and often more importantly, the initial conditions from measurements are not necessarily consistent with the model equations. Therefore, data assimilation is utilized in order to provide useful initial conditions [17].

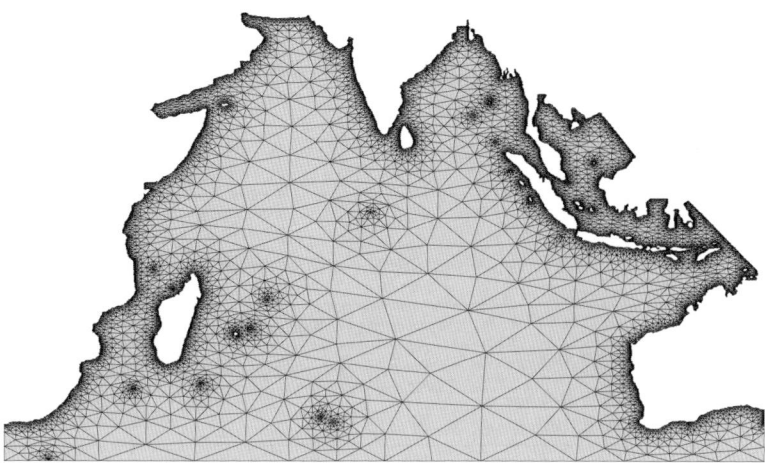

Figure 1. Computational mesh for Indian Ocean tsunami simulation demonstrating the complex boundary geometry

Similarly, it is often very difficult to derive boundary conditions for simulations. An example is the basal boundary for glacial flow problems. Since remote sensing techniques for deriving roughness properties below the glacier are limited in accuracy and resolution, these properties need to be derived by inverse methods from models [27].

Even if boundary conditions and initial conditions can be defined with good accuracy and the set of equations is comparably simple – as is the case for tsunami wave propagation simulations – the complex boundary geometry may pose significant challenges to the simulation accuracy (see Figure 1).

In summary, many geoscientific applications can be modeled by coupled systems of non-linear partial differential equations. However, major challenges are posed by setting up consistent and physically relevant initial and boundary conditions, in order to achieve well-posed equation sets.

3 Multiple scales

Assuming that such well-posed equation sets can be defined, their numerical solution is then the next challenge. The involved relevant scales (in time and in space) pose problems to the feasibility of numerical solution techniques.

An example is the tsunami simulation problem, mentioned already above. When considering the extent of the ocean a relevant length scale is of the order 10^8 m whereas inundation on land has typical length scales of the order of 10 m (see Table 1). Similar scale differences are valid for the time scales involved. In order to resolve the finest scales in discrete numerical approximations to the relevant equations, a discretization mesh of corresponding mesh spacing needs to be employed. When considering two-dimensional simulations, this results in of the order of $(10^7)^2 = 10^{14}$ mesh points. Considering the time scales an order of 10^4 time steps are involved in order to resolve the physical processes – not considering any numerical stability restrictions. Even for today's high performance computers this is a challenging size.

Table 1. Typical orders of time and length scales in tsunami propagation simulation

Regime	Length in m	Time in s
Domain	10^8	10^5
Deep ocean wave	10^5	10^3
Inundation	10	10

Table 2. Typical orders of time and length scales for atmospheric tracer transport

Regime	Length in m	Time in s
Global domain	10^8	10^6
Mesoscale mixing	10^4	10^3
Chemical reaction	1	1

A second example is given by atmospheric tracer transport. This is an important application in climate modeling, ecosystem simulation and disaster management (e.g., volcano ash dispersion). When looking at the latter, the domain is global, i.e. has a spatial extent of the order of 10^8 m, whereas the relevant length scale for chemical reactions is of the order of 1 m (see Table 2). Considering a three-dimensional setting with a vertical extent of the order of 10^5 m and looking at the time scales involved, this would mean an order of 10^6 time steps involving an order of 10^{23} unknowns. These are dimensions, completely inaccessible for today's computing infrastructures.

In order to cope with the vast extent of relevant scales and the inability of conventional numerical discretization methods to cope with these scales in a direct way, a number of strategies has been developed. An obvious first approach is to focus the investigation to a small selection of relevant scales and model these processes in approximate equation sets. This has been developed into some sophistication in all sections of geoscientific modeling. An example is the large number of simple-to-complex equation sets used in atmospheric modeling. In fact, the early success of numerical weather forecasting was partly based on the right decision to look at the large scale atmospheric circulation and solve for that purpose the barotropic vorticity equation [9]. Systematic scale analysis and subsequent derivation of scale-aware equation sets have been derived by Majda and Klein [20].

Numerical approaches for handling the scale gap are diverse. An early and common approach to simplify computations was to filter those (time) scales not relevant for the investigation. Implicit numerical methods were employed to slow down fast (e.g., acoustic) waves in atmospheric simulations [24]. An approach to deal with spatial scales is to use spectral expansion methods and to truncate the spectral expansion where appropriate (see, e.g., [16]). A typical approach to capture the behavior of small scale processes in large scale simulations is to employ averaging approaches or parameterizations. These act as diffusive processes (e.g., turbulence parameterizations) or source terms (e.g., radiation parameterizations) in the general large scale dynamical equation sets (some simple examples are described in [17]).

More recently, multi-scale numerical methods have been developed. One approach consists of multi-resolution methods, in which the computational discretization meshes are adapted to the situations involved. One such mesh is depicted in Figure 1, where the mesh is refined

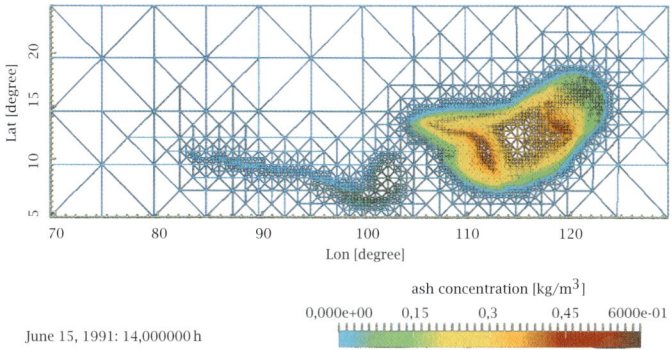

Figure 2. Adaptively refined grid in a two-dimensional prototypical simulation of the ash dispersion of volcano Mount Pinatubo (Philippines) eruption of 1991 (Figure taken from [12])

in shallow (coastal) areas in order to resolve tsunami wave shoaling and inundation. Adaptive mesh refinement has been successfully applied to several disciplines of geoscientific modeling. Examples of adaptive mesh refinement methods for atmospheric applications are given in [3]. In Figure 2 an adaptively refined mesh is shown that achieves high local resolution for resolving relevant processes in the ash cloud of a volcano ash dispersion simulation, while saving computational resources where no relevant tracer concentrations can be found.

In other fields of geosciences applications multi-scale variational methods have been developed that achieve an upscaling of sub-grid scale processes by modifying locally the discrete representation [1,14]. These methods achieve a discretely consistent upscaling of small scales and maintain physical properties like conservation, symmetries etc. while allowing for parameterized processes to modify the large scale. Several challenges remain in applying these methods to transient problems and problems exposing hyperbolic behavior.

In summary, the large extent of relevant scales is one of the major challenges – theoretically, as well as computationally – in geoscientific modeling. A number of approaches have been developed and it will be one of the main tasks for mathematical research in the future to solve the multi-scale interaction problem in a consistent way.

4 Uncertainty

As pointed out before, many geoscientific problems can be formulated in terms of partial differential equations. It has also already been mentioned that it is often a non-trivial task to acquire sufficiently accurate and reliable initial or boundary conditions. Additionally, many geoscientific phenomena are almost in some kind of balanced state, such that small perturbations to data can have large implications for the modeling results. Thus, uncertain data paired with highly sensitive (often non-linear) phenomena imply a large uncertainty in modeling results.

While a rich research area deals with uncertainty in a probabilistic sense (see, e.g., [10] for seismic hazards, [11] for tsunami hazard), this presentation will restrict itself to deterministic approaches to dealing with uncertainty. Therefore, a rough assessment of sources and types of uncertainty is necessary, and we will begin with the first.

Sources of uncertainty

It is obvious that inaccurate or inaccessible data is one of the predominant sources of uncertainty. Weather forecasting relies heavily on the accurate knowledge of the current state of the atmosphere. So, measurements are necessary. However, it is quite difficult and costly to gather all different types of data for all relevant points in space and time with sufficient accuracy. Even with remote sensing techniques a large uncertainty about in situ conditions remains. This becomes even more obvious for oceanographic data where deep ocean measurements are still hardly available with enough spatial and temporal resolution. Even more scarce is our knowledge about sub-surface geologic structures.

A second source of uncertainty comes from unknown or varying driving mechanisms. When considering climate change projections, human behavior and economic development cannot be reliably predicted. Therefore, only assumptions on such driving forces enter into the scenarios and pose uncertainty. Similarly, several internal feedback mechanisms in many geoscientific systems are not fully understood, so that their reaction to changing conditions and the interaction to the main dynamics is not completely clear.

Model uncertainty is another source. There are often a number of different assumptions leading to different models of the same phenomenon. Tsunami wave dispersion for example is very well represented by shallow water wave theory. But with good arguments one can also consider Boussinesq-type equations as the most appropriate model approach. Similarly, nonhydrostatic modeling approaches are valid approximations. And of course the most accurate (however not feasible in terms of computing resources and available initial conditions) approach would be a fully three-dimensional model employing the Navier-Stokes set of equations. The solution technique itself provides some source of uncertainty, however often considered minor in comparison to the data uncertainty.

Types of uncertainty

While uncertainties in data and driving mechanisms could be in principle diminished by higher investments in sensor networks and research, model uncertainty is hard to reduce, since it is inherent in the process of understanding nature by simplifying it to natural laws. Usually the first kind of uncertainties is called epistemic whereas the latter is called aleatoric [2].

Quantifying uncertainty

In order to quantify and possibly reduce epistemic uncertainty in deterministic modeling approaches, a number of methods have been developed. An important approach is a classical sensitivity analysis. By varying individual uncertain input parameters (e.g., initial values or boundary data) the sensitivity of the system to these variations is quantified. In mathematical terms this approach can be described by the following assumption: Let M : input $\mapsto$ output be the mapping established by the model. Let further $I = \{input_1, input_2, \ldots, input_n\}$ be a set of varying input parameters. Then $O = \{output_1, output_2, \ldots, output_n\}$ is the corresponding set of outputs. Let a metric $d(\cdot, \cdot) : V \times V \mapsto \mathbb{R}^+$ be given that measures the difference ("distance") between two elements of either set I or O. If we make some simplifying, yet not too restrictive assumptions (for example, we consider only a local situation and assume the behavior to be quasi-linear locally), we can determine the dependency of model results on input uncertainty (or the sensitivity) by the condition

$$\text{if } \max_{i,j} \left(d(\text{input}_i, \text{input}_j) \right) \leq \delta, \quad \text{then} \quad \max_{i,j} \left(d(\text{output}_i, \text{output}_j) \right) \leq \sigma \cdot \delta,$$

where δ is the variation in input parameters, and σ is the sensitivity parameter (or an amplification factor). This relation is justified by noting that we can (under further assumptions that we omit for educational reasons) express $\text{input}_k = \text{input}_l \oplus \delta$, where the signs $\oplus/\ominus$ denote a generalized addition/subtraction and we have used that $d(\text{input}_k, \text{input}_l) = \text{input}_k \ominus \text{input}_l$ in analogy to the distance in $\mathbb{R}$. Then the mapping induced by the model carries the input uncertainty to the output uncertainty by

$$M(\delta) = M(\text{input}_k \ominus \text{input}_l) = M(\text{input}_k) \ominus M(\text{input}_l)$$
$$= \text{output}_k \ominus M(\text{input}_k \oplus \delta) \leq \sigma \cdot \delta.$$

In this reasoning we have assumed linearity, and δ being the maximum of distances in the input set I. Note that this is a largely simplified notation intended for demonstration reasons. In real applications, it is a lot of effort to define the correct spaces for I and O, the metric, and linearity assumptions.

The sensitivity can be assessed by applying a number of model simulations and varying the input in a range that resembles the input uncertainty – a classical Monte-Carlo type approach. Observing that the sensitivity parameter resembles a differential by writing

$$\sigma = \frac{M(\delta)}{\delta} = \frac{M(\text{input}_k) \ominus M(\text{input}_k \oplus \delta)}{\text{input}_k \ominus (\text{input}_k \oplus \delta)},$$

another method for assessing the sensitivity is possible. If a derivative of M in the input parameter could be computed, the sensitivity on input uncertainty could also be quantified. Since in most cases it is impossible to compute a classical derivative, adjoint equations are derived and an adjoint model can yield the expected sensitivity parameter. Additionally, an adjoint method can be constrained by goal functionals, such that sensitivities with respect to certain desired objectives can be computed. With such a goal oriented adjoint method, one could study the influence of a mountain range or the vegetation cover on the precipitation in an area of interest. More interestingly, with an adjoint method one could study the range of variation in input, given a certain output. So, given measurements in a certain area, an adjoint method can yield the possible parameter ranges – e.g., spatial or temporal areas of influence – that create such output values.

5 A simple example for adjoint methods

A very simple such example is the adjoint to the linear advection (or transport) equation, which should be demonstrated here. It plays an important practical role in marine pollution monitoring and other areas of tracer transport problems. The main underlying question for this type of applications can be formulated as follows: *Given a (measured or observed) distribution of a (almost) passive tracer constituent, where is the source or area of release for this tracer?* This question is the inverse of the usual forward simulation problem, where initial conditions (a source) are given and the later dispersion of the tracer is computed, under the knowledge

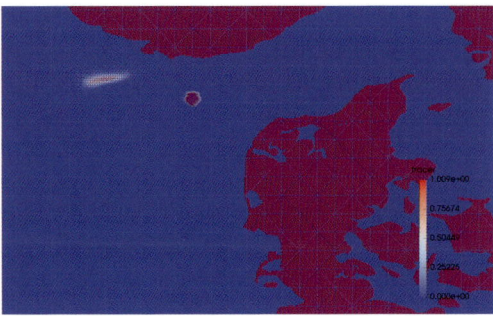 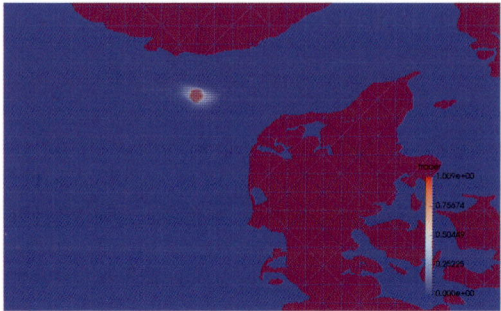

Figure 3.   A purely artificial demonstration of adjoint advection. A potentially measured pollution density (left panel) can be traced back to its source (right panel). The original source is depicted as a filled contour, whereas the measured/computed density is visualized on the adaptive mesh.

of a current or wind that disperses the constituent. So, in order to answer the question, a wind/current is assumed to be known. Furthermore, it needs to be assumed that – at least for a limited time interval – the underlying model equations are invertible, i.e. the physical processes are reversible. This is true for pure advection.

The advection equation is induced by the principle of mass conservation. Under a given velocity field $\mathbf{U} = (u_i(\mathbf{x}, t)) \in \mathbb{R}^d$ $(i = 1, \ldots, d)$, a tracer density distribution $p = p(\mathbf{x}, t) \in \mathbb{R}$, with $\mathbf{x} = (x_i) \in \mathbb{R}^d$, $t \in \mathbb{R}^+$ the spatial and time coordinates, can change its shape, but not its global integral. The (linear) advection equation can then be given by

$$\partial_t p + \nabla \cdot (\mathbf{U}p) = r$$

In this equation r represents a set of sources, sinks, or reactions. The two terms on the left hand side indicate that the change of p over time is balanced by the flux. We omit the formal derivation of the adjoint equation here and refer to the literature (e.g., [3]). However, it can be stated that the adjoint of the transport equation is the transport equation reversed. Intuitively this is quite obvious, since a constituent transported by a velocity field (without any other interaction) can be transported back by the negative velocity field.

Now, for practical application the adjoint operator can be used to trace back a constituent to its original source, assuming that the velocity field is known and that no additional sources, chemical reactions or diffusive processes have influenced the original dispersion. This is with good approximation true for oil spills caused by (illegal) bilge water deposition of ships for example [21]. Or certain pollutants that do not decay too quickly in water or air can be tracked by this approach.

An example application for this approach is depicted in Figure 3. This is a purely artificial example, in which an initial distribution of a pollutant in the Skagerrak region of the North Sea is tracked back to its source. Since the area affected is quite large, the modeling domain is taken to be the entire Skagerrak with surrounding seas. Near-realistic averaged current data were taken. Since the source is almost a point source, an adaptive mesh refinement strategy for solving the adjoint problem numerically helps to maintain the required accuracy without exhausting the computational resources. Such simple simulations can be performed within seconds on usual workstations.

6 Deterministic methods for geo-hazards

A number of different fields of modeling are involved in dealing with geo-hazards. Two have already been mentioned in this text: volcano hazards and oil spills. Many other application fields come to mind and it would go beyond the scope of this text to detail them all. It will be the aim of this section to conceptually describe common features of such simulation approaches as they occur in storm surge, hurricane, tsunami, wild fire, or pollutant dispersion modeling, to name only a few.

An important field of geo-hazard modeling lies in probabilistic approaches, which have proved to be a successful and useful tool for planning, preparedness, and risk analysis [2]. However, deterministic approaches are important tools in early warning situations as well as a basis for the probabilistic approaches. Therefore, we will discuss them here.

When societies deal with such hazards, simulation and modeling is likely in the center – often well hidden from even the experts in disaster management. In order to understand and plan for disaster management, computed scenarios are used. If personell needs training and agreed warning and mitigation procedures need testing, computed scenario data help to design realistic event settings. When the situation in an actual event needs to be evaluated, then often simulations help to derive a complete picture out of scarcely available data. And finally, in order to set up warning capacity, optimization of the data network and monitoring capability is supported by simulations and scenarios.

So, what are the common features of such simulations? In the above sections a number of application fields for geoscientific applications of applied mathematical methods have been introduced. There are a number of common features in these types of applications.

- Multiple scales that interact pose challenges to the numerical representation of geoscientific problems.
- Uncertainty in data and even in the proper mathematical description of processes have to be dealt with.
- In many cases inverse problem settings are to be answered.
- A large number of physically different sub-systems are coupled and interact.

If such methods are to be utilized to manage, plan or mitigate natural hazards, further requirements have to be considered.

- Accuracy requirements demand for highly complex set-ups.
- In case of emergency, efficient computations in very short time are required.

In order to highlight a few strategies for dealing with these requirements, the example of tsunami early warning will be explored. It is noted that similar approaches are taken and have been found relevant in other fields of geoscientific mathematical modeling. The tsunami example is just very educative.

When considering a tsunami event, there are several spatial and temporal scales involved, as described in Section 3. The extent of an ocean basin can be estimated by roughly of the order of $10^6 - 10^7$ m, the rupture area (i.e., the area of initial uplift caused by an earthquake) of a large scale tsunamogenic earthquake has often an extent of the order of $10^5 - 10^6$ m, the wave in deep ocean has an extent of approximately $10^3 - 10^4$ m, and local wave interaction at the coast or onshore requires a spatial resolution of down to the order of $10^1 - 10^2$ m. Since the vertical extent of the ocean is usually only of the order of 10^3 m and the wave height

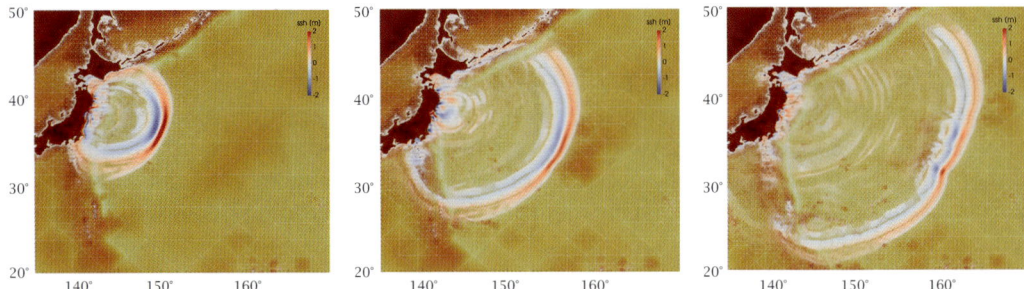

Figure 4.   Adaptive simulation of tsunami wave propagation for the 2011 Tohoku tsunami (taken from [26])

over mean zero is often less than approx. 10 m, it is in general easily justifiable to use two-dimensional modeling approaches. However, even with a two-dimensional grid of the highest local resolution covering the whole ocean would require up to the order of 10^{14} grid points, a challenging number even for today's super-computers.

The first step towards handling this complexity is to introduce non-uniform grids. Many groups have used nested or locally refined meshes [15]. A more advanced approach is to adaptively refine the mesh during computation and resolve small scale features of the wave propagation automatically by a suitable adaptation criterion [3, 18]. An adaptive simulation of the 2011 Tohoku Tsunami in Japan is depicted in Figure 4.

In an early warning system approach the uncertainty in input data needs to be taken into account. It turns out that in tsunami early warning, especially when the earthquake source lies close to the coast, large uncertainty is in the shape and the extent of the initial uplift, generating the tsunami. While the earthquake parameters (magnitude, and epicenter location) can be assessed with relatively high accuracy, these parameters alone do not allow for an accurate estimation of the source. One way to overcome this problem is to use a forecast scheme that tries to minimize the uncertainty by combining several independent measurements. This approach is supported by a simple, but effective, uncertainty model [4]. While traditional methods employ a chain of models (from seismic parameters to initial source, from source to wave propagation, and from wave propagation to coastal impact), where in each of the steps an amplification of initial uncertainty causes a large uncertainty in the result – effectively leading to many false warnings – the approach taken in [4] relates different types of measurements to scenarios. Thus, only those scenarios in the intersection set of different measurements are possible representatives for the true situation.

Figure 5 demonstrates this approach to uncertainty. The left panel shows the traditional approach for tsunami warning based on seismic parameters. While the assessment of these parameters is relatively certain (only a small circle symbolizes the uncertainty), the number of possible sources resulting from these parameters is already bigger. Plugging these into a wave propagation model further amplifies the uncertainty, leading to a large number of possible forecast scenarios. In contrast the right panel demonstrates the uncertainty reduction by independent data sets. While in each of the data sets individual uncertainty amplification can be large, the combination of them all, allows for only a small sub-set of scenarios to match all parameters simultaneously. This approach therefore decreases uncertainty.

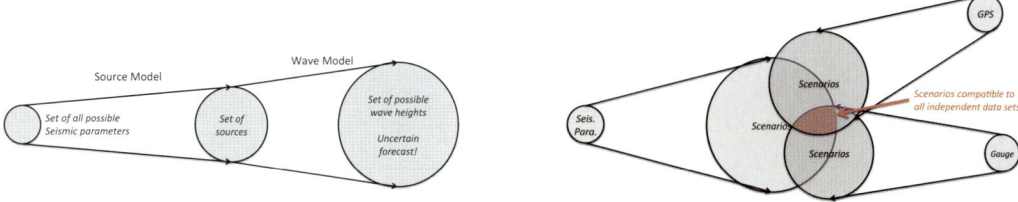

Figure 5.   Simple uncertainty model for near field tsunami warning. Uncertainty amplification in a chain of models (left) and uncertainty reduction through simultaneous utilization of independent parameters (right).

In order to improve computational efficiency, a number of approaches are taken. First, optimization on an algorithmic level, by introducing efficient data structures and utilizing locality-preserving ordering of grid unknowns is applied [5]. A second approach is to utilize highly efficient computing architectures like graphics processing units (GPUs) [8].

Finally, an important part of the modeling chain is validation. An even philosophically interesting question is how to validate simulations where solutions are not known, and in particular, the model development is driven by new knowledge and developing theories, which in turn creates further model development for verifying hypotheses, generated by the model. For the tsunami case, an agreed systematic approach to validation is taken. It is based on a hierarchy of simple to complex test cases, which start from simple tests on conservation properties and code correctness, over quasi one-dimensional analytical tests, cases involving laboratory measurement data, to the whole complexity of real events [25].

7 Concluding remarks

With the description in the previous sections, several examples of applied mathematical methods for geosciences applications have been demonstrated. It would have been too demanding to describe the details of an earth system model as a whole. Since in these complicated simulation tools, atmosphere, ocean, ice, biosphere and anthroposphere are represented with a number of complex interactions within and between these model components. Thus, in order to demonstrate the mathematical and conceptual challenges of multi-scale representation, uncertainty propagation, efficient computing, inverse modeling, and validation the approach was taken to select simplified settings in which to demonstrate the mathematics involved in solving geoscientific problems.

It is the author's belief that mathematics is necessary to solve some of the pressing problems related to geosciences modeling. In the recent past, tremendous progress has been made by mathematical methods for inverse problems, employing adjoint methods, advanced error estimation, and adaptive modeling techniques. A similar break-through is necessary in other fields of geoscientific research. In particular the multi-scale problem appears to be one that needs refined mathematical language and tools in order to be treated successfully. Therefore, a closer collaboration of interdisciplinary teams involving applied and theoretical scientists with a deep understanding and creativity in mathematical methods will be beneficial for future societies.

Acknowledgement.   The author gratefully acknowledges support through the Cluster of Excellence 'CliSAP' (EXC177), University of Hamburg, funded by the German Science Foundation (DFG), and through the ASCETE project, funded by Volkswagen Foundation.

References

[1] T. ARBOGAST, Analysis of a two-scale locally conservative subgrid upscaling for elliptic problems, *SIAM J. Numer. Anal.* **42** (2004), 576-598.

[2] T. BEDFORD and R. COOK, *Probabilistic Risk Analysis: Foundations and Methods*, Cambridge Univ. Press, Cambridge, 2001.

[3] J. BEHRENS, *Adaptive Atmospheric Modeling - Key Techniques in Grid Generation, Data Structures, and Numerical Operations With Applications*, Lecture notes in Computational Science and Engineering **54**, Springer, Heidelberg, 2006.

[4] J. BEHRENS, A. ANDROSOV, A. Y. BABEYKO, S. HARIG, F. KLASCHKA, and L. MENTRUP, A new multi-sensor approach to simulation assisted tsunami early warning, *Nat. Hazards Earth Syst. Sci.* **10** (2010) 1085-1100.

[5] J. BEHRENS and M. BADER, Efficiency considerations in triangular adaptive mesh refinement, *Phil. Trans. R. Soc. A* **367** (2009) 4577-4589.

[6] J. BEHRENS and F. DIAS, New computational methods in tsunami science, *Phil. Trans. R. Soc. A* **373** (2015) 20140382.

[7] V. BJERKNES, Das Problem der Wettervorhersage, betrachtet vom Standpunkte der Mechanik und der Physik (The problem of weather prediction, considered from the viewpoints of mechanics and physics). *Meteorol. Z.* **21** (1904) 1-7 (translated and edited by Volken E. and S. Brönnimann, *Meteorol. Z.* **18** (2009) 663--667).

[8] C. E. CASTRO1, J. BEHRENS and C. PELTIES, Optimization of the ADER-DG method in GPU applied to linear hyperbolic PDEs, *Int. J. Numer. Meth. Fluids* **81(4)** (2015) 195-219.

[9] J. G. CHARNEY, R. FJÖRTOFT, and J. VON NEUMANN, Numerical integration of the baroropic vorticity equation. *Tellus* **2** (1950) 237-254.

[10] C. A. CORNELL, Engineering seismic risk analysis, *Bull. Seismol. Soc. Am.*, **58** (1968) 1583-1606.

[11] E. L. GEIST and T. PARSONS, Probabilistic analysis of tsunami hazards, *Natural Hazards*, **37** (2006) 277-314.

[12] E. GERWING, An adaptive semi-Lagrangian advection model for volcanic emissions, Master Thesis, University of Hamburg, 2015.

[13] A. E. GILL, *Atmosphere-Ocean Dynamics*, International Geophysics Series, vol. 30, Academic Press, London, 1982.

[14] I. G. GRAHAM ET AL. (eds.), *Numerical Analysis of Multiscale Problems*, Lecture Notes in Computational Science and Engineering **83**, Springer, Heidelberg, 2012.

[15] S. HARIG, CHAERONI, W. S. PRANOWO and J. BEHRENS, Tsunami simulations on several scales – Comparison of approaches with unstructured meshes and nested grids, *Ocean Dynamics*, **58** (2008) 429-440.

[16] R. JAKOB-CHIEN, J. J. HACK and D. L. WILLIAMSON, Spectral transform solutions to the shallow water test set, *Jour. Comp. Phys.* **119** (1995), 164-187.

[17] E. KALNAY, *Atmospheric Modeling, Data Assimilation and Predictability.* Cambridge Univ. Press, Cambridge, 2002.

[18] R. J. LEVEQUE, D. L. GEORGE and M. J. BERGER, Tsunami modelling with adaptively refined finite volume methods, *Acta Numerica* **20** (2011) 211-289.

[19] E. N. LORENZ, Deterministic nonperiodic flow, *Jou. Atmos. Sciences* **20** (1963) 130-141.

[20] A. J. MAJDA and R. KLEIN, Systematic multiscale models for the tropics, *J. Atmos. Sci.* **60** (2003), 393-408.

[21] S. MASSMANN, F. JANSSEN, T. BRÜNING, E. KLEINE, H. KOMO, I. MENZENHAUER-SCHUMACHER and S. DICK, An operational oil drift forecasting system for German coastal waters, *Die Küste* **81** (2014), 255-271.

[22] R. ORIVE, M. L. OSETE, J. I. DÍAZ and J. FERNÁNDEZ, Introduction to Mathematics and Geosciences: Global and Local Perspectives, Volume I, *Pure Appl. Geophys.* **172** (2015) 1-5.

[23] L. F. RICHARDSON, *Weather Prediction by Numerical Process*. Cambridge Univ. Press, Cambridge, 1922.

[24] A. ROBERT, A stable numerical integration scheme for the primitive meteorological equations, *Atmosphere-Ocean* **19** (1981), 35-46.

[25] C. E. SYNOLAKIS, E. N. BERNARD, V. V. TITOV, U. KÂNOĞLU and F. I. GONZÁLEZ, Validation and verification of tsunami numerical models, *Pure Appl. Geophys.* **165** (2008) 2197-2228.

[26] S. VATER and J. BEHRENS, Validation of an adaptive triangular discontinuous Galerkin shallow water model for the 2011 Tohoku tsunami, *Geophys. Res. Abstr.* **18** (2016) EGU2016-9170.

[27] N. WILKENS, J. BEHRENS, T. KLEINER, D. RIPPIN, M. RÜCKAMP and A. HUMBERT, Thermal structure and basal sliding parametrisation at Pine Island Glacier – a 3-D full-Stokes model study, *The Cryosphere* **9** (2015) 675-690.

Authors

David H. Bailey – Lawrence Berkeley National Laboratory (retired), Berkeley, CA 94720, and University of California, Davis, Davis, CA 95616, USA ⟨david@davidhbailey.com⟩

Jörn Behrens – Department of Mathematics, Universität Hamburg, Bundesstraße 55, 20146 Hamburg, Germany ⟨joern.behrens@uni-hamburg.de⟩

Jean Bertoin – Institut für Mathematik, Universität Zürich, Winterthurerstrasse 190, 8057 Zürich, Switzerland ⟨jean.bertoin@math.uzh.ch⟩

Albrecht Beutelspacher – Mathematikum Gießen e. V., Liebigstraße 8, 35390 Gießen, Germany ⟨albrecht.beutelspacher@mathematikum.de⟩

Holger Boche – Lehrstuhl für Theoretische Informationstechnik, Technische Universität München, 80290 München, Germany ⟨boche@tum.de⟩

Jonathan Borwein – Centre for Computer Assisted Research Mathematics and its Applications (CARMA), University of Newcastle, Callaghan, NSW 2308, Australia ⟨jon.borwein@gmail.com⟩

Claus Diem – Mathematical Institute, University of Leipzig, Leipzig, Germany ⟨diem@math.uni-leipzig.de⟩

Jürg Fröhlich – Theoretical Physics, HIT K42.3, ETH Zurich, 8093 Zurich, Switzerland ⟨juerg@phys.ethz.ch⟩

Werner Kirsch – Fakultät für Mathematik und Informatik, FernUniversität in Hagen, 58084 Hagen, Germany ⟨werner.kirsch@fernuni-hagen.de⟩

Helmut Neunzert – Fraunhofer Institute for Industrial Mathematics ITWM, Fraunhofer-Platz 1, 67663 Kaiserslautern, Germany ⟨helmut.neunzert@itwm.fraunhofer.de⟩

Helmut Pottmann – Institute of Discrete Mathematics and Geometry, Technische Universität Wien, Wiedner Hauptstraße 8–10/104, 1040 Wien, Austria ⟨pottmann@geometrie.tuwien.ac.at⟩

Christiane Rousseau – Department of Mathematics and Statistics, University of Montreal, C.P. 61288 succ. Centre-ville, Montreal (Quebec), H3C 3J7, Canada ⟨rousseac@dms.umontreal.ca⟩

Walter Schachermayer – Fakultät für Mathematik, Universität Wien, Oskar-Morgenstern-Platz 1, 1090 Wien, Austria, and Institute for Theoretical Studies, ETH Zurich, Switzerland ⟨walter.schachermayer@univie.ac.at⟩

George G. Szpiro – 2109 Broadway, New York, NY, 10023, USA ⟨georgeszpiro@gmail.com⟩

Ezra Tampubolon – Lehrstuhl für Theoretische Informationstechnik, Technische Universität München, 80290 München, Germany ⟨ezra.tampubolon@tum.de⟩

Aad van der Vaart – Mathematical Institute, Leiden University, PO Box 9512, 2300 RA Leiden, Netherlands ⟨avdvaart@math.leidenuniv.nl⟩

Johannes Wallner – Institute of Geometry, Technische Universität Graz, Kopernikusgasse 24, 8010 Graz, Austria ⟨j.wallner@tugraz.at⟩

Wessel van Wieringen – VU University Medical Center, Amsterdam, Netherlands ⟨w.vanwieringen@vumc.nl⟩

Ofer Zeitouni – Faculty of Mathematics, Weizmann Institute of Science, POB 26, Rehovot 76100, Israel ⟨ofer.zeitouni@weizmann.ac.il⟩

Index